GPS, GLONASS, Galileo, and BeiDou for Mobile Devices

From Instant to Precise Positioning

Get up to speed on all existing GNSS with this practical guide. Covering everything from GPS, GLONASS, Galileo, and BeiDou orbits and signals to multi-GNSS receiver design, AGPS, network RTK systems, and VRS, you will understand the complete global range of mobile positioning systems. Step-by-step algorithms and practical methods provide the tools you need to develop current mobile systems, whilst coverage of cutting-edge techniques, such as the instant positioning method, gives you a head start in unlocking the potential of future mobile positioning. Whether you are an engineer or a business manager working in the mobile device industry, a student or a researcher, this is your ideal guide to GNSS.

Ivan G. Petrovski leads the development of GNSS applications at iP-Solutions, Japan. He has over 25 years' worth of research and development experience in the GNSS field, and has previously led GNSS-related R&D for DX Antenna, GNSS Technologies Inc., and the Institute of Advanced Satellite Positioning at TUMSAT. He has academic experience working as an associate professor with MAI and as a guest professor with TUMSAT. As an engineer he has developed RTK software, the algorithms and software for indoor and outdoor positioning with pseudolites, in addition to instant positioning algorithms, real-time GNSS software receivers, and the GNSS DIF recorder and RF signal simulator.

GPS, GLONASS, Galileo, and BeiDou for Mobile Devices

IVAN G. PETROVSKI
iP-Solutions, Tokyo

CAMBRIDGE
UNIVERSITY PRESS

CAMBRIDGE
UNIVERSITY PRESS

University Printing House, Cambridge CB2 8BS, United Kingdom

Cambridge University Press is part of the University of Cambridge.

It furthers the University's mission by disseminating knowledge in the pursuit of education, learning and research at the highest international levels of excellence.

www.cambridge.org
Information on this title: www.cambridge.org/9781107035843

First published 2014

Printed in the United Kingdom by TJ International Ltd. Padstow Cornwall

A catalog record for this publication is available from the British Library

Library of Congress Cataloging in Publication data
Petrovski, Ivan G., 1962–
GPS, GLONASS, Galileo, and BeiDou for mobile devices : from instant to precise positioning / Ivan G. Petrovski, iP-Solutions, Tokyo.
pages cm
ISBN 978-1-107-03584-3 (Hardback)
1. GPS receivers–Design and construction. 2. Mobile communication systems.
3. Global Positioning System. 4. Galileo satellite navigation system.
5. Artificial satellites in navigation. I. Title.
TK6565.D5P45 2014
910.285–dc23 2013032821

ISBN 978-1-107-03584-3 Hardback

Additional resources for this publication at www.cambridge.org/9781107035843

Contents

Foreword by Glen Gibbons

The Law of Unintended Consequences – case in point: the mandate for enhanced 911 (E911) services for mobile devices.

In 1996, the Federal Communications Commission (FCC), a US regulatory agency with broad powers over telecom providers and services, issued its first report, order, and notice of proposed rule-making (NPRM) regarding E911. The FCC action sought to require wireless telephone companies to be able to report automatically the location of an emergency caller.

This initiative proposed to bring to the rapidly expanding wireless space a requirement previously established for wireline providers. Unlike telephones at fixed locations, the whereabouts of a mobile phone was not readily known in emergencies, often not even to the callers themselves.

In issuing the NPRM, however, the Global Positioning System – let alone other GNSS systems – was not on the minds of the FCC commissioners. Indeed, if FCC staff and officials were aware of GPS in 1996, which had only reached full operational capability the year before, it was probably as the precise timing technology providing synchronization to wireline and some wireless networks.

No, in 1996, the FCC assumed that the positioning solutions for E911 would arise from within the networks themselves, using such techniques as angle of arrival (AOA), time of arrival (TOA), and time difference of arrival (TDOA).

This is where the unintended consequences enter in, and why the AOA of Dr. Petrovski's latest book – *GPS, GLONASS, Galileo, and BeiDou for Mobile Devices: From Instant to Precise Positioning* – is a particularly resonant and relevant point of entry.

With E911, the FCC was focused on helping first responders (e.g. police, ambulance, and fire department personnel) reach people in need quickly. The agency probably assumed that, as with many other regulatory mandates, the telecom industry would resist or at least seek to delay the imposition of the new rules. In fact, the full accuracy and reporting conditions of E911 only took effect in December 2012, nearly 16 years after the original FCC report and order.

For our purposes, however, we can leave the prolonged story of E911's implementation, filled as it is with bureaucratic woes and bewildering discoveries about the nature of emergency services' communications infrastructure and processes. Long before E911 arrived, a much more remarkable thing happened: GNSS capabilities began appearing voluntarily on almost every cell phone sold into the US market. Moreover, GNSS has spread to many other mobile devices, such as cameras and notepad computers, not intended primarily for communications.

What happened?

Well, to answer that question we need to return briefly to 1996 and consider the state of the GPS industry – GPS being the only practically accessible GNSS technology at the time. By the mid 1990s, GPS had become well accepted within certain professional and commercial sectors: surveying and mapping (including geographic information systems), along with the previously mentioned timing applications, and – in the wake of the first Persian Gulf war – the military community.

What GNSS did not have in 1996 was a mass market. In large part, this situation stemmed from the US policy of "Selective Availability" or SA – an effort to maintain an advantage for US and allied military forces. This measure imposed a timing dither on the open GPS C/A code signal that degraded the real-time positioning accuracy to "no worse than" 100 meters 95 percent of the time, typically in the range of 60 to 70 meters.

Aside from a few fisherman and recreational boaters, who were happy to use something that would provide them with a better accuracy than the quarter-mile or so then possible with Loran-C, nothing approaching a consumer market had emerged yet for GPS. (Marine users' counterparts in general aviation also found that a GPS receiver – even one that cost thousands of dollars – was still a lot cheaper and a lot more accurate than other types of navigation aids.)

At the time, visionaries and pioneers in a newly named, but underpopulated, world of location-based services (LBS) were looking to the automotive industry for the platform that would lift GPS into the popular imagination. For Americans, cars are the quintessential consumer product. With millions of them sold every year, automobiles seemed the ideal vehicle (no pun intended) for carrying GPS into a mass market.

In addition to SA, however, a couple of things turned out to be wrong with that paradigm. First, the complex network of interlocking hardware and software vendors and LBS providers had not sorted themselves out sufficiently to have found a successful business model. Second, demand for navigation systems among car owners, who do 90 percent of their driving on familiar local routes, remained unencouraging – it suffered from the "So what?" factor. And, more importantly, the US auto industry with a typical seven-year design cycle and price sensitivities, was distinctly uninterested in a largely untested technology that would add hundreds of dollars to the price of a new car.

The FCC's E911 mandate, however, changed that calculus. Mobile phone manufacturers, who had a much shorter design cycle, began thinking about how they might meet the federal mandate with a handset-based solution, rather than a network-based one. And they began looking more closely at the opportunity to build in money-making applications that took advantage of the regulatory mandate for location – a classic lemons-into-lemonade scenario.

By the time Selective Availability was removed in 2000, a substantial and ultimately irresistible momentum had begun for adding GNSS capabilities to mobile devices. Fast-forward to the recent Lightsquared controversy in which the massive installed base of GNSS receivers – driven largely by GPS in mobile phones – created a major obstacle to rollout of a high-powered wireless broadband system in frequencies adjacent to GNSS.

Thus, mobile phones had popularized GNSS with a speed and widespread adoption that vehicle telematics could not.

One further aside: the popularization of affordable, precise location brought about by GNSS has actually fed back and rejuvenated interest in other positioning technologies. Because, as we well know, GNSS does not work everywhere that people live, work, and play, but now that they have experienced its utility and even come to depend on it, people want the capability everywhere and all the time.

So, product developers and service providers have redoubled their efforts to find tools and methods to provide ubiquitous positioning – not merely among the network-based solutions mentioned earlier, but also in exploiting such things as WiFi, magnetic fields, inertial sensors, signals of opportunity, and even echoes. Sometimes alone, but more often in combination with GNSS in mobile devices.

So, about this book. . . .

First off, it could not be more timely. GPS, of course, is well established – there are easily more than a billion receivers in operation around the world, both as standalone tools in the hands of, for example, geocachers or integrated into other devices and systems.

At the same time, after 16 years of decline, restoration, and modernization, Russia's GLONASS system has returned to full operation and appears to be here to stay, despite recent troubles in its space program. Europe's Galileo, finally escaping a long series of political misadventures and technical travails, is on the threshold of a rapid build-out of the constellation – with a public interface control document (ICD) and Galileo-capable receivers already available.

China's GNSS program, BeiDou (actually, BeiDou-2, reflecting the second phase of a satnav program that began in 2000) has shown the fastest evolution – launching its second-generation program with a 2007 launch and following up over the next five years placing another 15 spacecraft into orbit, including two dual-satellite launches. That led to the declaration of regional service and the publication of a B1 civil signal ICD in December 2012.

With these systemic developments and the worldwide adoption of GNSS by all categories of users in mind, I believe that we have arrived at one of those axial ages of technology: a turning point at which a door opens into a world of myriad possibilities, in which familiar endeavors take on new forms and entirely novel ways of working and living appear.

Dr. Petrovski's discussion begins, as so many previous publications rightfully have, with the GNSS infrastructures themselves: the ground control and monitoring segment, the satellite constellations, and the signals. Dr. Petrovski continues by taking up the design of GNSS receivers, both hardware and software, before leading the reader into the application realm of mobile positioning – as my preface to these remarks should make clear, what I consider the pivotal development of GNSS for the general population.

As he has done in previous works, Dr. Petrovski brings his long-established expertise in the GNSS field to bear. The scope of the text's discussion is wide ranging but detailed, thorough without losing perspective and a sense of proportion, sober and informed yet piquant at points, reflecting the personality of the author. In the penultimate Chapter 10, "Trends, opportunities, and prospects," Dr. Petrovski wraps up his examination of GNSS with a thoughtful reflection on prospects for advancing receiver technology in the future. The final chapter of the book – on testing mobile devices – is essentially a practical "how-to" appendix for which the preceding discussion has well prepared the reader.

Let me now step away from the contents of Dr. Petrovski's book again. I must leave to others with far more expertise in the engineering and scientific disciplines the task of assessing, interpreting, and communicating the lessons provided within its covers.

Instead, I want to place this publishing project in a wider context, to highlight the book's purpose against a backdrop of social and historical factors that have informed GNSS technology even as it has altered some of our most quotidian human activities.

Such matters are easy to underestimate or overlook, because GPS had barely reached the public consciousness before it began to disappear back into the context of daily life, increasingly invisible within its applications and the unconscious expectations of location users – like light switches or water faucets, noticeable only by its absence.

In human affairs, we may describe some things as game-changing events, or watershed moments, or even sea changes. But such characterizations are inadequate to describe the pervasive and far-flung effects of GNSS. The changes brought about by this still-young technology are more on the scale of plate tectonics or gravity. And this is not mere hyperbole, not when we are speaking of a technology that is used to measure the wobble in the Earth's rotation itself.

Indeed, had it not been appropriated by the environmental sciences, we would speak of GNSS in terms of global change to capture its true dimensions. Because, since the advent of GNSS, the world has changed – and it almost certainly is not changing back.

Of course, the intersection of GNSS with the routines of average citizens does not have the weightiness of its applications in geodesy, weather monitoring, or spacecraft navigation. Rather, for almost all of us, GNSS manifests itself on a human, even mundane, scale. Nonetheless, we can no more think of people abandoning their location-based applications – traffic guidance, photo-georeferencing, rendesvouzing, friend finding, ATM finding, cinema finding, whatever finding – than we can imagine them giving up the Internet or social media.

So, what are some of the hallmarks of this technology that has so transformed the world? Let me suggest just a few.

GNSS are incremental and cumulative For all of their transformative effects, GNSS build on earlier endeavors, discoveries, and achievements. In fact, it was in the context of GNSS that I first heard the expression, "We stand on the shoulders of giants." Thus, Galileo and BeiDou have largely recapitulated GPS and GLONASS while bringing their own advances and novel contributions. The Soviet Union's GLONASS almost certainly reflected a security-related response to GPS, only using a different orbital configuration and adding an FDMA element to the spreading-code design of GPS's CDMA.

But the lineage of cause and effect, of research and practice, goes back much further. Most immediately, GPS built on earlier Defense Department initiatives, including the Navy's TRANSIT (Doppler) system and Timation program, the Air Force 621B Project, the Army's SECOR satellites, and OMEGA.

Looking further back, however, the Earth's first artificial moon, Sputnik, sent aloft by the Soviet Union in 1957, inspired researchers at Johns Hopkins Applied Physics Lab to use the satellite's radio signal to determine its orbital location by means of Doppler measurements. And soon they speculated that the reverse was possible: to determine a receiver's position on the ground if the satellite's orbital location was known. That, in turn, led to efforts to replicate in space the radionavigation capabilities seen in such terrestrial systems as Loran and DECCA.

If we broaden the bandwidth on our scan of history for GNSS antecedents, however, we will ultimately assemble a pantheon of pioneers in physics, astronomy, math, and engineering – from Einstein, to Kepler and Newton, Tycho Brahe, Copernicus, and on back to Archimedes and Euclid.

GNSS clearly is a "disruptive" technology, a term much in vogue. In the words of that familiar repository of knowledge, Wikipedia, "A disruptive innovation is an innovation that helps create a new market and value network, and eventually goes on to disrupt an existing market and value network (over a few years or decades), displacing an earlier technology."

Oddly enough, Wikipedia does not yet include GNSS in its list of these innovations – where such inventions as telephones, automobiles, and personal computers appear – but GNSS certainly fits the bill.

Moreover, GNSS has put a previously inconceivable capability into the hands of millions. Before GPS, positioning was not a term of art known to the general public. Yes, people were familiar with such concepts as navigation, with mapping, with surveying. But these were understood to be in the bailiwick of professionals, of people who trained for years with expensive, specialized equipment, and not a matter of concern for the rest of us.

How that view has changed! The quotation from Ralph Waldo Emerson on page xix with which Dr. Petrovski introduces this book is as true today for GNSS as it was for the disruptive technologies of the nineteenth century.

A GNSS is a strategic national or regional asset Undertaking a GNSS reflects the relative weight and aspirations of nations around the world. In the bipolar world of the 1970s, that meant the United States and the Soviet Union. In the twenty-first century, Europe and China have joined the club, with India and Japan adding regional satnav systems. These roles have been enshrined by membership in the UN-sanctioned International Committee on GNSS.

Consider this: as Europe was exploring the possibility of developing its own GNSS, it posed a variety of benefits – commercial opportunities, a boost to the manufacturing economy, education and employment of high-tech professionals, a civil alternative to "military" systems such as GPS and GLONASS. But the real argument that persisted through the years of studies, debates, and indecision, and ultimately persuaded the European Union to launch Galileo, was the desire for political sovereignty and control over a critical infrastructure.

GNSS are complexly global and universal By this, I do not mean the obvious physical reach of the system infrastructures themselves. Here, "global" refers to the scale and "universal" to the degree of penetration that GNSS technology has already achieved into so many areas that cross national and ethnic boundaries: transportation, communications, economics, politics, sociology, law, and on and on.

For example, a truly global technology like GNSS intrinsically points toward a global market. And along with such an expansive geography of business comes a panoply of related activities: intellectual property rights, trade issues (subsidies, "dumping," taxes, and tariffs), compatibility and interoperability of technologies, international rule-setting and adjudication of disputes, and so forth. In the meantime, location-based advertising has emerged as one of the fastest-growing revenue streams associated with mobile devices.

Another example: Use of GNSS by law enforcement agencies has become a contentious issue in many countries, having already reached the US Supreme Court. At the same time, concern about its potential misuse by private individuals is rising. In fact, by one measure we might even say that a technology has truly "arrived" when it begins being put to undesirable uses. That appears to be the situation with GNSS, where we see such misapplications of the technology as virtual stalking, warrantless surveillance, road toll avoidance, deceiving truck dispatchers, and manipulating stock markets.

At the same time, GNSS-inspired position/location has become part of the common ground of human experience. It has entered into our language and our mental maps – our sense of where we come from, where we are, where we are going. GNSS and associated technologies orient us to the world and its inhabitants around us – both known and unknown.

So, thank you Ivan for giving us new insights and access into use of the remarkable technology and tool of GNSS.

About this book

As far as GNSS is concerned there are four main areas: mobile applications, geodesy, geophysical applications, and INS-aided navigation. This book covers the first topic, and the last two topics are covered in the book written together with Dr. Toshiaki Tsujii and published by Cambridge University Press, *Digital Satellite Navigation and Geophysics*.

This book is different from *Digital Satellite Navigation and Geophysics* in one other respect. In this book I take a pragmatic approach, providing algorithms and methods, in comparison with an holistic approach in *Digital Satellite Navigation and Geophysics*, where we tried to explain how GNSS is related to other manifestations of the physical world and physical science, and to the nature of things.

Acknowledgments

I would like to thank colleagues and friends without whom this work wouldn't be possible: Dr. Toshiaki Tsujii from JAXA, Mr. Ken Satoh from Amtechs Corporation. Prof. Harumasa Hojo and Prof. Emeritus Akio Yasuda from Tokyo University of Marine Science and Technology, Mr. Andrew Addy from Spirent Communications, and Dr. Takuji Ebinuma from Tokyo University.

I would also like to thank Mr. H. Torimoto from GNSS Technologies Inc. for his support during the years I was working with him.

I would like to thank my wife Tanya for her support and for encouraging me to share my knowledge and experience in this book.

Figures 8.3 and 9.1 were created by Natalia I. Petrovskaia, BA (Cantab), MPhil (Cantab), PhD (Cantab).

Credits for cover image and Figure 2.29: ESA– J. Huart; for Figure 10.1: ESA– S. Corvaja.

Abbreviations and acronyms

ADC	analog-to-digital converter
AGPS (AGNSS)	assisted GPS (GNSS)
AMP	accelerated massive parallelism
API	application programming interface
ARAMIS™	adaptive receiver applied for monitoring of ionospheric scintillation
ASC	audio sub-channel
ASIC	application-specific integrated circuits
BDT	BeiDou Navigation Satellite System time
BOC	binary offset carrier
bps	bits per second
BPSK	binary phase shift keying
C/A	coarse/acquisition
CDGPS (CDGNSS)	carrier differential GPS (GNSS)
CDMA	code division multiple access
CEP	circular error probable
CGCS2000	China geodetic coordinate system 2000
CODE	Center for Orbit Determination in Europe
CPLD	complex programmable logic device
CPU	central processing unit
CUDA	computer unified device architecture
DFF	D-type flip-flop
DFT	digital Fourier transform
DGPS (DGNSS)	differential GPS (GNSS)
DIF	digitized intermediate frequency
DLL	delay-locked loop
DOP	dilution of precision
ECEF	Earth-centered, Earth-fixed
ECI	Earth-centered inertial
ECO	Earth-centered orbital
EGNOS	European Geostationary Navigation Overlay Service
EMC	electromagnetic compatibility
EOP	Earth orientation parameters

FCC	Federal Communications Commission
FDMA	frequency division multiple access
FFT	fast Fourier transform
FIFO	First In First Out
FLL	frequency-locked loop
FPGA	field-programmable gate array
FTF	fundamental time frame
GATE	Galileo test and development environment
GEO	geostationary Earth orbit
GIS	geographic information system
GLONASS	Global Navigation Satellite System
GMT	Greenwich Mean Time
GNSS	Global Navigation Satellite Systems
GPS	Global Positioning System
GPU	graphic processing unit
GSI	Geographical Survey Institute (Japan)
GST	Galileo system time
GUI	graphical user interface
HEO	highly elliptical orbit
HIO	highly inclined orbit
ICAO	International Civil Aviation Organization
ICD	Interface Control Document
ICRF	International Celestial Reference Frame
IERS	International Earth Rotation and Reference Frames Service
IF	intermediate frequency
IGP	ionospheric grid point
IGS	International GNSS Service
IGSO	inclined geosynchronous satellite orbit
IMES	indoor messaging system
INS	inertial navigation system
IONEX	ionosphere exchange format
IP	intellectual property
IPP	ionospheric pierce point
ISC	inter-signal correction
ITRF	International Terrestrial Reference Frame
ITU	International Telecommunication Union
JAXA	Japanese Aerospace Exploration Agency
JPL	Jet Propulsion Laboratory
LAAS	local area augmentation system
LEO	low Earth orbit
LNA	low-noise amplifier
LO	local oscillator
LOS	line-of-sight (between a receiver and a satellite)
LSE	least-squares estimation

LTO	long-term orbit
L2C	GPS civil signal on L2 frequency of L-band
LUT	lookup table
MEO	medium Earth orbit
MODIP	modified dip latitude
Mps	megasamples per second
MSAS	Multi-functional Satellite Augmentation System
NAV	navigation
NCO	numerically controlled oscillator
OCXO	oven-controlled crystal oscillator
OEM	original equipment manufacturer
PDF	probability density function
PLL	phase-locked loop
PPP	precise point positioning
PRN	pseudorandom noise
QPSK	quadrature phase shift keying
QZSS	Quasi-Zenith Satellite System
RAIM	receiver autonomous integrity monitoring
RF	radio frequency
RHCP	right-hand circularly polarized
RINEX	Receiver Independent Exchange Format
RMS	root mean square
RPS	record and playback system
RTCM	Radio Technical Commission for Maritime Services
RTK	real-time kinematic
RTOS	real-time operating systems
SA	selective availability
SARP	Standards and Recommended Practices
SAW	surface acoustic wave
SBAS	space-based augmentation system
SDCM	system of differential correction and monitoring
SDR	software-defined radio
SEP	spherical error probable
SIMD	single instruction on multiple data
SMA	sub-miniature version A
SNR	signal-to-noise ratio
SoL	safety of life
SOW	seconds of week
SP	standard precision
SV	space vehicle
SVID	space vehicle identification number
SVN	space vehicle number
TCXO	temperature-compensated crystal oscillator
TEC	total electron content

TOT	time of transmission
TOW	time of week
TTA	time to alert
TTFF	time to first fix
2DRMS	twice distance RMS
URA	user range accuracy (GPS)
UTC	Coordinated Universal Time
VCXO	voltage-controlled crystal oscillator
VLBI	Very Long Baseline Interference
VRS	virtual reference station
WAAS	Wide Area Augmentation System
WGS-84	World Geodetic System 1984

Definitions

Code phase measurements	Ambiguous code phase measurements are measured by receiver baseband processor. These measurements do not require any information to be decoded from navigation message.
Observables	Observables are formed using receiver raw measurements in various combinations. Pseudorange observables are just pseudoranges.
Pseudoranges or pseudorange observables	Pseudoranges are created from code phase measurements properly aligned with each other (removing time skew) and made unambiguous using time mark either from navigation message or external sources.
Receiver measurements	Receiver measurements usually refer to antenna coordinate measurements, when observables are processed in the receiver navigation processor.
Receiver raw measurements	Receiver code phase and carrier phase raw measurements are unambiguous pseudoranges and ambiguous carrier phase measurements.

Part I

GNSS: orbits, signals, and methods

1 GNSS ground and space segments

Global Navigation Satellite Systems (GNSS) at the time of writing comprise four systems, two of which are fully operational and two of which are on their way (see Table 1.1). A brief history of GNSS is given in Chapter 10 along with a timeline of application development and prospects for this development, especially concerning mobile applications.

Each GNSS comprises a constellation of satellites, called a space segment, and a ground segment (Figure 1.1). The main idea behind GNSS is to measure distances between a satellite and a user located on the surface of the Earth or in a lower atmosphere. Satellite coordinates can be calculated at any moment of time. The information that allows the calculation of satellite position is uploaded to and then broadcast from satellites to the user. The ground segment is responsible for determining satellite orbits, which it then uploads to the satellites, and also for defining the coordinate frame and time frame in which satellite and user positions are estimated.

Having received the information on satellite orbits and measured distances to the satellites, a user can calculate receiver position as an intersection of four spheres in a four-dimensional space-time continuum. If the receiver clock is perfectly synchronized with the satellite time frame, only three satellites would be required to determine receiver position in three-dimensional space (Figure 1.2).

This chapter describes the GNSS space segment and provides the information required to understand the process of calculating satellite coordinates. Algorithms used to calculate satellite positions using various orbit parameters for all existing GNSS are given. The first operational GNSS were the American Global Positioning System (GPS) and the Russian GLObal NAvigation Satellites System (GLONASS), or GLObal'naia NAvigacionnaia Sputnikovaia Systema in Russian. Next came those GNSS that are not yet operational: the European system, Galileo, named after Galileo Galilei (1564–1642), the Italian astronomer and philosopher; and the most recent, the Chinese BeiDou, named for the Chinese pronunciation of the Big Dipper constellation.

1.1 Ground segment and coordinate reference frames

The ground segment comprises tracking stations and facilities, which provide coordinate and clock reference frames, calculate satellite coordinates, and upload this information to satellites.

Table 1.1. GNSS as of 2013

	System			
	GPS	GLONASS	Galileo	BeiDou
Country	USA	Russia	EU	China
Signal from space availability (operational satellites)	1973	1982	2011	2010
Start of operation (positioning service worldwide)	1993	1995	2019 (expected)	2020 (expected)

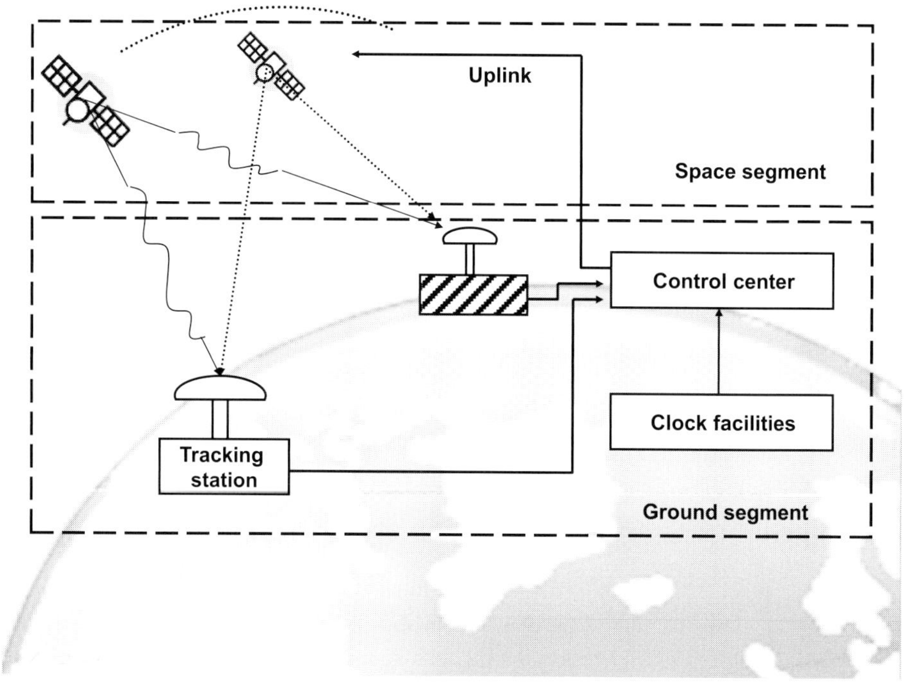

Figure 1.1 GNSS space and ground segments.

The functions of the ground segment can be summarized as follows.

(1) To establish a coordinate frame, in which satellite and user positions are calculated.
(2) To establish a time scale, which, along with the coordinate frame, defines the complete four-dimensional space-time continuum, in which satellite and user positions are calculated.
(3) To collect satellite measurements.
(4) To calculate satellite orbits.
(5) To monitor the satellite signal.
(6) To upload information on satellite orbits and health to the satellites.

A satellite's orbit is defined in a coordinate system, which combines three-dimensional coordinate frame and a time frame. First, we define two essential coordinate frames: Earth-centered inertial (ECI) and Earth-centered, Earth-fixed (ECEF) (see Figure 1.3).

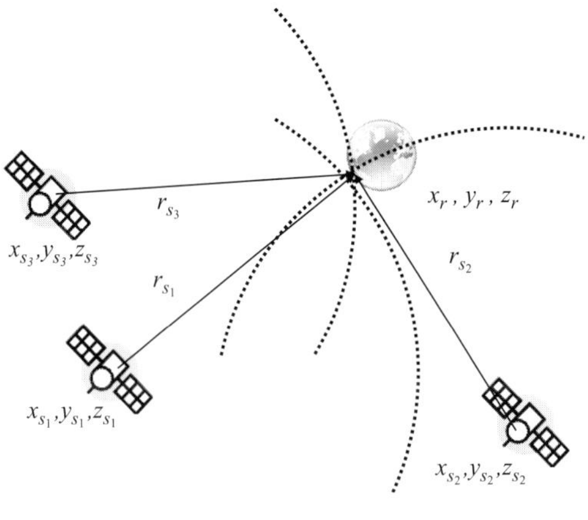

Figure 1.2 Positioning with satellites.

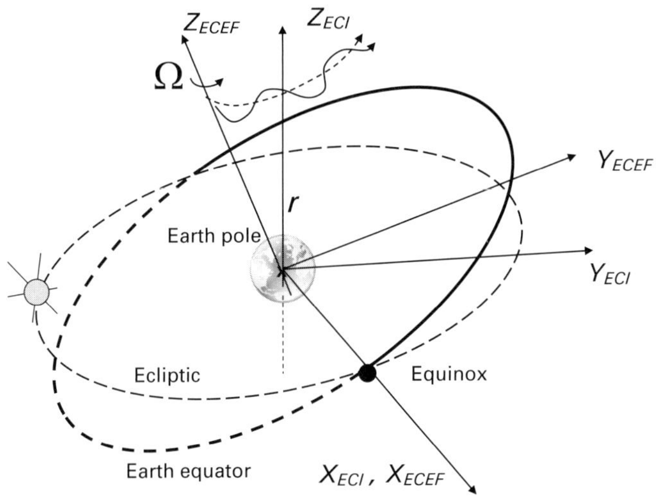

Figure 1.3 ECI and ECEF coordinate frames.

The ECI coordinate frame is a Cartesian coordinate frame with an origin placed at the Earth's center of mass and axes fixed relative to distant stars. The *z*-axis coincides with Earth's spin axis and the *x*-axis is defined by the direction from the Earth to the Sun on the first day of spring, when the Sun crosses the Earth's equatorial plane.

This point of intersection between the Sun's trajectory (ecliptic) and the Earth's equatorial plane is called the vernal equinox, or sometimes the First Point of Aries, as it was named thousands of years ago when the vernal equinox was in the zodiacal constellation of the Ram (Aries). The Aries zodiacal symbol ♈ is still used to mark the vernal equinox. The vernal equinox, defined as the intersection line of the equatorial and the ecliptic planes, is not fixed in space due to precession and nutation. This vernal equinox precession has a period of 26 000 years, which results in an ECI precession rate of 0.014° per year. This drift makes it necessary to reference coordinate frames to a certain date. The vernal equinox coordinates are then adjusted to the epoch of interest through precession and nutation transformations, which are given as sequences of rotations.

In order to make this precise transformation between the ECI and the ECEF, one can apply Earth orientation parameters (EOP), which are freely available from the International GNSS Service (IGS). A navigation message for an L2C GPS signal also contains EOP. The coordinate transformation of a satellite position from ECEF to ECI is described by a series of rotations as follows:

$$\overline{X}_{ECEF} = [R_{PM}][R_{ER}][R_N][R_P]\overline{X}_{ECI}, \tag{1.1}$$

where $[R_{PM}]$, $[R_{ER}]$, $[R_N]$, $[R_P]$ are the rotation matrices for polar motion, Earth rotation, nutation, and precession, respectively. For most mobile applications, all these movements, with the exception of the Earth daily rotation, can be safely neglected. However, when we consider tasks that require higher accuracy, including orbit determination by a ground network, it becomes necessary to account for all movements, including polar motion, nutation, and precession.

The ECEF frame is a Cartesian coordinate frame with its origin placed at the Earth's center of mass. The z-axis coincides with the Earth's spin axis, and the x-axis goes through the Greenwich meridian. The ECEF frame rotates with the Earth.

The satellite orbit parameters are defined in a tracking network, which is effectively an ECEF frame. However, as described in Section 1.2, the mathematical presentation of an orbit in the inertial frame is much simpler (see Figure 1.4). However, in order to be useful for a user located on the Earth's surface, satellite coordinates in the inertial frame must be transformed back to the ECEF frame.

A satellite navigational message embedded in the transmitted signal provides a user with orbital parameters. These parameters allow the user to calculate the satellite's position in the ECEF frame. Keplerian parameters given in a navigation message for almanac and ephemeris are not strictly speaking Keplerian parameters for the satellite orbit. These parameters are defined in the ECEF frame by control segments by fitting these parameters to a set of measurements from a control segment reference station network. The orbital parameters are calculated not in the inertial frame, which is fixed relative to stars, but in a special "non-rotating" with the Earth ECEF frame. This frame is different from the ECI frame due to small rotations caused by polar motion, nutation, and precession.

As a result of using this modified inertial frame, we can transfer satellite coordinates to the ECEF frame simply by multiplying them by a matrix that describes only the Earth's rotation. This provides us with satellite coordinates accurate enough for mobile applications.

For mobile applications, the relative motion of ECEF in relation to ECI can be confined to a rotation around the z-axis with the Earth's angular velocity, i.e.

$$\overline{X}_{ECEF} = \begin{bmatrix} \cos(\omega_E \cdot t) & \sin(\omega_E \cdot t) & 0 \\ -\sin(\omega_E \cdot t) & \cos(\omega_E \cdot t) & 0 \\ 0 & 0 & 1 \end{bmatrix} \times \overline{X}_{ECI}, \tag{1.2}$$

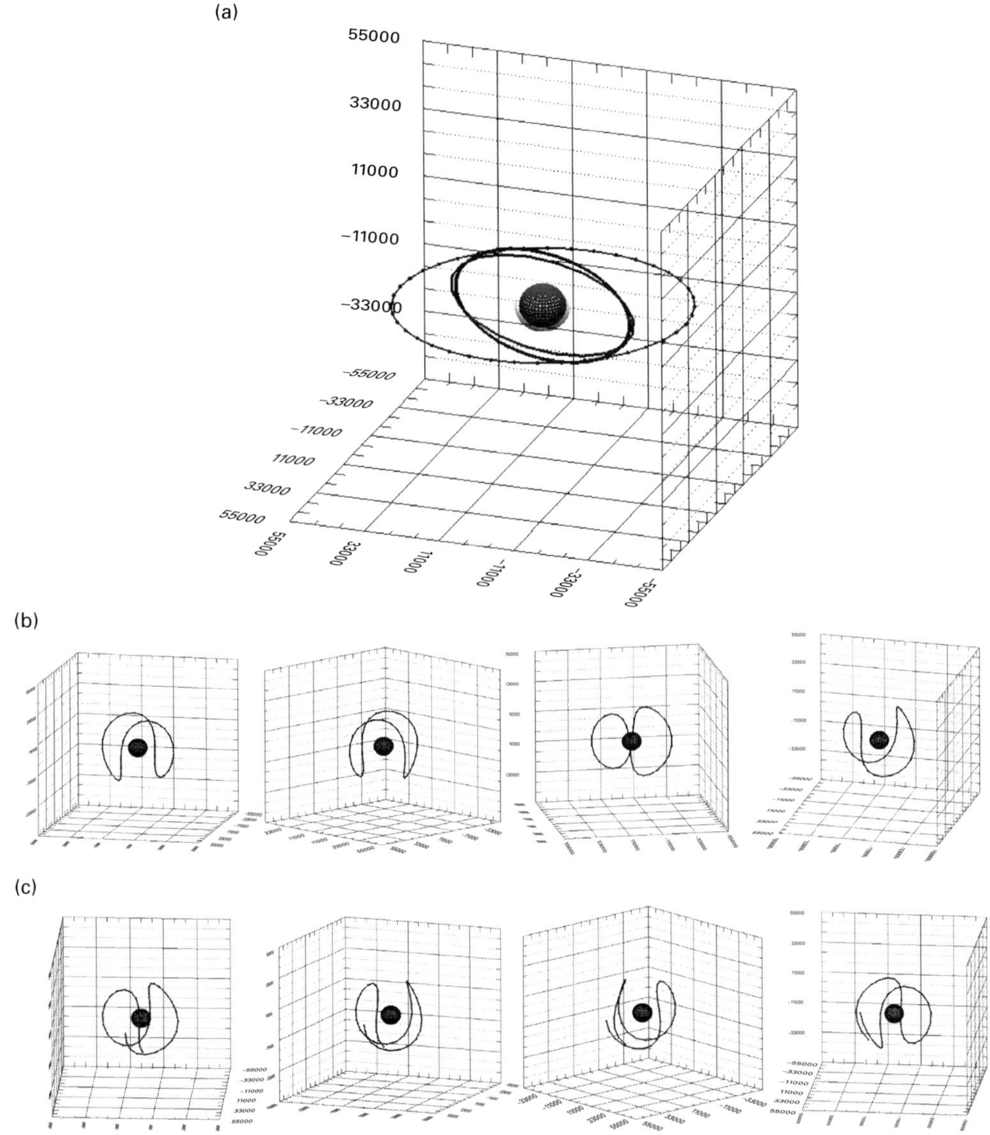

Figure 1.4 (a) GPS, GLONASS, and GEO orbits in ECI; (b) GPS orbit in ECEF; (c) GLONASS orbit in ECEF.

where ω_E is the angular velocity of the Earth's rotation.

In matrix form,

$$\overline{X}_{ECEF} = [R_{\omega_E}] \times \overline{X}_{ECI}, \tag{1.3}$$

where the rotation matrix $[R_{\omega_E}]$ satisfies the following conditions:

$$[R_{\omega_E}(\omega_E)]^{-1} = [R_{\omega_E}\omega_E]^T = [R_{\omega_E}(-\omega_E)]. \tag{1.4}$$

Correspondingly,

$$\overline{X}_{ECI} = \begin{bmatrix} \cos(\omega_E \cdot t) & -\sin(\omega_E \cdot t) & 0 \\ \sin(\omega_E \cdot t) & \cos(\omega_E \cdot t) & 0 \\ 0 & 0 & 1 \end{bmatrix} \times \overline{X}_{ECEF}. \tag{1.5}$$

In order to specify an inertial frame, we need to specify a reference epoch for both the equator and the equinox. The J2000.0 reference system is one of the most common, used, for example, by the Center for Orbit Determination in Europe (CODE) Analysis Center. More than one realization of the ECI frame exists, for example the International Celestial Reference Frame (ICRF), which is determined using a catalog of extragalactic stars based on Very Long Baseline Interference (VLBI) observations. The International Terrestrial Reference Frame (ITRF) is an ECEF frame, which corresponds to the ICRF. Although there are various realizations of the ITRF, we are particularly interested in the IGS realization, which is based on GNSS observations.

The IGS ITRF consists of coordinates and velocities of a set of globally distributed tracking stations, which are in fact GNSS receivers, for specific epochs. One can say that the ITRF frame is fixed to a network of IGS tracking stations on the Earth's surface.

In a similar way, unless ITRF is used, a ground segment defines a particular realization of the ECEF frame for its system. The problem is that the ground segment should rely on its own tracking stations in order to ensure integrity of the system. Therefore, the ground segment tracking stations ultimately define the underlying coordinate frame for each system, because a control segment reference network, which defines these coordinates, is, in a manner of speaking, the ECEF frame itself.

GPS uses the WGS-84 coordinate frame. The main difference between the WGS-84 and the ITRF is that the WGS-84 may only be realized by users with a resolution of about one meter in geocentric position (because of the quality of the broadcast orbits and satellite clocks). The ITRF may be employed with centimeter accuracy if IGS orbits and ITRF coordinates of the IGS sites are included in the processing. The two systems are therefore consistent at about the one-meter level.

Another very important issue concerning the ground segment is that the accuracy of orbit determination depends on the network distribution. This is in effect similar to the GNSS dilution of precision (DOP) factor [1], which is considered in detail in Chapter 3. Figure 1.5 demonstrates this effect. The achievable accuracy for a regional network is much less than for a global one. The effect can be mathematically described

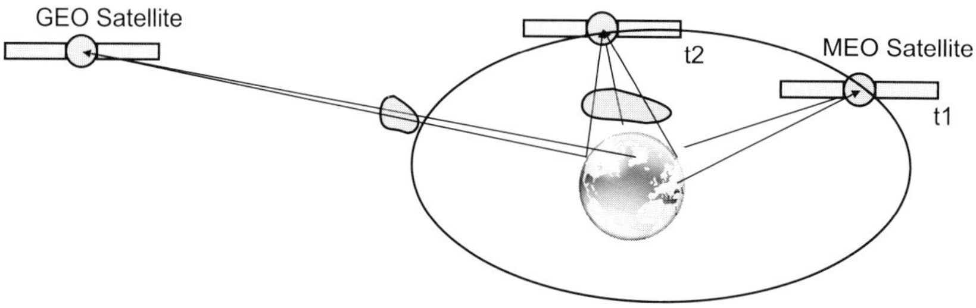

Figure 1.5 Ground segment geometry.

by reversed DOP, which is considered in detail in Chapter 3. At two moments in time, t_1 and t_2, the GPS satellite position can be measured with a much wider angle and by different stations, whereas geostationary satellites can see only the same stations because its position in relation to the Earth is fixed and the angle between the lines of sight to these stations is much narrower. In Chapter 3 we consider the satellite to station DOP, whereas here the effect is described by the station to satellite DOP. The satellite orbit stability and ground network geometry define not only the accuracy of ephemeris determination, but also how often satellite maneuvers are required. For example, for GPS maneuvers are required once a year, whereas geostationary satellites may require them once a week.

The DOP factor, however, is not the only effect that comes from the station network distribution. A globally distributed network allows for continuous satellite tracking over the whole orbit, whereas a regional network can track only part of the orbit. GNSS orbits operate at a distance of about 20 000 km, whereas the Earth radius is about 6000 km. Network density is a factor which affects accuracy significantly. To compensate for the regional character of the network, one can add additional constraints, which may include time synchronization [2] or inter-satellite measurements. Such constraints significantly improve accuracy and the time required for data collection.

The ground networks therefore define slightly different frames, unless they are using ITRF and correspondingly the same satellite tracking stations. Despite this, for mobile applications the difference is negligible. Even for much more demanding geodetic applications, various GNSS coordinate frames can be easily unified. The time frame, however, may differ significantly for various GNSS. We will look at these time scales in Section 1.2.

The GLONASS ground network has a regional character in comparison to GPS, which affects its achievable accuracy. Figure 1.6 shows a schematic distribution of GPS and GLONASS ground segment networks. In 2013, Russia had 19 ground network tracking stations, and in addition there are correction and monitoring stations. Three of these stations are located in Antarctica; others are in Russia, Belarus, and Ukraine. The Russian Space Agency has estimated that at least 40 tracking stations would be required to increase GLONASS's accuracy tenfold. There are negotiations between Russia, the USA, China, and some other countries to extend the monitoring network. However, not all of these stations can be considered as part of the control segment because of the integrity issue.

The first correction and monitoring station outside of the former Soviet Union territory was established in February 2013 in Brazil. The purpose of this station is to provide differential corrections for GLONASS and signal monitoring [3]. This station should significantly improve GLONASS's accuracy, especially in those parts of the world that previously had no tracking stations.

The IGS ground network has a global distribution and very high density. It provides the highest available accuracy of orbit estimation. However, it cannot guarantee real-time integrity. In post-processing mode, if invalid measurements occur or the quality of the measurements is not sufficient, for example due to problems with antennas, the environment, or interference, then such measurements can be caught and removed.

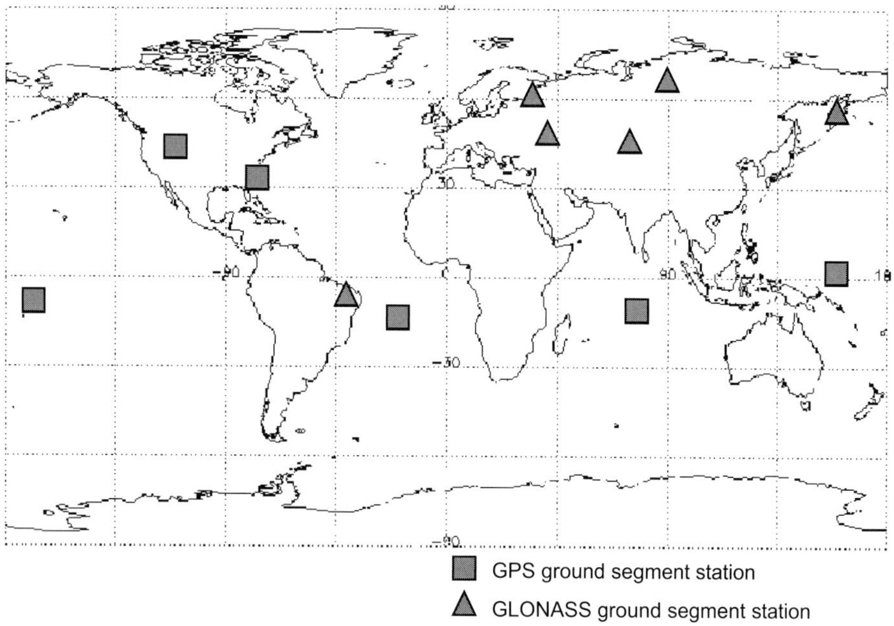

Figure 1.6 Ground segment networks for GPS (after [1]) and GLONASS (after [3]).

1.2 Space segment and time references

1.2.1 GPS time and calendar time

Satellite system time frames are connected to specific world or country time frames. In relation to time scales, we should mention Coordinated Universal Time (UTC). UTC is the main internationally accepted time standard, which has specific realizations for each GNSS within the country that owns that specific GNSS. For example, GPS time is referenced to UTC as it is maintained by the U.S. Naval Observatory (UTC (USNO)). Galileo is referenced to UTC as it is provided by the Galileo Time Service Provider based on information from European UTC laboratories.

All GNSS time scales differ from their corresponding UTC. For example, GPS time is continuous, whereas UTC is periodically adjusted to compensate for leap seconds. Also, the GNSS time scale may drift in relation to UTC. Control segments maintain their scales so the difference to the respective UTC realization is kept within certain limits. For GPS it is within one microsecond.

In terms of mobile applications, we can consider Greenwich Mean Time (GMT) instead of UTC. For practical use we express GMT via the Gregorian calendar. For mobile applications, we may for now reduce all considerations regarding various time frames to the following: connecting all GNSS time frames to GPS time and then converting GPS time to calendar time. It is important to remember that we also need to include leap second correction in the conversion of GPS time to calendar time.

GPS time is measured in GPS weeks, which are counted from January 6, 1980. One GPS week comprises 604 800 seconds, equivalent to the usual calendar week. The GPS time cycle is 1024 weeks (19.6249 years), which was originally defined by the number of allocated bits in a broadcast navigation message. The current cycle started at midnight on Saturday August 21, 1999. This corresponds to the Julian date JD = 2451412.5, the time scale used in astronomy.

The Gregorian calendar date is usually found from the GPS week and seconds through the Julian date, where the start of GPS time in Julian date is given. Examples of algorithms used to convert the Julian date to the Gregorian date are given in [4] and [5].

1.2.2 Other GNSS time scales

The GNSS time scale can be defined in various ways. It can be maintained by the master clock, as for the case of GLONASS (in its Central Synchronizer facilities); on the contrary, it can be defined as an average over the ensemble of clocks, like in the case of GPS. An ensemble also includes satellite clocks. Galileo also has the option to include both ground clocks and satellite clocks in the ensemble of clocks that maintains Galileo time, which is steered to UTC within 50 ns for 95% of the time, modulo 1 s.

Galileo and BeiDou calculate system time in a similar manner to GPS, counting it in weeks and seconds of week. GLONASS time is different and therefore we consider GLONASS time in more detail here.

In contrast to GPS, GLONASS system time is not continuous and is corrected along with UTC. GLONASS time in relation to its UTC realization can be expressed as follows:

$$\Delta_{GLONASS/UTC} = \Delta_{UTC(RUS)/UTC} + \delta_{GLONASS/UTC(RUS)} \tag{1.6}$$

and

$$\delta_{GLONASS/UTC(RUS)} = \tau_C + \tau_n(t_b) - \gamma_n(t_b)(t - t_b), \tag{1.7}$$

where τ_C is the GLONASS time scale correction to UTC(RUS), given for a calendar day number within a four-year period beginning with the leap year, t_b is the reference epoch within a day in Moscow time (UTC(RUS) + 3 hr), $\gamma_n(t_b)$ is the frequency offset for satellite n at epoch t_b, and $\tau_n(t_b)$ is the correction for the satellite clock,

$$\tau_n(t_b) = t_c(t_b) - t_n(t_b), \tag{1.8}$$

where $t_c(t_b)$ is GLONASS time at time t_b, and $t_n(t_b)$ is the satellite clock reading at time t_b.

The navigation message contains parameters that are required to transfer GLONASS time to UTC and GPS time. We consider these parameters (where they are in the navigation message and how to use them) in Chapter 2.

1.2.3 Onboard clock error

In order to make a positioning, the satellites and user receiver should be in the same time frame. Normally, user clocks are not synchronized precisely enough with the satellite time scale for any positioning task. Satellites are equipped with very stable and very expensive atomic clocks. Even those clocks would not be stable enough on their own, and the ground segment constantly measures their drift and keeps them synchronized with the system time scale within a specified allowance. The time scales of individual satellites in the constellation are synchronized not only physically, but also analytically. The physical clocks on board satellites are normally not corrected within some margin. Instead, a satellite broadcasts corrections, which must be applied in order to compensate for the shift between system time and satellite time. These corrections include drift of the satellite clock and its derivatives, predicted for some period of time (usually 12 or 24 hours). These parameters are then broadcast to the user within the satellite navigation message. This is still not good enough for precision applications or orbit determination. In those cases, the satellite clock errors are estimated along with unknown receiver antenna coordinates and receiver clock error.

Clock errors are impossible to distinguish from orbit error, and therefore they directly affect the positioning accuracy. Ephemeris information in a broadcast navigation message includes data from the satellite clock. Using the broadcast clock drift parameters, the clock error is calculated as follows:

$$\Delta t_{S_i} = (k_{2_i} dt + k_{1_i}) dt + k_{0_i}, \tag{1.9}$$

where $k_{1_i}, k_{2_i}, k_{3_i}$ are clock corrections transmitted in a navigation message for the ith satellite, and dt is the difference between the transmission epoch and the reference time for which corrections were calculated.

Satellite clocks are affected by very fine effects. For example, the theory of relativity should be taken into account; the relativistic effect on a satellite clock due to the eccentricity of the satellite orbit can be calculated using satellite position and velocity [5]:

$$\delta_{rel} = \frac{2}{c^2} r_s(T_s) \dot{r}_s(T_s). \tag{1.10}$$

For offline assisted GNSS (AGNSS) application, it is necessary to be able to predict satellite clock error. Satellites with less predictable clocks may significantly degrade the quality of position solution. One may want to use predicted orbits and clocks for various real-time applications, in particular when a broadcast navigation message is unavailable. For navigational purposes, clock values can be interpolated from predicted ephemerides. However, clocks are much more difficult to predict. Satellite orbits are very predictable, especially for GPS, for which even the most unpredictable factor, solar radiation pressure, is still well modeled. Clocks have less stable parameters, for example clock drift can change significantly over a short period. In such cases, a special modification of an integrity-monitoring algorithm in a receiver may be able to exclude the clock with the maximum drift.

Clock extrapolation can be used to predict clock behavior. A clock can be modeled, for example, as the sum of a polynomial, an exponentially correlated frequency noise, and random walk. Then the parameters of this model can be extrapolated. However, if we want to use predicted clocks for an extended period of time, we may encounter some difficulties. This is the point at which the specifics of onboard clocks, and to be more exact the predictability of their drift, come to play a very significant role in mobile applications. In this case, it is important how these clocks are designed. In addition, designing onboard clocks as an ensemble of less precise clocks is more advantageous from the mobile applications point of view. If one clock in the ensemble has a larger drift, the effect will be lessened by the other clocks, whereas, even with redundancy, an increase in drift in more precise clocks has a more severe impact.

1.3 Satellite motion description using Keplerian parameters

This section introduces orbital parameters, which are used by GNSS to describe satellite motion.

All GNSS satellites are moving around the Earth in elliptical orbits due to the Earth's gravitational force. Satellites are also experiencing effects from other forces. For the purposes of describing satellite movement, one can consider a satellite as a point mass moving around the Earth. A satellite's orbital motion can be derived by integration of non-linear differential equations:

$$\vec{a}(\vec{r},\vec{v},t) = \ddot{\vec{r}}(\vec{r},\dot{\vec{r}},t) = \vec{a}_g(\vec{r}) + \vec{a}_d(\vec{r},\vec{v},t) + \vec{a}_B(\vec{r},t) + \vec{a}_{SRP}(\vec{r},t), \qquad (1.11)$$

where $\vec{a}_g(\vec{r})$ is the acceleration caused by the Earth's gravitational field, $\vec{a}_d(\vec{r},\vec{v},t)$ is the acceleration caused by atmospheric drag, $\vec{a}_B(\vec{r},t)$ is the acceleration caused by the Sun, Moon, and planets, and $\vec{a}_{SRP}(\vec{r},t)$ is the acceleration caused by the solar radiation pressure.

The value of these forces depends on the satellite orbit. For more details on how satellite motion is derived from the laws of Newton, see [6]. The approximate values of these forces are given in Table 1.2 (after [7]–[9]) for a typical GNSS satellite on a medium Earth orbit (MEO), which is a typical type of orbit for a GNSS satellite.

If we simplify a satellite's movement by considering only the Earth's central gravitational force, we can describe the satellite's trajectory using only geometrical considerations and Kepler's laws. (For detailed derivations, see [6].) In this case, the GNSS satellite orbit would have an ideal elliptical shape in ECI (see Figure 1.4). The orbit shape in ECEF is more complex; it gives us the orbit view that we see from the surface of the Earth.

Table 1.2. Approximate values of forces affecting a GNSS satellite on MEO[a]

Term	Perturbation	Magnitude (km/s^2)
$\vec{a}_g^{(0)}(\vec{r})$	Central gravitational field	10^{-3}
$\vec{a}_g^{(2)}(\vec{r})$	Earth oblateness	10^{-7}
$\vec{a}_B(\vec{r},t)$	Sun, Moon, planets	10^{-8}
$\vec{a}_g^{(i)}(\vec{r}), i \geq 3$	Higher harmonics of the Earth's geopotential	10^{-9}
$\vec{a}_{SRP}(\vec{r},t)$	Solar radiation pressure	10^{-10}
$\vec{a}_d(\vec{r},\vec{v},t)$	Atmospheric drag	~0

[a] After [7]–[9].

For geodetic applications, where millimeter-level accuracy is required, the complete equations (1.11) should be considered. For most mobile applications, a model based on Kepler's laws is accurate enough. GLONASS, however, presents an exception by implementing a dynamical model that includes consideration of Earth's oblateness and gravitational forces due to the Sun and the Moon. Consideration of these forces allows GLONASS to compensate for regional distributions of the ground network stations (Figure 1.6).

Kepler's laws as they are applicable to a MEO satellite [6]

Kepler's first law A satellite is moving around the Earth in an elliptical orbit, with the Earth's center of mass collocated with one of the foci.

Kepler's second law A line joining the centers of mass of a satellite and the Earth sweeps out equal areas in equal intervals of time.

Kepler's third law The sum of the masses of the Earth and a satellite multiplied by their period of mutual revolution is proportional to the cube of the mean distance between them:

$$(m + M_E)T_o^2 = \frac{4\pi^2 a^3}{G},$$

where m is the mass of the satellite, M_E is the mass of the Earth, T_o is the period of mutual revolution, G is Newton's gravitational constant, and a is the mean distance between the Earth and the satellite.

In this section we consider a satellite to be a point mass in a central gravitational field. A satellite can in general have an orbit of one of three types: elliptic, parabolic, or hyperbolic. All GNSS satellites are on elliptical orbits. An elliptical satellite orbit always has the Earth in one of its foci.

A satellite position at any point of time or *epoch* can be derived from six Keplerian parameters that describe the satellite's orbit and location on that orbit (Figure 1.7). Three Keplerian parameters, the inclination, the longitude of the ascending node, and the argument of the perigee, describe the orbit position in relation to the Earth. Two Keplerian parameters, the semi-major axis and the eccentricity, describe the orbit size and shape. The sixth parameter, the time of perigee passage, describes the satellite position in the orbit at the specific (reference) epoch. Knowing these parameters allows one to calculate a satellite's position at any given epoch by using a time difference between the given epoch and the reference epoch. Table 1.3 shows the set of six Keplerian parameters, their notations, and descriptions.

The six Keplerian parameters constitute satellite almanac data, which are transmitted for each satellite as part of a navigation message. These Keplerian parameters describe an ideal orbit. Due to various forces (Table 1.2) and the irregular shape of the Earth's gravitational field, a satellite's orbit cannot be precisely described by an ellipse. Therefore the Keplerian parameters are accurate enough only to describe a satellite's position for a relatively short period. Over longer time scales, they can be used to find an approximate satellite position to assist a receiver in satellite search and acquisition. Knowing the approximate satellite position, even with errors on the order of kilometers,

Table 1.3. Keplerian parameters

Parameter	Defines	Notation	Definition
Semi-major axis	orbit size	a	half the sum of the perigee[a] and apogee[b] distances
Eccentricity	shape of ellipse in terms of its circularity	e	
Inclination	orbit orientation in relation to the Earth	i	vertical tilt of orbit with respect to equatorial plane
Longitude of the ascending node[c]		Ω	angle in equatorial plane between the directions to the vernal equinox[d] and to the ascending node[e,f]
Argument of perigee		ω	angle in orbital plane between the ascending node and the perigee in the direction of the motion of the satellite
Time of perigee passage	satellite position in orbit	t_p	an epoch when the satellite is at the perigee point

[a] The perigee is a satellite's closest point in its orbit to the Earth.
[b] The apogee is a satellite's furthest point in its orbit to the Earth.
[c] Or right ascension of the ascending node.
[d] The vernal equinox is the point of intersection between the Sun's trajectory (ecliptic) and the Earth's equatorial plane.
[e] The ascending node is the intersection of the satellite's orbit with the equatorial plane in a northerly direction.
[f] The longitude of the ascending node is measured counterclockwise when viewed from the north side of the equatorial plane.

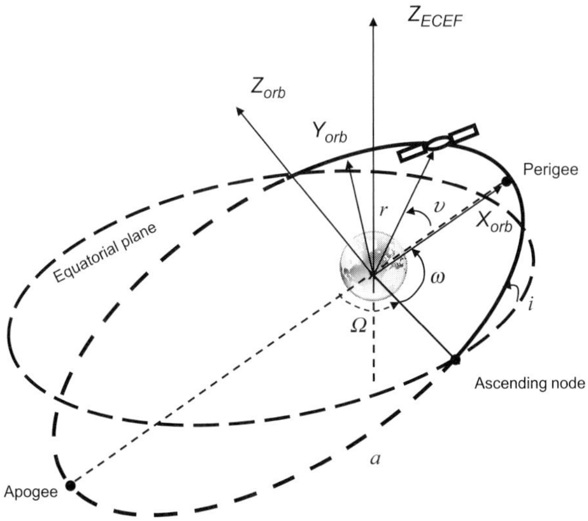

Figure 1.7 Keplerian parameters for elliptical orbit.

the receiver can estimate which satellite to search for and the approximate frequency of the search, i.e. the Doppler frequency shift, which is defined by satellite velocity in relation to the receiver, How to use the approximate satellite position information is described in detail in Chapter 8.

```
******** Week 713 almanac for PRN-01 ********
ID:                              01
Health:                          000
Eccentricity:                    0.1785278320E-002
Time of Applicability(s):        405504.0000
Orbital Inclination(rad):        0.9600886146
Rate of Right Ascen(r/s):        -0.7977475151E-008
SQRT(A)  (m 1/2):                5153.641113
Right Ascen at Week(rad):        -0.8392485290E+000
Argument of Perigee(rad):        0.281532662
Mean Anom(rad):                  0.1941663521E+001
Af0(s):                          0.2002716064E-004
Af1(s/s):                        0.3637978807E-011
week:                            713
```

Figure 1.8 An example of Keplerian parameters for a satellite; an extract from a YUMA file.

An almanac can also be used for mission planning, and for this purpose can be valid for months. Mission planning was an important consideration for geodetic tasks years ago when a number of satellites and geometry was not always satisfactory for precise positioning.

YUMA standard for Keplerian parameters

The GPS almanac given in YUMA format is provided by the U.S. Coast Guard (http://www.navcen.uscg.gov). The format gives orbital parameters in standard Keplerian format (see an example of a file in YUMA format in Figure 1.8).

ID: __ // PRN of the SVN

Health:__ // 000 = usable

Eccentricity: __ // This shows the amount of the orbit deviation from circular (orbit). It is the distance between the foci divided by the length of the semi-major axis (our orbits are very circular).

Time of Applicability: __ // The number of seconds in the orbit during which the almanac was generated. A type of time tag.

Orbital Inclination: __ // The angle at which the SV orbit meets the equator (GPS is at approx. 55°). Roughly, the SV's orbit will not rise above approx. 55 degrees latitude. The number is part of an equation: $\# = \pi/180 =$ true inclination.

Rate of Right Ascension: __ // Rate of change in the measurement of the angle of right ascension as defined in the Right Ascension mnemonic.

SQRT(A) Square Root of Semi-Major Axis: __ // This is defined as the measurement from the center of the orbit to either the point of apogee or the point of perigee.

Right Ascension at Time of Almanac (TOA): __ // Geographic longitude of the ascending node of the orbit plane at the weekly epoch.

Argument of Perigee: __ // An angular measurement along the orbital path measured from the ascending node to the point of perigee, measured in the direction of the SV's motion.

Mean Anomaly: __ // Angle (arc) traveled past the longitude of ascending node (the value can change from 0 to either 180° or −180°). If the value exceeds 180°, subtract 360° to find the mean anomaly. When the SV has passed perigee and is

Continued

Continued

heading towards apogee, the mean anomaly is positive. After the point of apogee, the mean anomaly value will be negative to the point of perigee.

Af(0): __ // SV clock bias in seconds.

Af(1): __ // SV clock drift in seconds per second.

week: __ // GPS week (0000–1024), every seven days since August 22, 1999.

1.4 Algorithm for satellite position calculation using standard Keplerian parameters

By knowing the Keplerian parameters, one can calculate a satellite's position at *any epoch*. Such parameters, for example, are transmitted as almanac data from the GPS, Galileo, and BeiDou satellites. The algorithm is basically the same for these systems [10]–[12]. Keplerian parameters are given in a different form for GLONASS. In this sense, we define the forms of the Keplerian parameters as they are specified in Table 1.3 as standard. The satellite position calculation algorithm for GLONASS is also significantly different, and is presented in the following sections.

From Kepler's third law (see the boxed text on p. 14), the period of Keplerian orbit can be found from the following equation:

$$T_o = 2\pi\sqrt{\frac{a^3}{\mu}}, \tag{1.12}$$

where a is the mean distance between the Earth's and the satellite's centers of mass, and μ is the Earth's gravitational constant. The Earth's gravitational constant is defined as follows:

$$\mu \triangleq GM_E, \tag{1.13}$$

where M_E is the mass of the Earth and G is Newton's gravitational constant.

For GPS the Earth's gravitational constant is defined in the GPS Interface Control Document (ICD) [10] as

$$\mu = 3.986005 \cdot 10^{14}\ [\mathrm{m}^3/\mathrm{s}^2]. \tag{1.14}$$

For GLONASS the Earth's gravitational constant is defined in the GLONASS ICD [13] as

$$\mu = 398600.4418 \cdot 10^9 = 3.986004.418 \cdot 10^{14}\ [\mathrm{m}^3/\mathrm{s}^2]. \tag{1.15}$$

A satellite movement along an idealized circular orbit can be characterized by a constant angular velocity value called the mean motion, given by

$$n = \sqrt{\frac{\mu}{a^3}}. \tag{1.16}$$

The angle showing the position of the satellite on the orbit is called the mean anomaly. At any given epoch t, the mean anomaly can be found by using the mean motion and time elapsed from a reference epoch, in this case an epoch of perigee passage:

$$M = n(t - t_p). \tag{1.17}$$

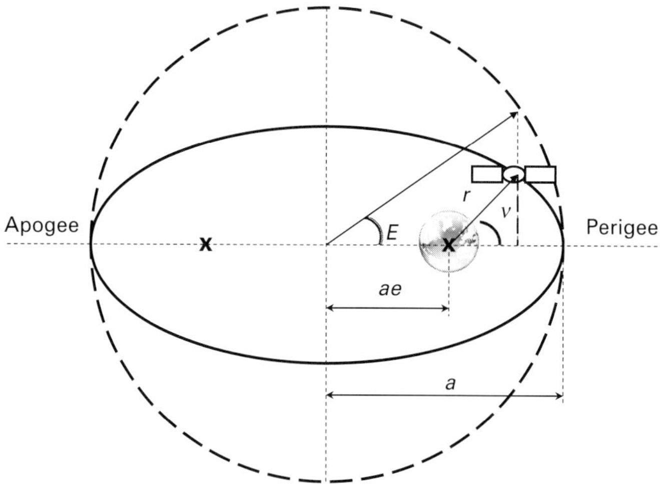

Figure 1.9 True anomaly and eccentric anomaly.

The mean anomaly is an abstraction, which changes linearly with time, because it was derived for a circular orbit with zero eccentricity. Earth satellites are moving in elliptical orbits and their movement along the orbit is not uniform. To define an angle that describes the satellite position in an elliptical orbit, we introduce the eccentric anomaly E (see Figure 1.9). The eccentric anomaly corresponds to the mean anomaly position of the abstract satellite, which moves on a corresponding circular orbit. The movement of this abstract satellite is non-uniform (unlike the uniform movement of a satellite in an elliptical orbit). In accordance with Kepler's second law, the satellite velocity reaches a maximum at perigee and minimum at apogee. The eccentric anomaly is defined by considering the mean anomaly at a given epoch using the Kepler equation as follows:

$$M = E - e \cdot \sin E, \tag{1.18}$$

where e is the orbit eccentricity.

The solution of this equation yields the position of the satellite. The equation is non-linear and can be solved either analytically by expansion to a Fourier series or Bessel function series, or numerically. Newton's method of iterations uses the following initial approximation for the eccentric anomaly:

$$E_0 = M + \frac{e \cdot \sin(M)}{1 - \sin(M + e) + \sin(M)}, \tag{1.19}$$

with sequential iterations defined as

$$E_{n+1} = E_n - \frac{E_n - e \cdot \sin(E_n) - M}{1 - e \cdot \cos(E_n)}, \tag{1.20}$$

where $n = 1,2,\ldots$. Newton's method of iterations usually converges within a few iterations.

As the final step, the real satellite position on the elliptical orbit is defined by an angle called the true anomaly. It can be derived from the eccentric anomaly as follows:

$$\upsilon = \cos^{-1}[(e - \cos E)/(e \cos E - 1)], \tag{1.21}$$

where υ is defined in the same quadrant with E.

In order to find the satellite coordinates, we introduce an intermediate coordinate frame, called the Earth-centered orbital (ECO) coordinate frame. The axes X_{orb} and Y_{orb} are located in the orbital plane, where the axis X_{orb} points to orbit's perigee and axes Y_{orb} and Z_{orb} are as depicted in Figure 1.7. Satellite coordinates in the orbital frame can be calculated as follows:

$$\overline{X}_{orb} = \begin{bmatrix} x_{orb} \\ y_{orb} \\ z_{orb} \end{bmatrix} = \begin{bmatrix} r \cos(\upsilon) \\ r \sin(\upsilon) \\ 0 \end{bmatrix}, \tag{1.22}$$

where r is the distance to the satellite, and it can be expressed from the ellipse geometry as

$$r = a \cdot (1 - e \cos E). \tag{1.23}$$

Then the following transformation to the ECI frame can be described by sequential matrix multiplication with rotation matrices:

$$\overline{X}_{ECI} = [R_3(-\Omega)][R_1(-i)][R_3(-\omega)]\overline{X}_{orb}. \tag{1.24}$$

A rotation matrix $[R(\alpha)]$ defines a rotation of the orthogonal frame in relation to one of its axes by an angle α. The rotation matrices are generally defined as follows. Note that two rotations are around the same axis, and rotation matrix $[R_2(\alpha)]$ is not used in this case:

$$[R_1(\alpha)] = \begin{bmatrix} 1 & 0 & 0 \\ 0 & \cos(\alpha) & \sin(\alpha) \\ 0 & -\sin(\alpha) & \cos(\alpha) \end{bmatrix}, \tag{1.25}$$

$$[R_2(\alpha)] = \begin{bmatrix} \cos(\alpha) & 0 & -\sin(\alpha) \\ 0 & 1 & 0 \\ \sin(\alpha) & 0 & \cos(\alpha) \end{bmatrix}, \tag{1.26}$$

$$[R_3(\alpha)] = \begin{bmatrix} \cos(\alpha) & \sin(\alpha) & 0 \\ -\sin(\alpha) & \cos(\alpha) & 0 \\ 0 & 0 & 1 \end{bmatrix}. \tag{1.27}$$

Multiplication of a vector by a rotation matrix does not change the vector magnitude. Correspondingly, the satellite coordinates in the ECI frame can be calculated as follows:

$$\overline{X}_{ECI} = \begin{bmatrix} x_{ECI} \\ y_{ECI} \\ z_{ECI} \end{bmatrix} = \begin{bmatrix} r(\cos(\upsilon + \omega) \cdot \cos\Omega - \sin(\upsilon + \omega) \cdot \sin\Omega \cdot \cos i) \\ r(\cos(\upsilon + \omega) \cdot \sin\Omega + \sin(\upsilon + \omega) \cdot \cos\Omega \cdot \cos i) \\ r \cdot \sin(\upsilon + \omega) \cdot \sin i \end{bmatrix} \tag{1.28}$$

The argument of the latitude, i.e. the angle between the ascending node and the position of the satellite at the current epoch, can be calculated as a sum of the argument of the perigee and the true anomaly:

$$u = \upsilon + \omega. \tag{1.29}$$

The argument of the latitude can be used as the sixth Keplerian parameter.

The satellite velocities can be calculated as follows [10]:

$$\overline{V}_{ECI} = \begin{bmatrix} V_{x_{ECI}} \\ V_{y_{ECI}} \\ V_{z_{ECI}} \end{bmatrix} = \begin{bmatrix} V_r(\cos u \cdot \cos\Omega - \sin u \cdot \sin\Omega \cdot \cos i) - V_u(\sin u \cdot \cos\Omega + \cos u \cdot \sin\Omega \cdot \cos i) \\ V_r(\cos u \cdot \sin\Omega + \sin u \cdot \cos\Omega \cdot \cos i) - V_u(\sin u \cdot \sin\Omega + \cos u \cdot \cos\Omega \cdot \cos i) \\ V_r \sin u \cdot \sin i + V_u \cos u \cdot \sin i \end{bmatrix}, \tag{1.30}$$

where

$$V_r = \sqrt{\frac{\mu}{a}} \cdot \frac{e \cdot \sin\upsilon}{\sqrt{1-e^2}}, \tag{1.31}$$

$$V_u = \sqrt{\frac{\mu}{a}} \cdot \frac{1 + e \cdot \cos\upsilon}{\sqrt{1-e^2}}. \tag{1.32}$$

1.5 Theoretical background for the spherical harmonics of the Earth's geopotential

Algorithms for GLONASS satellite position calculations use not only the central component of the Earth's gravitational field, but also its higher harmonics. An acceleration $\vec{a}_g(\vec{r})$ in Newton's equations (1.11) can be expressed as a gradient of a potential energy of the orbit ∇U,

$$\vec{a}_g(\vec{r}) = -\nabla U, \tag{1.33}$$

where

$$\nabla U = \frac{\mu}{r^3}\vec{r} - \nabla R \tag{1.34}$$

and R is a disturbing potential. The first term in the equation is the central component of the Earth's gravitational field, which would be the only component for an ideal spherical body.

The geopotential function U can be expanded in a series of spherical harmonics as a function of coordinates [14]:

$$U(r, \lambda, \varphi) = \sum_{l=0}^{\infty} \sum_{m=0}^{l} (r/R_e)^{l+1} P_{lm} \sin\varphi (C_{lm} \cos m\lambda + S_{lm} \sin m\lambda), \tag{1.35}$$

where P_{lm} are Legendre associated functions, R_e is the Earth's radius, λ is latitude, φ is longitude, and r is a distance from the center of the Earth to the point where the geopotential is calculated. The harmonics are called *zonal* for $m = 0$, *sectorial* for $m = 1$, and *tesserial* for $m \neq 0$, $m \neq 1$. Figure 1.10 shows the zonal harmonics, which are those accounted for in GLONASS orbit calculations.

A geopotential model of the Earth is defined as a set of coefficients in the series expansion of (1.35). The zonal coefficients are denoted as follows:

$$J_l \triangleq C_{l0}. \tag{1.36}$$

The J_0 term represents the spherical distribution of a potential changed by $1/r$ law:

$$J_0 \triangleq 1. \tag{1.37}$$

The coordinate frame is chosen to go through the center of mass, therefore

$$J_1 \triangleq 0. \tag{1.38}$$

The J_2 term represents the equatorial bulge mass distribution, or oblateness. It is the most significant term, and it is accounted for in all algorithms for GLONASS satellite position calculations, as recommended by the GLONASS ICD. This term causes the right ascension of the ascending node and the argument of the perigee to rotate through several degrees per day. The terms of higher order than J_2 are small. The geopotential model may include from 100 to 1000 terms. Table 1.4 shows an example of the first four terms for a Goddard Earth Model 10b [15].

The spherical harmonics model (1.35) can also account for elastic properties of the Earth. In that case, coefficients should be expressed as functions of time.

Table 1.4. First four most significant geopotential terms

J_0	J_1	J_2	J_3	J_4
1	0	0.00108263	−0.00000254	−0.00000161

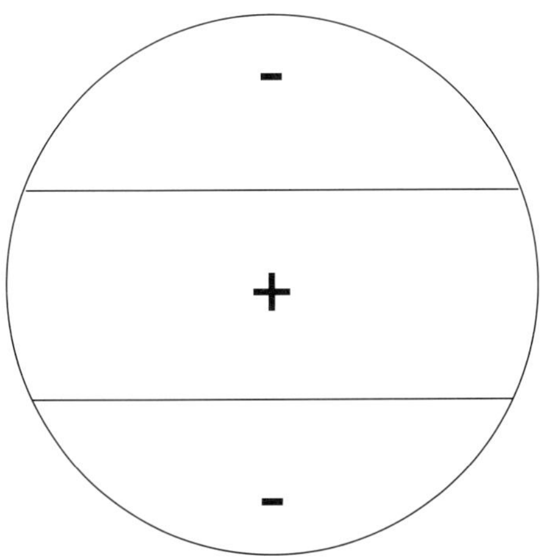

Figure 1.10 Zonal harmonics of the Earth's geopotential.

1.6 Algorithm for transformation of GLONASS almanac parameters into standard Keplerian parameters

The algorithms for satellite position calculation are similar for GPS, Galileo, and BeiDou, but different for GLONASS [13]. The main difference is that the GLONASS almanac takes into account the second zonal harmonic of the Earth's gravitational field, which accounts for the Earth's oblateness, whereas other GNSS almanacs use only the central gravitational force. This section looks at the algorithm used to calculate corrections to the Keplerian parameters from the second zonal harmonic. Almanac parameters are specified in the ITRF at reference epoch t_λ, which corresponds to a first passage of the ascending node (hence the subscript λ).

The GLONASS almanac consists of the following parameters (we omit the satellite index for clarity):

N_A = a day of the reference epoch within a four-year period calculated from the previous leap year;
λ = longitude of ascending node at reference epoch in radians;
t_λ = a reference epoch within a day, in seconds;
Δi = correction to the mean value of inclination at the reference epoch[1];
ΔT = correction to the mean value of the Draconic period[2] at the reference epoch;
$\Delta T'$ = rate of orbital period change;
e = orbit eccentricity;
ω = argument of perigee at the reference epoch;
≜ Draconic orbit period is defined as the time between subsequent nodal crossing by a satellite.

Satellite orbit coordinates are calculated in the ECI frame.

(1) Calculate the semi-major axis a using an iterative method of successive approximations:

$$p^{(n)} = a^{(n)} \cdot (1 - e^2), n = 0, 1, 2, \ldots, \tag{1.39}$$

$$T_{osc}^{(n+1)} = T_{dr} \cdot \left\{ 1 + \frac{3}{2} \cdot C_{20} \left(\frac{a_e}{p^{(n)}} \right)^2 \left[\left(2 - \frac{5}{2} \cdot \sin^2 i \right) \cdot \frac{(1 - e^2)^{3/2}}{(1 + e \cdot \cos \omega)^2} + \frac{(1 + e \cdot \cos \upsilon)}{1 - e^2} \right] \right\}^{-1}, \tag{1.40}$$

where T_{osc} is the osculating Draconic period, T_{dr} is the Draconic period, $\nu = -\omega$, $i = i_{cp} + \Delta i$, $T_{dr} = T_{cp} + \Delta T$, and

$$a^{(n+1)} = \sqrt[3]{\left(\frac{T_{osc}^{(n+1)}}{2\pi} \right)^2 \cdot \mu}. \tag{1.41}$$

[1] The mean value of inclination is 63°.
[2] The mean value of Draconic period is 43 200 s

Table 1.5. Geodetic constants

	Earth rotation rate (s^{-1})	Gravitational constant (km^3/s^2)
GPS: WGS-84	$7.2921151467.10^{-5}$	398600.50
GLONASS: PZ-90	$7.29221150.10^{-5}$	398600.4418
Galileo	$7.2921151467.10^{-5}$	398600.4418
BeiDou: CGCS2000	$7.29221150.10^{-5}$	398600.4418

An initial value for the semi-major axis is

$$a^{(0)} = \sqrt[3]{\left(\frac{T_{dr}}{2\pi}\right)^2 \cdot \mu}. \tag{1.42}$$

The algorithm iterates until the following condition is satisfied:

$$|a^{(n-1)} - a^{(n)}| < 10^{-3} \text{ [km]}. \tag{1.43}$$

Three iterations are usually enough to meet this condition.

(2) Calculate the time of the ascending node (t_{λ_k}) for the orbital period containing the reference epoch and the corresponding longitude of the ascending node (λ_k).

The values for the Earth's rotation rate and gravitational constant may be slightly different, but the difference is negligible for mobile applications (Table 1.5).

The coefficient of the second zonal harmonic of Earth's geopotential

$$C_{20} = -1082,62575 \cdot 10^{-6}; \tag{1.44}$$

Earth's equatorial radius

$$a_e = 6378.136 \quad \text{[km]}; \tag{1.45}$$

Earth rotation rate

$$\omega_E = 0.7292115 \cdot 10^{-4} \quad [\text{s}^{-1}]; \tag{1.46}$$

gravitational constant

$$\mu = 398600.4418 \quad [\text{km}^3/\text{s}^2]; \tag{1.47}$$

sidereal time for the reference epoch

$$S = S_0 + \omega_E \cdot (t_{\lambda_k} - 10800) \quad \text{[rad]}, \tag{1.48}$$

where S_0 is the true sidereal time at midnight GMT on day N_0 of reference epoch; the longitude of the ascending node is given by

$$\Omega = \lambda_k + S, \tag{1.49}$$

$$n = \frac{2\pi}{T_{dr}}, \tag{1.50}$$

$$\Omega' = \frac{3}{2} C_{20} \cdot n \cdot \left(\frac{a_e}{a}\right)^2 \cdot \cos i \cdot (1 - e^2)^{-2}, \tag{1.51}$$

longitude of the ascending node

$$\lambda_k = \lambda + (\Omega' - \omega_E) \cdot (T_{dr} \cdot W + \Delta T' \cdot W^2), \tag{1.52}$$

$$t^* = t_i - t_\lambda + 86400 \cdot (N_0 - N_A), \tag{1.53}$$

$$W = \text{int}(W_k) = \text{int}\left(\frac{t^*}{T_{dr}}\right). \tag{1.54}$$

The time of the ascending node is calculated as follows:

$$t_{\lambda_k} = t_\lambda + T_{dr} \cdot W + \Delta T' \cdot W^2, \tag{1.55}$$

and it should not exceed a 24-hour period, i.e.

$$t_{\lambda_k} = \text{mod}(t_{\lambda_k}, 86400). \tag{1.56}$$

(3) Calculate corrections to orbital parameters at epoch t_i due to Earth's geopotential zonal harmonic.

The GLONASS ICD provides a slightly different calculation for the eccentric anomaly than the GPS ICD:

$$tg\frac{E}{2} = \sqrt{\frac{1-e}{1+e}} tg\frac{\upsilon}{2}. \tag{1.57}$$

The mean anomaly is calculated in the same way as for GPS:

$$M = E - e \cdot \sin E, \tag{1.58}$$

$$\bar{\lambda} = M + \omega, \tag{1.59}$$

Denote $h = e \sin \omega$, $l = e \cos \omega$, $J = -\frac{3}{2} C_{20}$, $a = a^{(n)}$. Then

$$\begin{aligned} \delta a &= \delta a^{(2)} - \delta a^{(1)}, \\ \delta h &= \delta h^{(2)} - \delta h^{(1)}, \\ \delta l &= \delta l^{(2)} - \delta l^{(1)}, \\ \delta \Omega &= \delta \Omega^{(2)} - \delta \Omega^{(1)}, \\ \delta i &= \delta i^{(2)} - \delta i^{(1)}, \\ \delta \lambda^* &= \delta \bar{\lambda}^{(2)} - \delta \bar{\lambda}^{(1)}, \end{aligned} \tag{1.60}$$

where parameters δa, δh, δl, $\delta \Omega$, δi, $\delta \lambda^*$ are calculated as follows for epoch m, which takes two instants:

$$\frac{\delta a^{(m)}}{a} = 2J\left(\frac{a_e}{a}\right)^2 \left(1 - \frac{3}{2}\sin^2 i\right)(l \cdot \cos\bar{\lambda} - h \cdot \sin\bar{\lambda}) + J\left(\frac{a_e}{a}\right)^2 \sin^2 i \cdot$$
$$\left(\frac{1}{2}h \cdot \sin\bar{\lambda} - \frac{1}{2}l \cdot \cos\bar{\lambda} + \cos 2\bar{\lambda} + \frac{7}{2}l \cdot \cos 3\bar{\lambda} + \frac{7}{2}h \cdot \sin 3\bar{\lambda}\right),$$

$$\delta h^{(m)} = J\left(\frac{a_e}{a}\right)^2\left(1-\frac{3}{2}\sin^2 i\right)\left[l\cdot n\cdot\tau + \sin\bar{\lambda} + \frac{3}{2}l\cdot\sin 2\bar{\lambda} - \frac{3}{2}h\cdot\cos 2\bar{\lambda}\right]$$
$$-\frac{1}{4}J\left(\frac{a_e}{a}\right)^2\sin^2 i\left[\sin\bar{\lambda} - \frac{7}{3}\cdot\sin 3\bar{\lambda} + 5l\cdot\sin 2\bar{\lambda} - \frac{17}{2}l\cdot\sin 4\bar{\lambda} + \frac{17}{2}h\cdot\cos 4\bar{\lambda} + h\cdot\cos 2\bar{\lambda}\right]$$
$$+J\left(\frac{a_e}{a}\right)^2\cos^2 i\left(l\cdot n\cdot\tau - \frac{1}{2}l\cdot\sin 2\bar{\lambda}\right),$$

$$\delta l^{(m)} = J\left(\frac{a_e}{a}\right)^2\left(1-\frac{3}{2}\sin^2 i\right)\left[-h\cdot n\cdot\tau + \cos\bar{\lambda} + \frac{3}{2}l\cdot\cos 2\bar{\lambda} + \frac{3}{2}h\cdot\sin 2\bar{\lambda}\right]$$
$$-\frac{1}{4}J\left(\frac{a_e}{a}\right)^2\sin^2 i\left[-\cos\bar{\lambda} - \frac{7}{3}\cdot\cos 3\bar{\lambda} - 5h\cdot\sin 2\bar{\lambda} - \frac{17}{2}l\cdot\cos 4\bar{\lambda} - \frac{17}{2}h\cdot\sin 4\bar{\lambda} + l\cdot\cos 2\bar{\lambda}\right]$$
$$+J\left(\frac{a_e}{a}\right)^2\cos^2 i\left(-h\cdot n\cdot\tau + \frac{1}{2}h\cdot\sin 2\bar{\lambda}\right),$$

$$\delta\Omega^{(m)} = -J\left(\frac{a_e}{a}\right)^2\cos i\left(n\cdot\tau + \frac{7}{2}l\cdot\sin\bar{\lambda} - \frac{5}{2}h\cdot\cos\bar{\lambda} - \frac{1}{2}\sin 2\bar{\lambda} - \frac{7}{6}l\cdot\sin 3\bar{\lambda} + \frac{7}{6}h\cdot\cos 3\bar{\lambda}\right),$$

$$\delta i^{(m)} = \frac{1}{2}J\left(\frac{a_e}{a}\right)^2\sin i\cdot\cos i \cdot\left(-l\cdot\cos\bar{\lambda} + h\cdot\sin\bar{\lambda} + \cos 2\bar{\lambda} + \frac{7}{3}l\cdot\cos 3\bar{\lambda} + \frac{7}{3}h\cdot\sin 3\bar{\lambda}\right), \qquad (1.61)$$

$$\delta i^{(m)} = \frac{1}{2}J\left(\frac{a_e}{a}\right)^2\sin i\cdot\cos i\left(-l\cdot\cos\bar{\lambda} + h\cdot\sin\bar{\lambda} + \cos 2\bar{\lambda} + \frac{7}{3}l\cdot\cos 3\bar{\lambda} + \frac{7}{3}h\cdot\sin 3\bar{\lambda}\right),$$

$$\delta\bar{\lambda}^{(m)} = 2J\left(\frac{a_e}{a}\right)^2\left(1-\frac{3}{2}\sin^2 i\right)\left(n\cdot\tau + \frac{7}{4}l\cdot\sin\bar{\lambda} - \frac{7}{4}h\cdot\cos\bar{\lambda}\right) + 3J\left(\frac{a_e}{a}\right)^2\sin^2 i$$
$$\cdot\left(-\frac{7}{24}h\cdot\cos\bar{\lambda} - \frac{7}{24}l\cdot\sin\bar{\lambda} - \frac{49}{72}h\cdot\cos 3\bar{\lambda} + \frac{49}{72}l\cdot\sin 3\bar{\lambda} + \frac{1}{4}\sin 2\bar{\lambda}\right) + J\left(\frac{a_e}{a}\right)^2\cos^2 i$$
$$\cdot\left(n\cdot\tau + \frac{7}{2}l\cdot\sin\bar{\lambda} - \frac{5}{2}h\cdot\cos\bar{\lambda} - \frac{1}{2}\sin 2\bar{\lambda} - \frac{7}{6}l\cdot\sin 3\bar{\lambda} + \frac{7}{6}h\cdot\cos 3\bar{\lambda}\right).$$

The perturbed orbital parameters at the reference epoch are calculated as follows:

$$h_i = h + \delta h, \qquad (1.62)$$

$$l_i = l + \delta l, \qquad (1.63)$$

$$\varepsilon_i = \sqrt{h_i^2 + l_i^2}, \qquad (1.64)$$

$$\omega_i = \begin{cases} \arctan\left(\frac{h_i}{l_i}\right), & \text{if}(\varepsilon_i \neq 0 \text{ and } l_i \neq 0) \\ 0, & \text{if}(\varepsilon_i = 0) \\ \frac{\pi}{2}, & \text{if}(\varepsilon_i \neq 0 \text{ and } h_i = \varepsilon_i) \\ -\frac{\pi}{2}, & \text{if}(\varepsilon_i \neq 0 \text{ and } h_i = -\varepsilon_i) \end{cases} \tag{1.65}$$

$$a_i = a + \delta a, \tag{1.66}$$

$$i_i = i + \delta i, \tag{1.67}$$

$$\Omega_i = \Omega + \delta\Omega, \tag{1.68}$$

$$\lambda^* = M + \omega + n \cdot (t_i - t_\lambda) + \delta\lambda^* \tag{1.69}$$

$$M_i = \lambda^* - \omega_i. \tag{1.70}$$

Finally, coordinates and velocity are calculated as in Section 1.4.

1.7 Medium Earth GNSS orbits

Once we have mastered the almanac parameters, we can look at the GNSS orbits in detail. Today, GNSS satellites are all located on orbits that can be classified according to three main groups, depending on the value of orbit altitude, in terms of Keplerian parameters represented by the semi-major axis.

Medium Earth orbit (MEO) satellites have low eccentricity, and their orbits are quite close to circular orbits (see Figure 1.11). An MEO has an altitude in the range

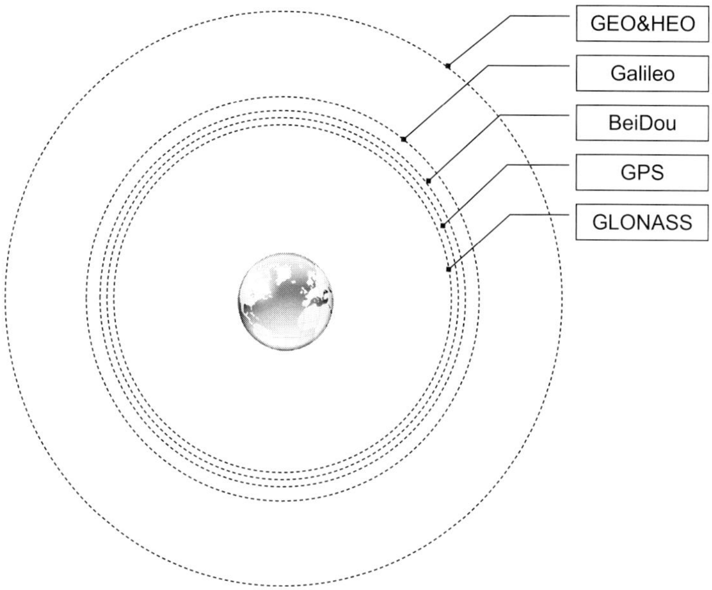

Figure 1.11 Medium Earth orbits.

Table 1.6. GNSS orbits

	GNSS								
	GPS		GLONASS		Galileo		BeiDou		
Parameter	MEO	GEO–WAAS	MEO	GEO–Luch	MEO	GEO–EGNOS	MEO	HEO	GEO
Number of satellites	32	3	24 + 3 spares	3	$27^a + 3^a$ spares	3	27^a	3	5
Position	N/A	INMARSAT 4F3 (98°W) LM RPS-1 (133°W) LM RPS-2 (107.3°W)	N/A	Luch-5A (16°W), Luch-5b (95°E), Luch-4 (167°E)	N/A	INMARSAT 3F2 (15.5°W) ARTEMIS (21.5°E) INMARSAT 4F2 (25°E)	N/A	3 × 118°E 2 × 95°E	144.5°E, 84°E, 160°E, 59°E, 80°E
Semi-major axis (km)	26 560	35 856	25 500	35 856	29 601	35 856	~27 900	~36 000	35 856
Inclination	~55°	~0°	~64.8°	~0°	~56°	~0°	~55°	~55°	~0°

[a] Planned.

19 000 (GLONASS) to 24 000 (Galileo) km. Table 1.6 lists the orbit parameters for GNSS MEO.

A revolution period for MEO satellites defined by Kepler's third law is given by

$$(m + M_E)T_o^2 = \frac{4\pi^2 a^3}{G}, \tag{1.71}$$

and is approximately 12 hours.

The period T_o is called the sidereal period to stress that it is calculated with respect to the inertial frame, and it is defined as the time interval between two epochs of a satellite passing a specific orbit point, for example the perigee. It is different from the synodic period, which is defined as the time interval between two epochs of satellite footprint passing the same meridian. The difference is due to the Earth's rotation. For GPS, a sidereal period is 12 hr, which is one-half of a sidereal day. A sidereal day is 4 min shorter than a calendar day; therefore a GPS satellite reappears at the same place on the sky after 11 hr 58 min. The GLONASS satellite orbit is lower and the sidereal period is correspondingly shorter. For GLONASS, a sidereal period is 8/17 of a sidereal day. This means that the GLONASS satellite will reappear at the same place on the sky every 8 days, after it has made 17 revolutions of the Earth. The ground tracks for GPS and GLONASS satellites are shown in Figure 1.12.

GLONASS orbits are less prone to the influence of the irregularities of the Earth's gravitational field. This allows GLONASS to maintain good accuracy, despite a regional ground station monitoring network [16],[17].

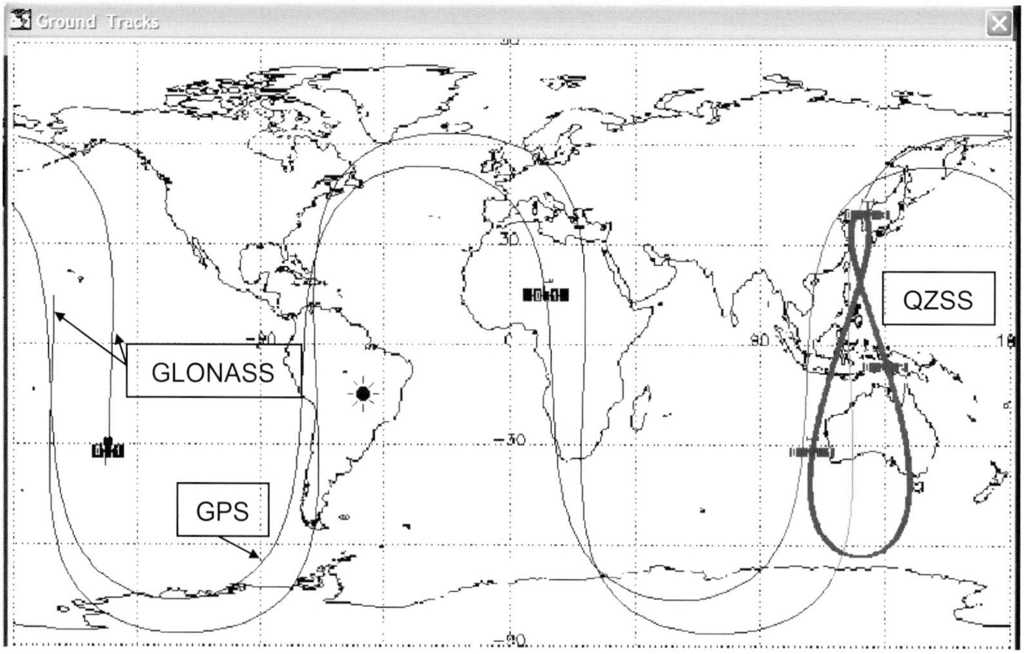

Figure 1.12 GNSS ground tracks.

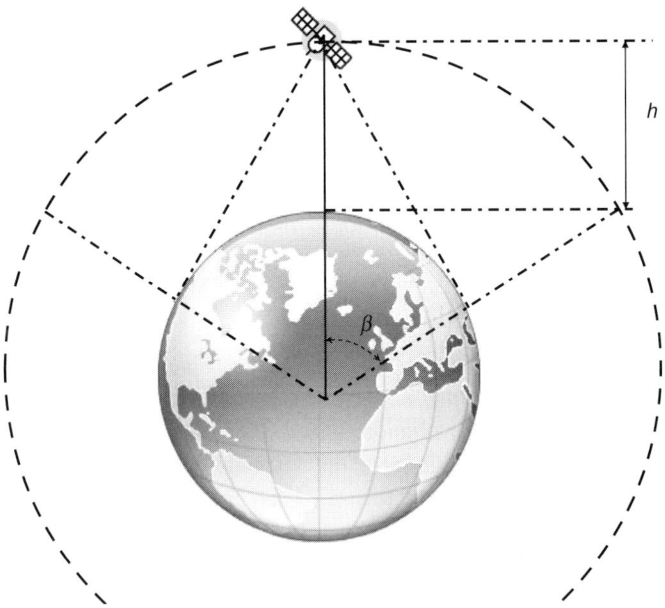

Figure 1.13 Area of satellite visibility.

The area of satellite visibility, i.e. the area on the surface of the Earth from which a satellite is visible, is proportional to a visibility angle given by (Figure 1.13)

$$\beta = \arccos\left(\frac{\cos(\alpha)}{1 + h/R_E}\right) - \alpha, \tag{1.72}$$

where α is the satellite elevation angle, h is the satellite altitude, and R_E is the radius of the Earth. In relation to the entire surface of the Earth, the area of visibility can be found to be

$$S \approx \sin^2\left(\frac{\beta^2}{2}\right). \tag{1.73}$$

This defines a visibility period, which is the period of time during which the satellite is visible to a static user.

1.8 GEO and HEO for SBAS

1.8.1 GEO

Clarke or geostationary Earth orbit (GEO) satellites have an altitude of 35 856 km. At this altitude GEO satellites are rotating synchronously with the Earth. GEO satellites also have low eccentricity, and their orbits are quite close to circular orbits (see Figure 1.4 and

Figure 1.11). GEO satellites are restricted to strictly defined slots and should undergo periodical orbit corrections to keep their position.

GNSS also widely use GEO satellites to provide regional corrections and integrity service checks as the so-called space-based augmentation systems (SBAS). Basically all navigation satellite systems on GEO can be classified as SBAS, which can be defined as a system to augment satellite navigation core systems (e.g. GPS and GLONASS) for the operation of civil aviation from en-route to approach. The requirements for SBAS are specified in the Standards and Recommended Practices (SARP) of the International Civil Aviation Organization (ICAO) [18]. The most important function of the SBAS is to provide integrity of satellite navigation, i.e. the ability to alert the users within a specified time, called the time to alert (TTA), if a problem occurs such that the navigation system signal should not be used for navigation solution. The SBAS include the American Wide Area Augmentation System (WAAS), the European Geostationary Navigation Overlay Service (EGNOS), the Japanese Multi-functional Satellite Augmentation System (MSAS), and the Russian Luch. BeiDou has combined satellites on MEO and GEO. This is similar to considering GPS and WAAS, GLONASS and Luch, or Galileo and EGNOS as being two parts of one country's satellite system. Russia is going to introduce regional satellite systems with three satellites in GEO. The GEO satellites are called Luch (meaning "Beam" in Russian). Their designated location is at an orbital slot at 16° west longitude (2011), 95° east longitude (2012), and 167° east longitude (2014). The coverage from GEO satellites excludes northern regions. Therefore it is planned to enhance coverage with HEO/HIO satellites, in particular using Molniya orbits (see Section 1.8.2). The system will also provide an Internet-based service.

The SBAS satellites are on geosynchronous orbits and therefore are regional by nature, as are highly elliptical orbit (HEO) satellites. We have, however, included SBAS in GEO under the heading of GNSS. We did this for the sake of unified classification, because BeiDou includes all MEO, HEO/HIO, and GEO satellites. So we either have to exclude GEO satellites from BeiDou here and put them into a separate chapter with WAAS, EGNOS, and Luch, or include those SBAS in the corresponding constellations. The difference between BeiDou GEO satellites and other SBAS GEO satellites is that all BeiDou GEO satellites use the same signals, and their ephemerides and almanacs are part of the BeiDou satellite signal. The WAAS and EGNOS transmit signals with different data rates and content, and WAAS satellites are not included in GPS navigation messages.

The SBAS have three main functions. The first is to provide corrections for global system users in order to improve the ranging service from global systems. The second function is to provide system integrity. To provide integrity means to guarantee that there is a certain time interval within which a user will be notified if there is any fault in GNSS signals that are used for positioning. This interval is set to six seconds for WAAS. The third function is to provide an additional ranging service.

1.8.2 HEO

Highly elliptical orbits are orbits with the same altitude as GEO, but with high eccentricity. Sometimes HEO are called Molniya type orbits [15], named after Russian communication satellites designed to cover mostly Siberian areas of the former USSR.

These types of orbits can also have lower eccentricity and high inclination, in which case they can be called highly inclined orbits (HIO) or inclined geosynchronous satellite orbits (IGSO). Satellites on HEO spend most of the time in the apogee area. This comes directly from Kepler's second law.

As we have already discussed, Kepler's second law states that a line joining a satellite and the Earth sweeps out equal areas in equal intervals of time. Therefore, due to the highly elliptical shape of the HEO, a satellite will linger in that part of the orbit with high altitude, as its velocity decreases when it travels far from Earth. This allows HEO satellites to spend most of their time over the desired region.

The first HEO satellite for navigation was the Quasi-Zenith Satellite System (QZSS), developed in Japan [19]. A QZSS satellite typically operates more than 12 hr a day with an elevation above 70°. This is why the satellites were called "quasi-zenith." The ground track of a QZSS satellite has the shape of an asymmetrical figure 8. This type of orbit is said to have an analemma shape. This shape is clearly visible in the ECEF frame (see Figure 1.14).

The first satellite of Japan's QZSS "Michibiki," was launched in September 2010. The total QZSS constellation should have three satellites, which can be used to augment

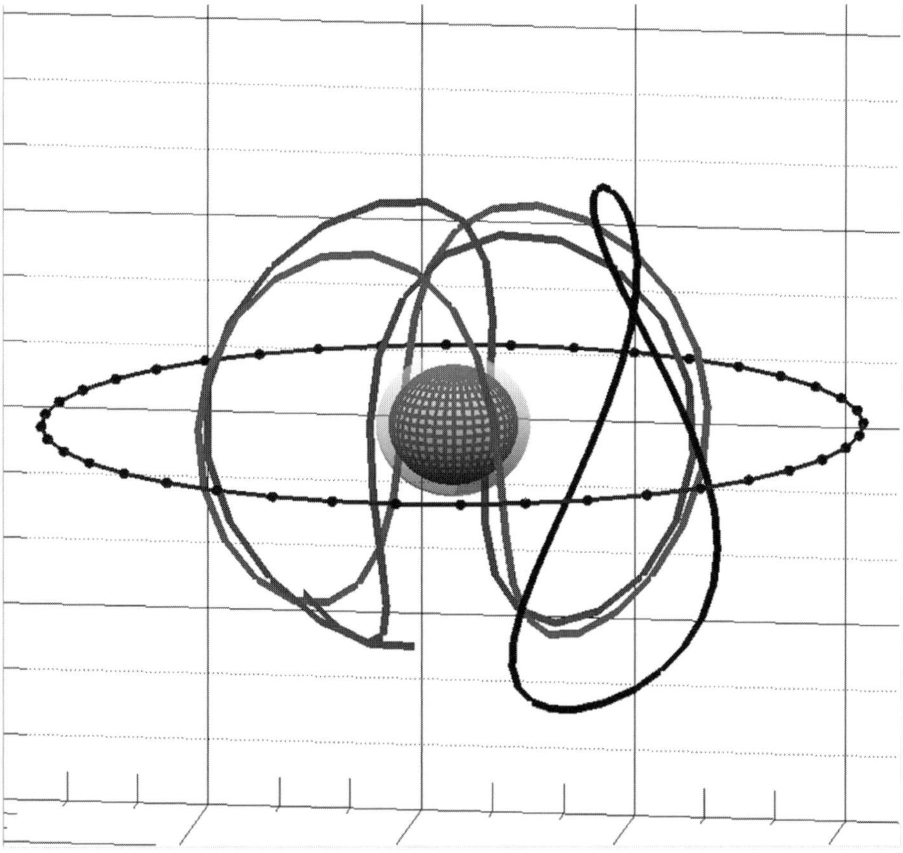

Figure 1.14 GPS, GLONASS, and QZSS orbits and geostationary belt with GEO satellite slots. From iP-Solutions ReGen simulator screenshot.

primary GNSS constellations and even to provide a basic positioning service on its own. The QZSS, however, may require more frequent satellite maneuvers than, for example GPS satellites for station keeping and collision avoidance.

The accuracy of the estimation of the ephemeris and the time to achieve the required accuracy may be greater problems for regional (GEO and HEO) than for global systems. There are fewer ground network stations available for such systems, and the system generally suffers from much poorer geometry. Also, such satellite orbits may be less stable in comparison to MEO. These satellites may in addition be subject to extra requirements, such as being confined in predefined slots in the geostationary satellite belt. These requirements in turn will result in more frequent maneuvers. The frequency of these maneuvers depends on the orbit parameters, i.e. on how stable the orbit is, on the requirements for station keeping, on the allowance for satellite maintenance, etc. In this respect, the above-mentioned Molniya orbits exhibit critical inclination, suppressing precession of the perigee, which makes them more stable.

An approximate estimation of the rate of the argument of the perigee rotation, ω, for HEO satellites can be estimated as follows [15],[20]:

$$\omega = \omega_0 + \frac{d\omega}{dt}(t - t_0). \tag{1.74}$$

Expressing ω through orbit parameters, we get the following:

$$\omega = \omega_0 + \frac{3}{2} J_2 M \left(\frac{R_E}{a}\right)^2 (1 - e^2)^{-2} (2 - 2.5 \sin^2 i)(t - t_0), \tag{1.75}$$

where a is the semi-major axis, i is inclination, R_E is the radius of the Earth, M is the mean motion, $t - t_0$ is a period of evaluation, e is eccentricity, and J_2 is a term relating to Earth's geopotential, caused by the oblateness of the Earth. For a QZSS analemma orbit with inclination 45° and eccentricity 0.1, the change in argument of the perigee per day would be on the order of 0.01°. This orbital drift would be eliminated for an inclination of 63.435°. An orbit with this inclination would have basically the same characteristics from the visibility point of view.

Navigation satellites cannot be used immediately after maneuvers, because the orbital parameters have been changed. Satellites can be used in the operational mode only if ephemerides have been successfully estimated with the specified accuracy. Because regional systems usually consist of just a few satellites, a non-operational satellite can significantly affect overall system characteristics. Therefore the ephemerides for a satellite which has executed a maneuver must be available as soon as possible. Table 1.7 lists some parameters for QZSS, BeiDou, and Molniya HEO satellites.

1.9 Algorithm for GPS, Galileo, and BeiDou for satellite position calculation using ephemeris in the form of osculating elements

In addition to an almanac, navigation messages contain satellite ephemeris data that describe satellite orbits much more accurately. The term *ephemeris* (plural ephemerides)

Table 1.7. HEO satellites

	Satellites		
Parameter	Molniya	QZSS	BeiDou
Application	communication	communication + navigation	navigation
Country	Soviet Union	Japan	China
Launched	1965	2010	2010
Orbit inclination	63.435°	45°	55.06°
Eccentricity	0.74105	0.1	0.0036

came from a Greek word meaning journal, and was subsequently used to describe a table of astronomical body coordinates versus time. Ephemerides can be written in the form of either osculating orbital elements (GPS, Galileo, and BeiDou) or Cartesian vectors (GLONASS, IGS).

Satellite movements for a specific epoch can be described in terms of six Keplerian parameters or equally well in terms of six coordinates (three coordinate and three velocity projections) in the ECI frame. This implies univocal correspondence between the description of an orbit in Descartes (Cartesian) coordinates $(X, Y, Z, \dot{X}, \dot{Y}, \dot{Z})$ and in Keplerian parameters $(a, e, i, \Omega, \omega, t_p)$ (see Table 1.3). Satellite coordinates and velocity can be precisely calculated by numerical integration of the vector equation given in (1.1). We can also calculate the Keplerian parameters that correspond to these sets of coordinates and velocities. These Keplerian parameters, called *osculating Keplerian parameters*, are different for each point at which we make the calculations. The whole orbit therefore can be viewed as an envelope of osculating parameters.

GPS, Galileo, and BeiDou use satellite ephemerides in the form of refined Keplerian parameters. Ideal elliptical orbits are perturbed by higher harmonics and irregularities due to the Earth's gravitational field, by gravitational forces from planets and the Moon, and by a force resulting from solar radiation. Therefore a particular ellipse with a certain set of Keplerian parameters can describe an orbit for only a small interval. The orbit then requires another set of parameters that fits that part of the orbit better. The osculating Keplerian parameters give the correct coordinates and velocities in the vicinity of a specific, reference epoch. The total orbit can then be presented as an envelope of osculating Keplerian orbits, each of which corresponds to satellite position at a specific epoch.

For GPS, a 12 hr satellite orbit is divided into six 2 hr segments [21]. Osculating Keplerian parameters are calculated for each segment. In order to account for changes in the Keplerian parameters from one segment to another, and to provide a more accurate orbit presentation, the Keplerian parameters in each segment are presented as a time series (see Table 1.8). The almanac parameters in GPS are defined in the same sense as those for the ephemeris, and indeed can be viewed as a subset of the ephemeris data [6]. This led to the idea of substituting the broadcast almanac with the predicted ephemeris, thus facilitating an offline assist mode [22].

The algorithm presented here for satellite position calculation follows the GPS ICD [10], the Galileo ICD [11], and the BeiDou ICD [12], with slightly modified notations

Table 1.8. Keplerian parameters presented as a time series in broadcast ephemeris

Main Keplerian parameter	Rate	Sine harmonic	Cosine harmonic
Mean anomaly, M	mean motion Δn difference from computed value	–	–
Eccentricity, e	–	–	–
Square root of semi-major axis, $\sqrt{a}$	–	correction term to the orbit radius, C_{RS}	correction term to the orbit radius, C_{RC}
Longitude of ascending node, Ω	rate of right ascension, $\dot{\Omega}$	–	–
Inclination, i	rate of inclination angle, $\dot{i}$	correction term to the inclination, C_{IS}	correction term to the inclination, C_{IC}
Argument of perigee, ω	–	correction term to the argument of latitude (u), C_{US}	correction term to the argument of latitude (u), C_{UC}

and order of presentation. The ephemerides for a satellite are calculated for a specific reference epoch (t_{OE}), called the *time of ephemeris*. A time interval (Δt) between the reference epoch and time of signal transmission is

$$\Delta t = t - t_{OE}. \tag{1.76}$$

A time count restarts each week, and the reference epoch is specified in the navigation message within a week. If a user encounters the end of week between the reference epoch and the current epoch, the end of week crossover is performed as follows:

$$\begin{cases} \text{if}\, \Delta t > 302\,400[\text{s}] \text{ then } \Delta t - \Delta t - 604\,800 \\ \text{else if } \Delta t < -302\,400[\text{s}] \text{ then } \Delta t = \Delta t + 604\,800. \end{cases} \tag{1.77}$$

Now we apply corrections from the navigation message as they are specified in the second, third, and fourth columns of Table 1.8 to the nominal values of the Keplerian parameters given in the first column. The corrected mean motion is calculated as

$$n = n_0 + \Delta n, \tag{1.78}$$

where the mean motion n_0 is calculated from (1.16).

The mean anomaly is calculated as follows:

$$M = M_0 + n \cdot \Delta t. \tag{1.79}$$

The Keplerian equation (1.18) may be solved interactively for the eccentric anomaly E. The true anomaly should be calculated using the following formula:

$$v = \tan^{-1}\left\{ \frac{\sqrt{1-e^2}\sin E/(1-e\cos E)}{(\cos E - e)/(1-e\cos E)} \right\}. \tag{1.80}$$

Neither the numerator nor the denominator can be simplified by reducing $(1 - e \cos E)$ because the arctangent should be defined using the signs of both arguments to determine the quadrant of the return value.

The corrected argument of the latitude is calculated corresponding to its definition (1.29) and by applying corrections as follows:

$$u = \upsilon + \omega + C_{US} \sin(2u_0) + C_{UC} \cos(2u_0). \tag{1.81}$$

The corrected radius is calculated as follows:

$$r = a(1 - e\cos E) + C_{RS} \sin(2u_0) + C_{RC} \cos(2u_0). \tag{1.82}$$

The corrected inclination is calculated as follows:

$$i = i_0 + C_{IS} \sin(2u_0) + C_{IC} \cos(2u_0) + \dot{i}\Delta t. \tag{1.83}$$

The satellite position in the orbital plane is given by the same equations as in (1.22).

All the parameters described earlier for the ECI frame are calculated in the ECEF frame because all GNSS ground segment network stations basically define the ECEF frame. Therefore, if we account only for Earth rotation, we obtain equations for satellite position in the ECEF instead of the ECI frame, as in (1.28), where we assumed that all Keplerian parameters are defined in the ECI frame. If that were the case, we would need also to account for pole rotation, precession, and nutation as well as rotation of the Earth. In general, we use these transformations only if we want to transfer coordinates to the ECI frame, for applications demanding a high accuracy.

The corrected longitude of the ascending node with Earth rotation taken into account is as follows:

$$\Omega = \Omega_0 + (\dot{\Omega} - \dot{\omega}_E)\Delta t - \dot{\omega}_E t_{OE}. \tag{1.84}$$

The corresponding satellite position in the ECEF frame can be calculated as follows:

$$\overline{X}_{ECEF} = \begin{bmatrix} x_{ECEF} \\ y_{ECEF} \\ z_{ECEF} \end{bmatrix} = \begin{bmatrix} r[\cos u \cos\Omega - \sin u \sin\Omega \cos i] \\ r[\cos u \sin\Omega + \sin u \cos\Omega \cos i] \\ r \sin u \sin i \end{bmatrix}, \tag{1.85}$$

which looks exactly the same as (1.28), except that all parameters are corrected and Earth rotation is hidden in the corrected longitude of ascending node.

1.10 Algorithm for GLONASS satellite position calculation using ephemerides in the form of Cartesian vectors

We have seen how one can describe satellite movement at any epoch in either Keplerian parameters (a, e, i, Ω, ω, t_p) or Descartes (Cartesian) coordinates ($X, Y, Z, \dot{X}, \dot{Y}, \dot{Z}$) with three coordinate and three velocity projections. The GLONASS navigation message uses this latter representation of ephemeris data in its navigation messages. Satellite coordinates and velocity at any epoch can be precisely calculated by numerical integration using these Cartesian coordinates known at certain points along the orbit. The relations between these two presentations are unequivocal, and Keplerian parameters can always be calculated from Cartesian coordinates and vice versa.

GLONASS and IGS provide orbit information in Descartes coordinates as tabular orbits. The satellite coordinates, which may include higher derivatives, are provided at equally spaced points, which are located on the boundaries of each orbital segment. We may need to use force models when working with tabular orbits. The constraints for these models depend on the interval between tabular points and the required orbit accuracy between them. It may also be necessary to use some force models if the interval between points is large, such as in the case of GLONASS broadcast orbits, or when it is necessary to define an orbit with high accuracy, such as in the case of geodetic applications.

The coordinates of a GLONASS satellite are estimated via a solution of the boundary value problem. The ephemerides in such a format are provided in the GLONASS navigation message. The satellite position in this case should be found using a fourth-order Runge–Kutta method, as recommended in the GLONASS ICD [13],[23]:

$$\begin{cases} \dfrac{dx_{ECEF}}{dt} = V_{x_{ECEF}}, \\ \dfrac{dy_{ECEF}}{dt} = V_{y_{ECEF}}, \\ \dfrac{dz_{ECEF}}{dt} = V_{z_{ECEF}}, \\ \dfrac{dV_{x_{ECEF}}}{dt} = -\dfrac{\mu}{r^3}x_{ECEF} - \dfrac{3}{2}J_0^2\dfrac{\mu a_E^2}{r^5}x_{ECEF}\left(1 - \dfrac{5z_{ECEF}^2}{r^2}\right) + \omega_E^2 x_{ECEF} + 2\omega_E V_{y_{ECEF}} + \ddot{x}_{ECEF}, \\ \dfrac{dV_{y_{ECEF}}}{dt} = -\dfrac{\mu}{r^3}y_{ECEF} - \dfrac{3}{2}J_0^2\dfrac{\mu a_E^2}{r^5}y_{ECEF}\left(1 - \dfrac{5z_{ECEF}^2}{r^2}\right) + \omega_E^2 y_{ECEF} + 2\omega_E V_{x_{ECEF}} + \ddot{y}_{ECEF}, \\ \dfrac{dV_{z_{ECEF}}}{dt} = -\dfrac{\mu}{r^3}z_{ECEF} - \dfrac{3}{2}J_0^2\dfrac{\mu a_E^2}{r^5}z_{ECEF}\left(1 - \dfrac{5z_{ECEF}^2}{r^2}\right) + \ddot{z}_{ECEF}, \end{cases} \tag{1.86}$$

where $r = \sqrt{x_{ECEF}^2 + y_{ECEF}^2 + z_{ECEF}^2}$.

In these equations, the ECEF frame is defined as PZ-90. The transformation between the main coordinate frames becomes very straightforward with a new edition of PZ-90, PZ-90.02. For mobile applications, GLONASS PZ-90 can be considered to coincide with GPS WGS-84 on a decimeter level. For high-accuracy applications, it is recommended that the transformation of PZ-90 to the ITRF2000 frame can be applied by shifting the center of origin by the displacement vector (−36 cm; 8 cm; 8 cm). However, for high-accuracy applications GLONASS and GPS coordinates should be initially calculated using precise ephemerides in the same ITRF realization.

1.11 Algorithm for GLONASS satellite position calculation accounting for lunar and solar gravitational perturbations

This section describes a more precise satellite position calculation algorithm, which also accounts for the gravitational effects of the Sun and the Moon [13]. In this case we consider a system of differential equations in the ECI frame because we need to account for the Sun and the Moon, the ephemerides for which are available in the inertial frame:

$$\begin{cases} \dfrac{dx_{ECI}}{dt} = V_{x_{ECI}}, \\ \dfrac{dy_{ECI}}{dt} = V_{y_{ECI}}, \\ \dfrac{dz_{ECI}}{dt} = V_{z_{ECI}}, \\ \dfrac{dV_{x_{ECI}}}{dt} = -\dfrac{\mu}{r^3} x_{ECI} - \dfrac{3}{2} J_0^2 \dfrac{\mu a_E^2}{r^5} x_{ECI} \left(1 - \dfrac{5 z_{ECI}^2}{r^2} \right) + j_{x_{ECI}}^{Sun} + j_{x_{ECI}}^{Moon}, \\ \dfrac{dV_{y_{ECI}}}{dt} = -\dfrac{\mu}{r^3} y_{ECI} - \dfrac{3}{2} J_0^2 \dfrac{\mu a_E^2}{r^5} y_{ECI} \left(1 - \dfrac{5 z_{ECI}^2}{r^2} \right) + j_{y_{ECI}}^{Sun} + j_{y_{ECI}}^{Moon}, \\ \dfrac{dV_{z_{ECI}}}{dt} = -\dfrac{\mu}{r^3} z_{ECI} - \dfrac{3}{2} J_0^2 \dfrac{\mu a_E^2}{r^5} z_{ECI} \left(1 - \dfrac{5 z_{ECI}^2}{r^2} \right) + j_{z_{ECI}}^{Sun} + j_{z_{ECI}}^{Moon}, \end{cases} \quad (1.87)$$

where $j_{x_{ECI}}^{Sun}, j_{y_{ECI}}^{Sun}, j_{z_{ECI}}^{Sun}$ are accelerations due to the solar gravitational force, and $j_{x_{ECI}}^{Moon}, j_{y_{ECI}}^{Moon}, j_{z_{ECI}}^{Moon}$ are accelerations due to the lunar gravitational force.

Note that, in comparison with (1.86), the velocity and acceleration components have disappeared because we are considering movement in the inertial frame. Those were the components that caused the additional curves to appear in the orbit representation in the ECEF frame (see Figure 1.4). After the coordinates are calculated, they are transferred to the ECEF frame using (1.5).

References

[1] J. J. Spilker Jr., Satellite constellation and geometric dilution of precision, in *Global Positioning System: Theory and Applications, Vol. I*, B. W. Parkinson and J. J. Spilker, Eds. Washington, D.C.: American Institute of Aeronautics and Astronautics Inc., 1996.

[2] I. Petrovski and U. Hugentobler, Analysis of impact of onboard time scale instrumental error accuracy of Earth satellite ephemerides estimation with one-way measurements, in *Proc. European Navigation Conf. ENC-2006*, Royal Institute of Navigation, Manchester, UK, May 8–10, 2006.

[3] S. Revnivykh, *GLONASS Status and Modernization, Proceedings of the 24th International Technical Meeting of The Satellite Division of the Institute of Navigation (ION GNSS 2011)*, Portland, OR, Sept. 21, 2011, pp. 839–854.

[4] E. Richards, Calendars, in *Explanatory Supplement to the Astronomical Almanac*, 3rd edn., S. E. Urban and P. K. Seidelmann, Eds. Mill Valley, CA: University Science Books, 2013, pp. 585–624.

[5] G. Xu, *GPS: Theory, Algorithms and Applications*, Berlin: Springer-Verlag, 2003.

[6] I. Petrovski and T. Tsujii, *Digital Satellite Navigation and Geophysics: A Practical Guide with GNSS Signal Simulator and Receiver Laboratory*. Cambridge: Cambridge University Press, 2012.

[7] O. Montenbruck and E. Gill, *Satellite Orbits, Models, Methods, Applications*. Berlin: Springer-Verlag, 2000.

[8] M. Capderou, *Satellites, Orbits and Missions*. Paris: Springer-Verlag France, 2005.

[9] G. Beutler, *Methods of Celestial Mechanics, Volume I: Physical, Mathematical, and Numerical Principles*. Berlin: Springer-Verlag, 2005.

[10] Navstar GPS Space Segment/Navigation User Segment Interfaces, GPS Interface Specification IS-GPS-200, Rev F. Global Positioning Systems Directorate, Sept. 2011.

[11] European GNSS (Galileo) Open Service Signal In Space Interface Control Document (OS SIS ICD), Issue *1.4*. European Union and European Space Agency, Sept. 2010.

[12] BeiDou Navigation Satellite System Signal In Space Interface Control Document. Open Service Signal B1I (Version 1.0). China Satellite Navigation Office, Dec. 2012.

[13] Global Navigation Satellite System GLONASS, Interface Control Document, Navigational radio signal in bands L1, L2, Edition 5.1. Russian Institute of Space Device Engineering, Moscow, 2008.

[14] W. M. Kaula, *Theory of Satellite Geodesy – Applications of Satellites to Geodesy*. Waltham, MA: Blaisdell Publishing Company, 1966.

[15] J. R. Wertz, *Mission Geometry: Orbit and Constellation Design and Management*. El Segundo, CA: Microcosm Press; Dordrecht: Kluwer Academic Publishers, 2001.

[16] D. Ineichen, G. Beutler, and U. Hugentobler, Sensitivity of GPS and GLONASS orbits with respect to resonant geopotential parameters, *Journal of Geodesy*, vol. 77, pp. 478–486, 2003.

[17] U. Hugentobler, *Astrometry and Satellite Orbits: Theoretical Considerations and Typical Applications*. Vol. 57 of Geodätisch-geophysikalische Arbeiten in der Schweiz, Schweizerische Geodätische Kommission, Institut für Geodäsie und Photogrammetrie, Eidg. Technische Hochschule Zürich, Zürich, Switzerland, 1998.

[18] ICAO, International Standards and Recommended Practices, Aeronautical Telecommunications, Annex 10 to the Convention on International Civil Aviation, vol. I, 2008.

[19] I. Petrovski, H. Kishimoto, T. Furukawa *et al.* QZSS – Japan's New Integrated Communication and Positioning Service for Mobile Users, *GPS World*, vol. 14, no. 6, pp. 24–29, 2003.

[20] W. Larson and J. Wertz, Eds., *Space Mission Analysis and Design*, 3rd edn. El Segundo, CA: Microcosm Press; Dordrecht: Kluwer Academic Publishers, 1999.

[21] J. J. Spilker Jr. and B. W. Parkinson, Overview of GPS operation and design, in *Global Positioning System: Theory and Applications, Vol. I*, B. W. Parkinson and J. J. Spilker, Eds. Washington, D.C.: American Institute of Aeronautics and Astronautics Inc., 1996.

[22] F. van Diggelen, *A-GPS: Assisted GPS, GNSS, and SBAS*. Boston, MA: Artech House, 2009.

[23] V. A. Bartenev, *et al.*, Russia's Global Navigation Satellite System, Arlington, VA, ANSER Doc. call no. M-U 42106–39, 1994.

2 GPS, GLONASS, Galileo, and BeiDou signals

A GNSS signal is transmitted from a satellite in order to measure the distance between the satellite and a receiver. There are four GNSS today: GPS, GLONASS, Galileo, and BeiDou. Each GNSS has its signal transmitted on specific radio frequencies. In this book we mainly consider single-frequency devices working with frequencies in the L1 band. The reasons for this are discussed in Chapter 10. The L1 signals for all current navigation satellite systems are shown in Table 2.1. Figure 2.1 shows how the signal frequency bands are distributed. The signals on these frequencies share some specific features, which we consider in this chapter. The frequencies for the GNSS L1 signal are slightly different, but this difference is not large enough to cause any variation in their error budget. For readers interested in the position of GNSS signals in the overall electromagnetic wave spectrum, the signal description from an electromagnetic theory point of view, and the fine aspects of signal propagation through the atmosphere, which may be derived from consideration of electromagnetic theory, such as signal scintillation theory and models, we recommend [1].

2.1 GNSS signals

The description of GNSS signals is given in the corresponding Interface Control Documents (ICD): for GPS L1/L2 in IS-GPS-200 [2], for GPS L1C in IS-GPS-800 [3], for GLONASS L1/L2 [4], for Galileo [5], and for BeiDou [6].

In this chapter we describe a GNSS signal in terms of a general model for academic purposes, in order to avoid giving the impression that every system should be learned from scratch. Section 2.1.1 gives a general description of a GNSS signal. It also describes specific features of new signals, including BOC modulation, pilot channel, and tiered code.

2.1.1 GNSS signals in general

2.1.1.1 CDMA method

All GNSS signals use the concept of a *spread spectrum*. The name implies that the signal's energy is spread over a wide bandwidth. A specific implementation is *code division multiple access* (CDMA), which is a basis not only for GNSS, but also for modern communication applications. In the case of GNSS, however, the spread-spectrum

Table 2.1. GNSS L1 frequencies

System	Owner	L1 signal		Central frequency [MHz]	Modulation	Bandwidth [MHz]	Signal power [dBW]
GPS	USA	L1 C/A	C/A	1575.42	BPSK	2.046	−158.5
		L1C	$L1C_P$	1575.42	TMBOC(6,1)	30.69	−158.25
			$L1C_D$		BOC(1,1)		−163
GLONASS	Russia	FDMA	SP	$1602 + k \times 0.5625$, $k = -7, \ldots, 0, \ldots, 6$	BPSK	1.022	−161
		CDMA	$L1OC_P$	1600.995	BOC(1,1)	N/A	N/A
			$L1OC_D$		BPSK		
Galileo	EU	E1	$E1_P$	1575.42	CBOC(6,1)	24.552	−157
			$E1_D$				
BeiDou	China	B1	B1	1561.098	BPSK	16	−163

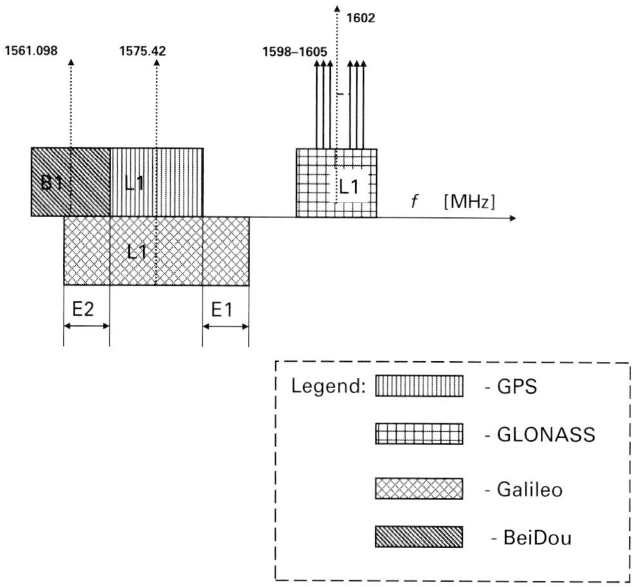

Figure 2.1 GNSS signal spectrums.

concept is much more important. A carrier frequency, on which a satellite signal is transmitted, is modulated by a pseudorandom code sequence or spread code. The pseudorandom code spreads a harmonic line spectrum into a wider frequency band, 2 MHz in the case of GPS and GLONASS (see Figure 2.2). What are the benefits of the spread spectrum?

(1) The pseudorandom code is used in a receiver to measure the distance to the satellite by comparing it with replica code generated inside the receiver. This application of spreading code is unique for GNSS. The code phase which is

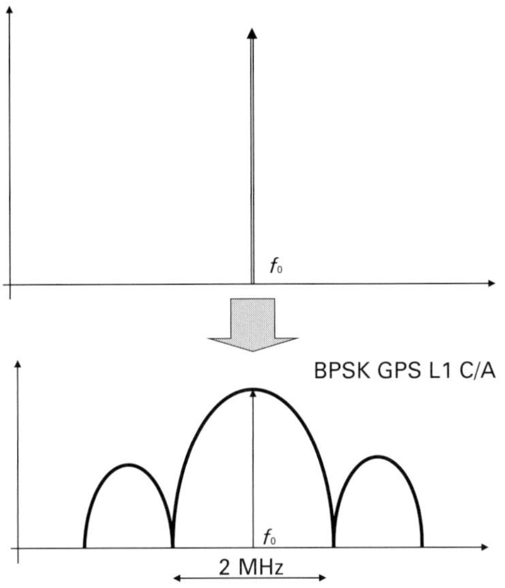

Figure 2.2 Spread spectrum vs. harmonics (carrier).

derived from this measurement is translated into a distance to the satellite. How the distance to a satellite is derived from the measured code phase is considered in detail later in this chapter.

(2) The spread-spectrum method allows a receiver to acquire and track a signal of power well under noise level.

GNSS satellites are located on medium Earth orbits (MEO) at a distance of about 20 000 km from a user receiver, or even further for the additional satellites located on geostationary Earth orbits (GEO). The energy of the signal obeys the inverse square law. This is because the signal power energy is spread equally over the spherical surface at the receiver distance. Therefore the power is spread over a spherical surface area, which is proportional to the square of the radius of the sphere. The satellite payload is minimized in order to make its delivery to an orbit cheaper, and therefore the transmission energy also should be minimized in order for the transmitter to consume less energy. As a result, the signal power near the Earth's surface may be well below noise level. Table 2.1 shows the GNSS signal power levels, as specified by the ICDs, at the user receiver antenna. This very low power signal can be detected because a receiver knows the signal pattern. It works on the same principle as a resonance. The incoming code has maximum impact on the receiver when its pattern coincides with a receiver replica pattern, i.e. the same pseudorandom code sequence is generated in the receiver.

(3) All GNSS except GLONASS transmit signals on the same frequency for all satellites. CDMA allows the receiver to distinguish between signals from different satellites which are using the same carrier frequency. This explains the name code division multiple access: it allows multiple signals to be accessed on the same frequency by distinguishing between them with different spread codes.

Each satellite within the GPS, Galileo, and BeiDou constellations is distinguished from every other from its own constellation by its pseudorandom code. Consequently, the pseudorandom code serves as a satellite ID. In addition to other code features, it allows the measurement of the distance to the satellite and makes the signal less prone to interference and noise.

For the case of the GLONASS L1 legacy signal, all satellites transmit the same code on slightly different frequencies, which means that the GLONASS frequency band occupies more space (see Figure 2.1). GLONASS can therefore be classified as having a frequency division multiple access (FDMA) system on top of CDMA, because GLONASS benefits from the first two above-mentioned advantages of CDMA.

2.1.1.2 GNSS signal structure

We can express a satellite signal in general as a combination of carrier, spread code, and data:

$$S_i = A_i \cdot B_i \cdot D_i, \tag{2.1}$$

where A_i is the carrier of the ith satellite signal, B_i is the spread code, and D_i is the navigation (NAV) data or NAV message. The $(\cdot)$ sign denotes the combination operation, which may differ from signal to signal.

The carrier can be expressed as follows:

$$A(t) = A_0 \sin(\omega_A t + \varphi). \tag{2.2}$$

For GPS, Galileo, and BeiDou, the signal (2.1) takes the form

$$S_i = A \cdot B_i \cdot D_i, \tag{2.3}$$

where A is a specific frequency for a given GNSS.

For GLONASS, the carrier frequencies are different for the satellites, but the spread code is the same on all frequencies:

$$S_i = A_i \cdot B \cdot D_i. \tag{2.4}$$

There are additional components which are used in specific GNSS to modify one of those three main components or make them more complex, for example

$$S_i = A_i \cdot A_i^{(2)} \cdot B_i \cdot B_i^{(2)} \cdot D_i \cdot D_i^{(2)}, \tag{2.5}$$

where $A_i^{(2)}$ is the carrier offset code, $D_i^{(2)}$ is the secondary code, and $D_i^{(2)}$ is an additional data layer, such as the GLONASS meander sequence, for example.

2.1.1.3 GNSS spread codes: past, present, and future

2.1.1.3.1 Shift register and memory codes

A general spread code can be seen as an infinite sequence of independent, identically distributed, random binary variables taking one of two values (0 or 1) with equal probability.

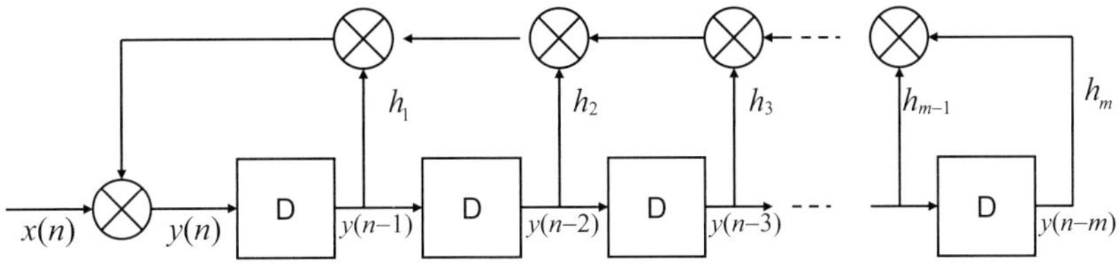

Figure 2.3 Linear feedback shift registers.

The code should behave randomly in order to facilitate all the aforementioned CDMA functions; on the other hand, it should be generated economically in both transmitter and receiver, and therefore it should in that sense be deterministic.

Originally, binary random codes were generated by linear feedback shift registers (see Figure 2.3). This type of code sequence is named pseudorandom noise (PRN) code. The shift register can be described as follows:

$$y(n) = \sum_{k=1}^{m} h_k y(n-k), \tag{2.6}$$

where m is the number of adders, and h_k can take either of two values, 1 or 0, where 0 would indicate an absence of the corresponding adder in the shift register.

The generated specific code sequence would be different for each initial state of the shift register $[y_0(n), y_0(n-1), \ldots, y_0(n-m)]$.

In polynomial form, the shift register (2.6) can be expressed as follows:

$$G(X) = \sum_{k=0}^{m} X^k \tag{2.7}$$

The essential qualities of the PRN sequences are defined by the correlation measures. Two infinite random sequences should be uncorrelated. The closer a given PRN sequence comes to a truly random sequence, the better the PRN sequence is. The correlation between two sequences $x(k)$ and $z(k)$ is described by the following expression:

$$R(n) = \frac{1}{N} \sum_{k=0}^{N-1} y(k) z(n+k). \tag{2.8}$$

Moreover, a random sequence would have very specific autocorrelation properties, i.e. a random code would correlate with its own replica at only one point. The autocorrelation function can be found as a result of multiplication of the sequence and its own shifted version:

$$R(n) = \frac{1}{N} \sum_{k=0}^{N-1} y(k) y(n+k), \tag{2.9}$$

where N is the sequence period.

The PRN codes are generated by a deterministic system and have finite length. A shift register can generate sequences with a period no greater than

$$N = 2m - 1. \tag{2.10}$$

Therefore, these sequences are sometimes called *maximal length* sequences or m-sequences. As a result of their deterministic nature, and consequently their finite length, they are not uncorrelated. Specific sequences have different correlation properties. In general, PRN codes are almost orthogonal to each other [7]; in other words, they are almost uncorrelated. The orthogonal codes also have an autocorrelation function, with a single correlation peak.

When we describe signals, we are always interested in how they are represented in the frequency domain, i.e. in the signal spectrum. The signal spectrum is a distribution of signal energy as a function of frequency. For this generally described signal, the spectrum of the carrier is a line at the carrier frequency $f_A = \omega_A/2\pi$. The spread code redistributes the energy over a wider frequency band (see Figure 2.2). The spectrum takes this particular shape because it is the shape of a pulse train signal, which can be represented as a superposition of harmonic signals with a more powerful harmonic at the central frequency (see Figure 2.4) and with other frequencies as follows:

$$A_{f_i} = \frac{\sin(f_i)}{f_i}. \tag{2.11}$$

The lines on the spectrum are separated by

$$\Delta f = \frac{1}{T}, \tag{2.12}$$

where T is the pulse period. The nulls of the spectrum are separated by an interval defined by the pulse width:

$$\Delta F = \frac{1}{\Delta T}, \tag{2.13}$$

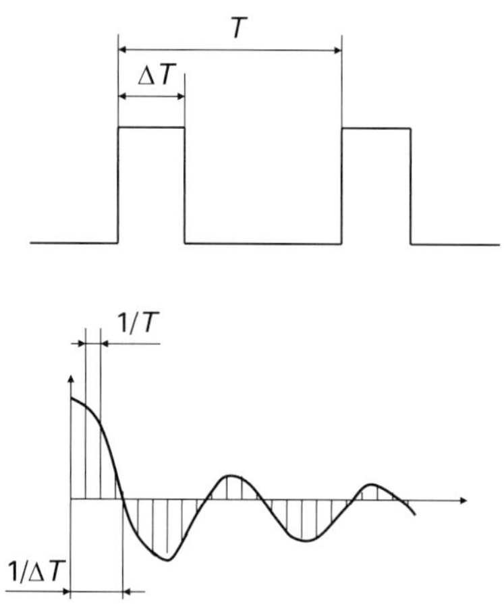

Figure 2.4 Pulse train spectrum.

where ΔT is the pulse width. For a GNSS signal designed as a binary sequence $\{0, 1\}$, $\Delta T = T/2$. The GNSS signal is similar to the pulse train spectrum centered on the carrier radio frequency (RF), and is usually depicted with no information given on the signal phase.

Today, sometimes the GNSS spread codes are designed to be kept in a memory instead of being generated by shift registers. For some codes this provides a more economical way of generating them. In general, this allows much more flexibility when changing a transmitted signal onboard a satellite on the fly. This last feature is closely related to a new GNSS paradigm, which we consider in Chapter 10.

2.1.1.3.2 *Strange attractor codes*

These codes are becoming closer to ideal random sequences for some features and moving further away from them for others. There are always trade-offs, some of which we consider further when describing particular signals.

Ideal code can be generated today using strange attractors and chaotic theory. Such codes have great potential and may be implemented in the future. The chaotic code concept has been studied for some time [8]. The idea behind such codes is that they are implemented as an output of a non-linear dynamic system.

Consider for example the Lorenz attractor, developed by Edward Lorenz in 1963 to describe a mathematical model for atmospheric convection [9],[10] which is described by the following non-linear equations:

$$\begin{aligned} \frac{dx}{dt} &= \sigma(y - x), \\ \frac{dy}{dt} &= \rho x - y - xz, \\ \frac{dz}{dt} &= xy - \beta y, \end{aligned} \tag{2.14}$$

where Lorenz specifies $\sigma = 10$, $\rho = 28$, $\beta = 8/3$. These values cause the system to have two stable solution points, between which the system oscillates chaotically.

The system of equations can be solved by numerical integration. The three-dimensional plots shown in Figure 2.5 demonstrate the solution of the system in three dimensions and its projection on the YZ plane. The solution goes to two states, which can be represented in a binary phase shift keying (BPSK) sequence as 0 and 1. The state cannot be predicted without specifying the equations and their initial conditions. The sequence goes to infinity, and does not repeat itself. If we have two initial values chosen to be very close to each other, after some period they still generate completely different behavior. Implementing code generation allows us to generate an infinite number of sequences, even for the basic Lorenz attractor, with ultimately minimum autocorrelation and cross-correlation properties. This type of behavior is called *chaotic* behavior. Figure 2.5 also shows the difference in system behavior as a function of initial conditions. The chaotic sequence can be defined by the following factors:

(1) a set of non-linear equations, for example the Lorenz attractor (2.14);
(2) parameter values;

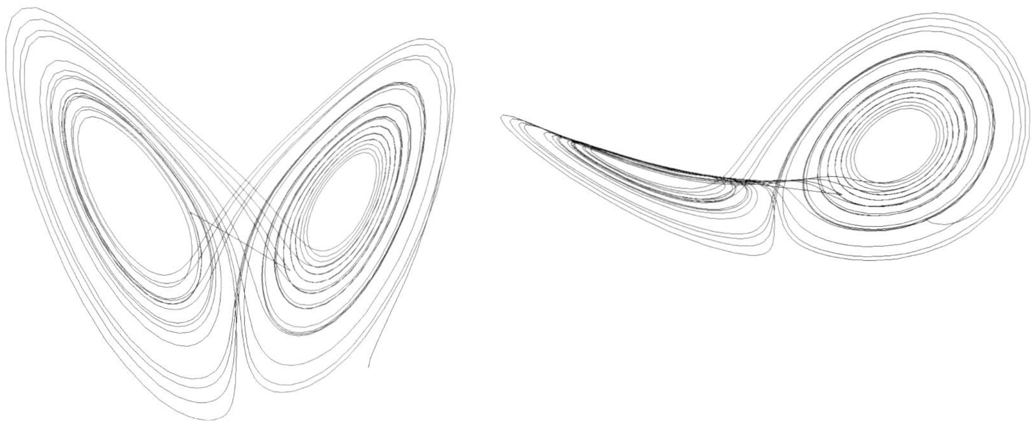

Figure 2.5 Lorenz strange attractors, which can be used to generate chaotic spread code sequences. (a) and (b) depict the same three-dimensional figure, but from different angles.

(3) initial conditions, for example with a variation on the order of 0.00001%;
(4) an independent parameter value (time).

Another strange attractor that can be used for the generation of chaotic binary codes is the Burke–Shaw attractor,

$$\begin{aligned}\frac{dx}{dt} &= -\sigma(y + x),\\ \frac{dy}{dt} &= -\sigma xz - y,\\ \frac{dz}{dt} &= \sigma xy + \beta,\end{aligned} \tag{2.15}$$

where $\sigma = 10$, $\beta = 4.272$.

The output of such a system looks like a random sequence. Potentially it has cross-correlation and autocorrelation properties indistinguishable from those of the ideal random sequence. The difference is that such a sequence is deterministic in that it is exactly the same each time, as long as the dynamic system and the initial conditions are the same. It also goes to infinity. This code would be seen as random, and would have infinite length, but at the same time it can be generated using a deterministic non-linear model.

Strange attractors can be used to generate quadrature phase shift keying (QPSK) sequences or signals for data and pilot channels. One such strange attractor, for example, is the four-dimensional Dadras strange attractor given in [11].

2.1.1.4 BOC modulation

Binary offset carrier (BOC) modulation is implemented by adding to a binary periodical {0, 1} sequence called a subcarrier [12]. In terms of our definitions (2.5), it can be expressed as follows:

$$A^{(2)}(t) = sign(\cos(\omega_{A^{(2)}} t)), \tag{2.16}$$

or alternatively as

$$A^{(2)}(t) = sign(\sin(\omega_{A^{(2)}} t)). \tag{2.17}$$

The subcarrier is generated synchronously with the code. BOC(n,m) is defined as the BOC signal with chipping code rate of 1023m chips per millisecond and a subcarrier frequency of 1.023n MHz. The subcarrier can be aligned with code or shifted on some phase.

Even though the term subcarrier is used, because it changes the position of the carrier frequency, it is actually a code sequence. Figure 2.6 shows the spectrum of the signal, modulated by BOC. The spread code changes the width of the signal spectrum band. The additional data sequence, such as for example GLONASS meander, does not affect the signal spectrum at all.

The signal energy of the BOC modulated signal is no longer concentrated around the carrier frequency, but instead is split. The main reasons behind the development of such signals were, on the one hand, the need to improve traditional GNSS signal properties for better resistance to multipath and interferences, and, on the other hand, the need for improved spectral sharing between various GNSS. For details on the comparison of the legacy GPS code with BOC in relation to multipath effects, see [1].

2.1.1.5 Data

Data (D_i in equation (2.1)) are encoded into the signal usually by a modulo-2 (logical exclusive OR, or XOR) operation. For GPS, the L1 coarse/acquisition (C/A) data rate is

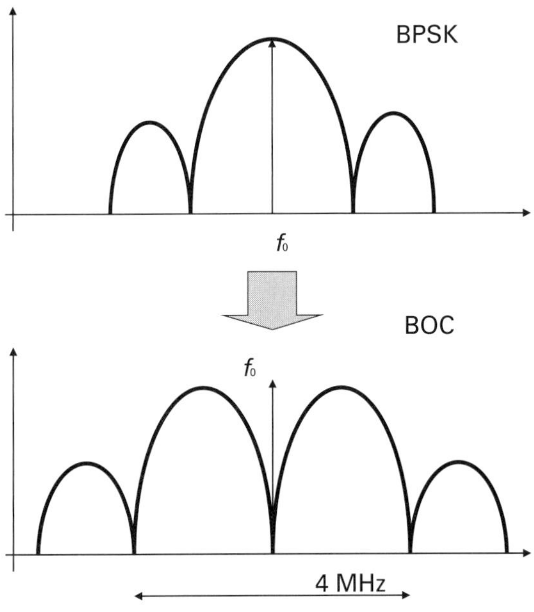

Figure 2.6 BOC spectrum.

50 bits per second (bps). The GPS L1 C/A code sequence is of 1 ms duration, and therefore each data bit contains 20 code repetitions. Navigation data bear information about satellite ephemeris, system time, satellite clock error estimate, satellite health status, and so on. The duration of a GPS L1 C/A navigation message is 12.5 min. GLONASS navigation data have the same data rate and an additional data layer called a meander. The duration of a GLONASS L1 navigation message is 2.5 min.

Apart from orbital information and corrections, navigation data also bear time marks, which allow one to create a range of measurements from measured code phases and pinpoint satellite location at the time of signal transmission.

2.1.1.6 Tiered code

Tiered code is a long spreading code sequence, which is constructed using two codes, the *primary* and *secondary* codes. The secondary code controls the repetitions of the primary code. The secondary code is in a way similar to a data sequence (Figure 2.7). The secondary code may remove primary code repetitions and change the minimal period of the resulting code. It generally makes primary code directly inaccessible for acquisition. In its most straightforward implementation, one secondary code chip contains one period of primary code. In this way, tiered codes are implemented for GPS L1C, Galileo E1.

During signal acquisition in the receiver, it is often difficult to acquire a signal during one sequence of the spread code. For the case of the GPS L1 C/A signal, which we consider in detail in Section 2.1.2.1, the length of the spread code is 1 ms. Due to low signal power, power loss in the atmosphere, receiver noise, interference, and so on, it is usual to acquire 10 or 20 ms of the signal and integrate. A change in data bit value sequence changes all the bits in the spread code, and, when it is summed with other instances of code, it yields zero instead of doubling the result. This generally limits coherent integration to the length of the data bit.

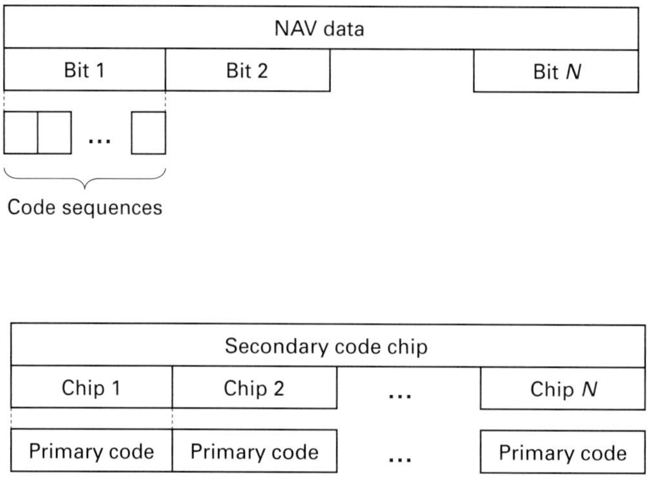

Figure 2.7 Data and tiered code.

This limitation arises because the navigation data are not known in advance. For secondary code there is a similar situation when the data content is known at the receiver, and the code sequence also becomes longer.

2.1.1.7 Pilot channel

Some of the signals (GPS L1C, L5, GLONASS L3, Galileo E1) will have two channels: one data channel and one pilot channel. The data channel contains the code and navigation message. The pilot channel contains a secondary code instead of a navigation message. The receiver knows the secondary code, and therefore, as we see in Chapter 3, can have a higher sensitivity due to the possibility of integrating the signal coherently over a longer interval.

The pilot and data channel signals can be combined using various methods. In general, we can express the total signal as follows:

$$S_i = S_i^{(p)} \cdot S_i^{(d)}, \tag{2.18}$$

where again the meaning of the $(\cdot)$ symbol may vary from signal to signal.

2.1.2 GPS L1 signals

GPS satellites transmit L1 on the central frequency 1575.42 MHz. The GPS L1 signals for civil users are the L1 C/A signal and the new L1C (L1 Civil) signal. Since 2012, the L1C signal has been transmitted by Japan's QZSS satellites in test mode along with the L1 C/A signal; QZSS signals are described in the QZSS ICD [13].

2.1.2.1 GPS L1 C/A signal

The general signal description (2.5) for the GPS L1 C/A signal takes a form similar to (2.3):

$$S_i = A \cdot B_i \cdot D_i, \tag{2.19}$$

where A is the carrier, B_i is the spread code, and D_i is the data. The code and data are added using the modulo-2 (XOR) operation:

$$S_i = A \cdot (B_i \wedge D_i). \tag{2.20}$$

We can use the XOR operation between the code and the data because both have a data width of 1 bit. This total signal is generated using BPSK.

The GPS L1 carrier has frequency $f_A = 1575.42$ MHz (see Table 2.1). The L1 C/A signal codes belong to the Gold codes family [14]. Gold codes are short, with a maximum length of 1023 chips. The GPS satellite uses a 10.23 MHz onboard satellite clock to generate signals on all frequencies. The L1 C/A code is of 1 ms duration. The chip rate is therefore 1023 chips/ms. The millisecond is the GPS C/A code length, and is therefore an essential unit in GPS signal processing algorithms. The minimum sequence was important because it simplifies an implementation in a receiver and makes acquisition easier by allowing integration of sequential acquisition results. On the contrary, most modern codes are rather long. Some new codes cannot be generated by shift registers and therefore can only be set in the memory.

Gold codes are not orthogonal, but they guarantee uniformly low cross-correlation properties with other Gold codes. This property is very important for GPS signals, because all satellites transmit on the same frequency. The autocorrelation function for these codes is not minimal for non-zero lag as it is for orthogonal codes. Figure 2.8 shows typical autocorrelation and cross-correlation functions for GPS Gold codes.

The C/A codes are generated at the rate of 1023 chips/ms. The shift registers, which are used for code generation, can be described by polynomials. GPS C/A codes are described as a product of two polynomials as follows:

$$\begin{aligned} G1(X) &= 1 + X^3 + X^{10}, \\ G2(X) &= 1 + X^2 + X^3 + X^6 + X^8 + X^9 + X^{10}. \end{aligned} \tag{2.21}$$

Figure 2.9 presents a GPS C/A code generator. The initial states of the polynomials and the phase between them define the code for a specific satellite. The initial states for both polynomials are set to the same value, i.e. 1:

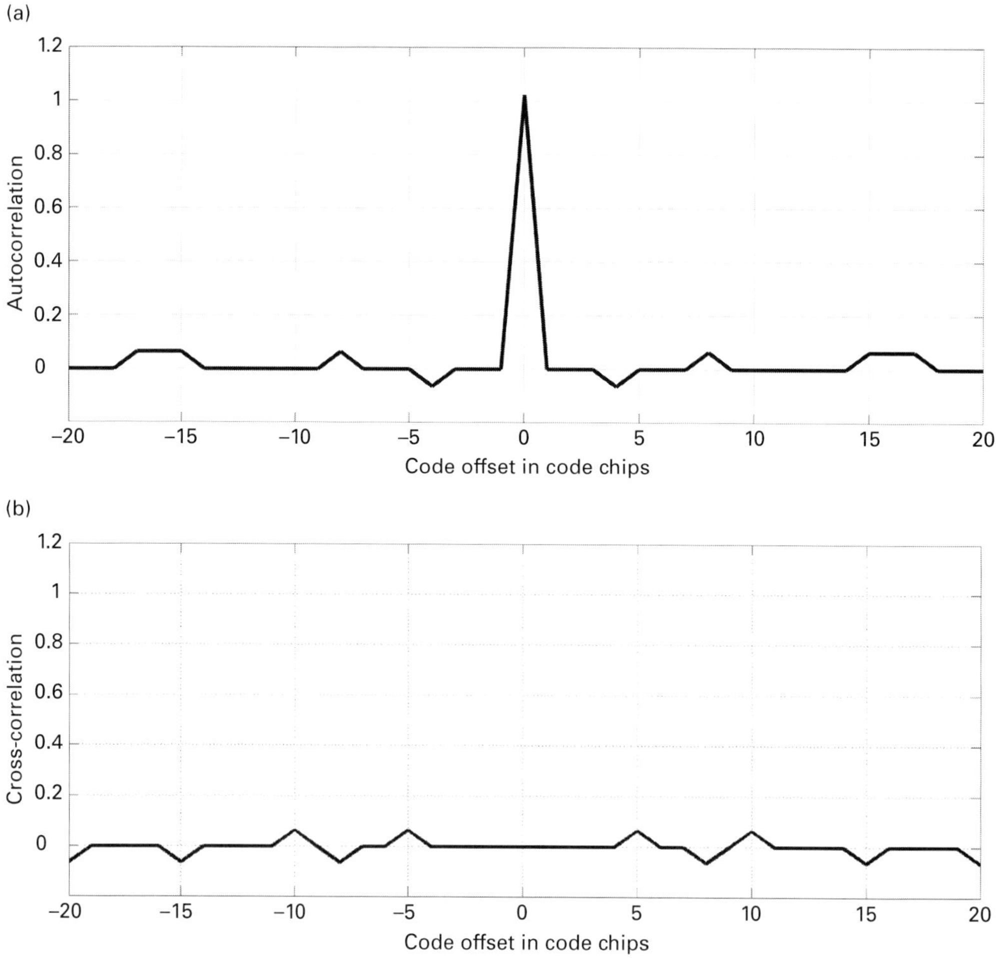

Figure 2.8 GPS C/A code autocorrelation and cross-correlation functions. After [1].

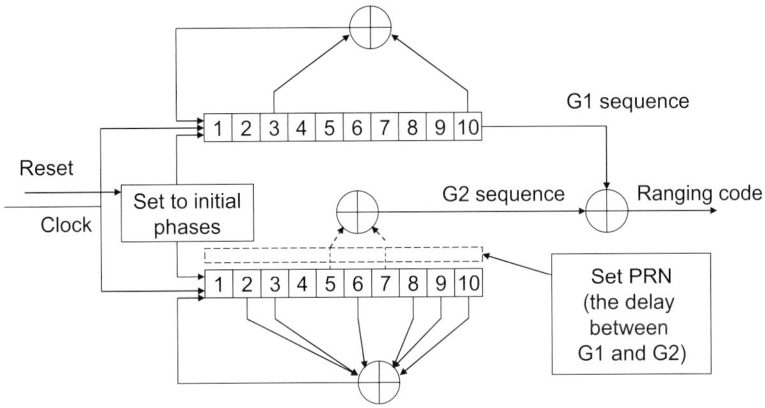

Figure 2.9 GPS C/A code generator. After [7].

$$G1_0 = G2_0 = \{1, 1, 1, 1, 1, 1, 1, 1, 1, 1\}. \tag{2.22}$$

Therefore, only the delay between the polynomials matters in this case, because the sequence repeats itself. The delay is set by choosing specific tap outputs and combining them with an exclusive OR operation, as shown in Figure 2.9. The delays are given in Table 2.2.

The L1 C/A signal is also transmitted by QZSS satellites. Table 2.2 also includes the information for QZSS, SBAS, and for pseudolites, which are considered in Chapter 8, Section 8.5.

The code is then combined with data, as previously described. The navigation message is of duration 12.5 min. Here, it is important that this is divided into subframes, each of 6 s duration. Each subframe contains a time mark, which indicates time of the subframe from the beginning of the week. Figure 2.10 shows C/A signal generation.

The L1 C/A signal is added to the precision P-code. The carrier for P-code has the same L1 frequency, but is shifted by $\pi/2$ relative to the C/A carrier. Two quadrature carrier components then form a QPSK signal. P-code is also transmitted on the L2 frequency, which allows its users to compensate for part of the propagation error. P-code signals are not very useful for mobile applications because civil users have no access to P-code and cannot create its replica in a receiver. This procedure of C/A and P-code summation does not affect C/A code, and can be disregarded by a civil user.

2.1.2.2 GPS L1C signal

The L1C signal contains pilot and data channels, and it can be expressed as follows. Both signals use BOC(f_s,1), which refers to (see Section 2.1.1.4) a subcarrier frequency $f_s \times 1.023$ MHz and a spreading code with chipping rate 1×1.023 MHz. The data channel uses BOC(1,1); the pilot channel uses time-multiplexed BOC(1,1) ($A^{(2)}$) and BOC(6,1) ($A^{(3)}$). The data and pilot channels are multiplexed in such a way that the data channel has 25% of the power and the pilot channel has 75%. The codes have a 10 ms period with a chipping rate of 1.023 Mbps, for a total length of 10 230 chips:

Table 2.2. PRN initial polynomial states for GPS, QZSS, and pseudolites

System	Satellite	PRN	Code delay chips
GPS	IIF	1	5
	IIR	2	6
	IIA	3	7
	IIA	4	8
	IIR-M	5	17
	IIA	6	18
	IIR-M	7	139
	IIA	8	140
	IIA	9	141
	IIA	10	251
	IIR	11	252
	IIR-M	12	254
	IIR	13	255
	IIR	14	256
	IIR-M	15	257
	IIR	16	258
	IIR-M	17	469
	IIR	18	470
	IIR	19	471
	IIR	20	472
	IIR	21	473
	IIR	22	474
	IIR	23	509
	IIF	24	512
	IIF	25	513
	IIA	26	514
	IIF	27	515
	IIR	28	516
	IIR-M	29	859
	IIA	30	860
	IIR-M	31	861
	IIA	32	862
QZSS	QZSS-1 MICHIBIKI	193	339
		194	208
		195	711
		196	189
		197	263
Pseudolites		33	863
		34, 37	950
		35	947
		36	948
WAAS	INMARSAT 4F3	133	603
	–	134	130
	LM RPS-1	135	359
	LM RPS-2	138	386

Table 2.2. *(cont.)*

System	Satellite	PRN	Code delay chips
SDCM	Luch-5A	125	235
	Luch-5B	140	456
	Luch-4	141	499
EGNOS	INMARSAT 3F2	120	145
	ARTEMIS	124	237
	INMARSAT 4F2	126	886

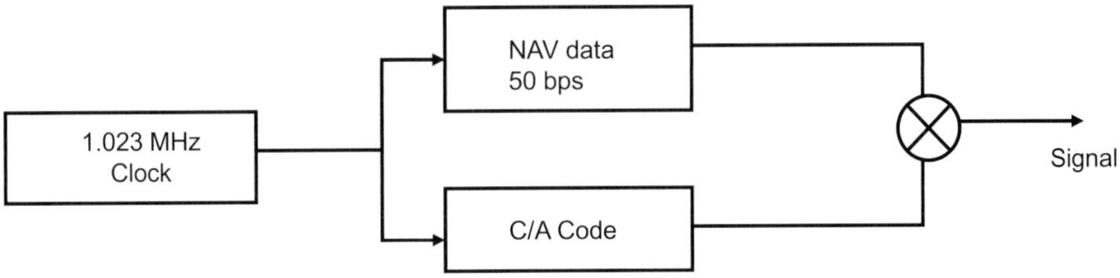

Figure 2.10 C/A signal generation.

$$
\begin{aligned}
S_i &= S_i^{(p)} \cdot S_i^{(d)}, \\
S_i^{(p)} &= A \cdot A^{(2)} \cdot A^{(3)} \cdot B_i \cdot B_i^{(2)}, \\
S_i^{(d)} &= A \cdot A^{(2)} \cdot B_i \cdot D_i.
\end{aligned} \tag{2.23}
$$

The pilot channel uses tiered code, i.e. a modulo-2 combination of primary and secondary codes $B_i \cdot B_i^{(2)}$. The secondary code is 18 s, and it has a rate of 100 bps. The total length is thus 1800 bits. The GPS L1C code autocorrelation function is shown in Figure 2.11.

The L1C navigation message, called CNAV-2, is 1800 bits in length and has a rate of 100 bps.

2.1.3 GLONASS L1 signals

GLONASS currently transmits two signals on L1, both FDMA on top of CDMA: a standard precision (SP) signal for civil users and a restricted high-precision signal. It is planned to transmit a pure CDMA signal. The satellites are promised to keep transmitting legacy signals indefinitely [15].

The fundamental clock frequency for GLONASS is 5.0 MHz, in comparison to 10.23 MHz for GPS, Galileo, and BeiDou. The signal description (2.5) for the GLONASS standard accuracy signal can be expressed as follows:

$$S_i = A_i \cdot (B \wedge D_i \wedge D^{(2)}), \tag{2.24}$$

where $D^{(2)}$ is an extra binary sequence known as the meander.

GLONASS satellites are distinguished by a shift in frequencies and not by different spreading codes. This is why GLONASS does not use Gold codes. The main advantage

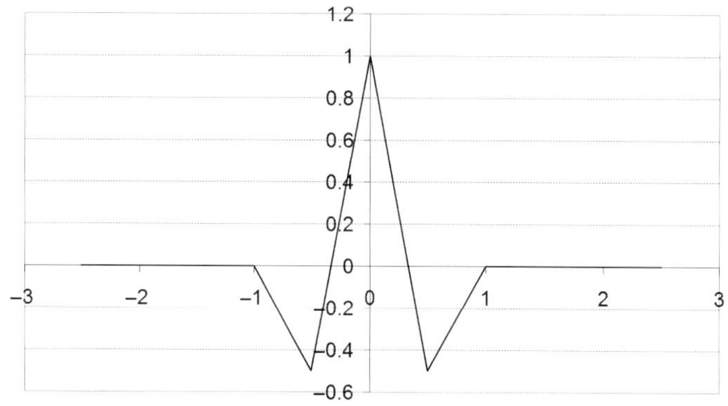

Figure 2.11 GPS L1-C code autocorrelation function.

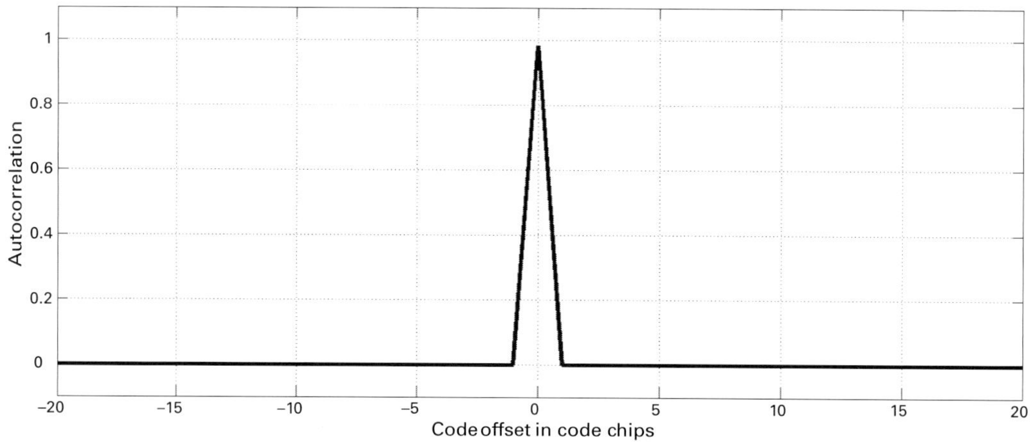

Figure 2.12 GLONASS code autocorrelation function. After [1].

of using Gold codes is the uniformly low cross-correlation properties with other Gold codes. This is not useful for GLONASS legacy codes, because each satellite transmits the code on its own frequency. The GLONASS civil signals on L1 and L2 frequencies implement an m-sequence with minimal auto-correlation properties. Figure 2.12 shows the GLONASS m-sequence autocorrelation function.

The GLONASS frequencies are defined for L1 as follows:

$$\Delta L_1 = 0.5625\,\text{MHz} \;\; L1_k = L1_0 + k\Delta L_1 \tag{2.25}$$

where $L1_0 = 1,602$ MHz, $\Delta L_1 = 0.5625$ MHz, and $k = -7,\ldots 0,\ldots 6$.

The number of frequency channels is 14. The satellites, which occupy opposite slots on the orbital plane, transmit on the same frequencies. Such satellites are called *antipodal* satellites.

Ideally, the GLONASS signal spectrum should be represented by a set of non-overlapped narrow beams for each frequency. The frequencies in fact overlap, which introduces some signal degradation on the order of 54 dB. Similarly, the GPS L1 C/A

Gold code experiences signal degradation in comparison with the pure orthogonal code sequences on the order of 21.6 dB.

On L1 and L2, GLONASS currently uses only one m-sequence for all satellites. A receiver distinguishes between signals from different GLONASS satellites by frequency. This feature sometimes leads to GLONASS being referred to as a FDMA (frequency division multiple access) system. Initially, FDMA was envisioned to provide more resistance against interference. The single-carrier system is less resistant to intentional or unintentional interference than a system in which carriers are spread over some range. In the new GPS signals, and planned Galileo and GLONASS signals, a similar resistance is achieved by modulating the signal with a subcarrier. Use of the FDMA term is, however, not exactly correct when applied to GLONASS, because GLONASS essentially uses spread code for two other reasons, namely to be able to work with low-power signals and to recover a pseudorange to a satellite. The spread code is an essential component of CDMA, but is not required for FDMA. Therefore, GLONASS can be more properly referred to as FDMA on top of CDMA.

Code is transmitted at 511 bits/ms and is generated by the following polynomial:

$$G(X) = 1 + X^5 + X^9. \tag{2.26}$$

The code period is 1 ms, the same as for GPS L1 C/A. Figure 2.13 shows the GLONASS code generator.

The navigation message is transmitted at the same rate as for the GPS, 50 bps, and has a length of 2.5 min. It is summed together with the meander code, which is a binary periodical sequence transmitted at a rate of 100 bps:

$$D^{(2)} = \{0, 1, 0, 1, 0, \ldots, 1, 0, 1, 0, 1, 0, \ldots\} \tag{2.27}$$

The meander is a secondary data message, because it does not change the minimum period of the resulting sequence $B \wedge D^{(2)}$. A secondary code makes the primary code directly inaccessible to acquisition due to the fact that it changes the minimum period of the resulting code.

The GLONASS navigation message is divided into subframes. One subframe of the GLONASS navigation message has duration 2 s, and has a time mark transmitted during the last 0.3 s (see Figure 2.14). The time mark is generated by the shift register defined by the following polynomial:

$$G(X) = 1 + X^3 + X^5, \tag{2.28}$$

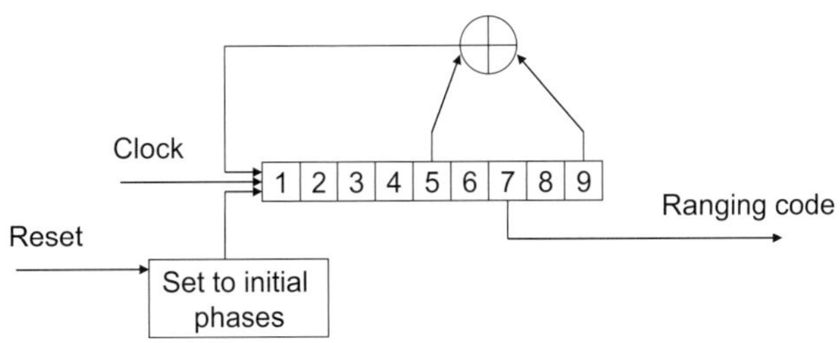

Figure 2.13 GLONASS code generator.

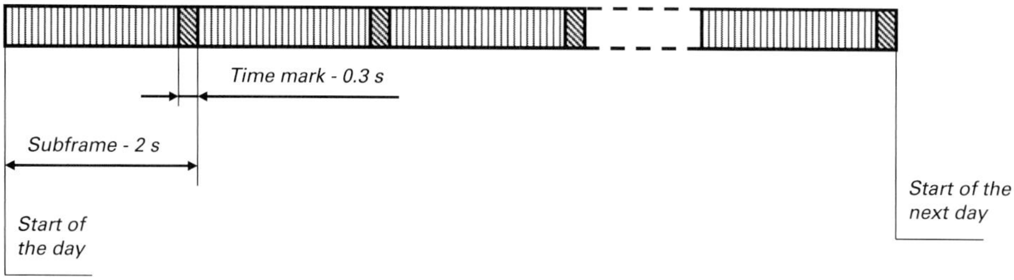

Figure 2.14 GLONASS time mark.

cut down to 30 symbols. The duration of each symbol is 10 ms. The time mark also serves a purpose similar to the preamble in the GPS navigation message. The truncated time mark sequence generated by (2.28) is

$$M = \{1, 1, 1, 1, 1, 0, 0, 0, 1, 1, 0, 1, 1, 1, 0, 1, 0, 1, 0, 0, 0, 0, 1, 0, 0, 1, 0, 1, 1, 0\}. \quad (2.29)$$

The GLONASS time mark does not bear time information explicitly, but allows time to be deduced if a receiver keeps time with 2 s accuracy.

With the exception of the meander sequence, the signal generation is similar to that of GPS C/A, as depicted in Figure 2.10.

2.1.4 Galileo signal

Galileo satellites transmit L1 (called E1 in the case of Galileo) signal on the central frequency 1575.42 MHz, the same as GPS (see Table 2.1).

The E1 signal contains pilot and data channels and can be expressed as follows:

$$S_i = S_i^{(p)} \cdot S_i^{(d)}. \quad (2.30)$$

The data and pilot channels both use CBOC modulation, which is multiplexed BOC (1,1) ($A^{(2)}$) and BOC(6,1) ($A^{(3)}$). Code A_i has a 1023 MHz chipping rate:

$$S_i^{(p)} = A \cdot A^{(2)} \cdot A^{(3)} \cdot B_i \cdot B^{(2)}, \quad (2.31)$$

$$S_i^{(d)} = A \cdot A^{(2)} \cdot A^{(3)} \cdot B_i \cdot D_i. \quad (2.32)$$

The data channel has a navigation message with 250 bps rate. The pilot $S_i^{(p)}$ channel is called E1-C, the data channel is called E1-B (see Figure 2.15). The combination sign ($\cdot$) in equation (2.31) refers to the modulation scheme, which combines both channels onto the same carrier component, with 50% power sharing. As in the case of GPS and GLONASS, the signal is then multiplexed with a restricted access signal. So we have

$$S_i = \frac{1}{3}\left\{\left[\sqrt{2} \cdot S_i^{(d)} - \sqrt{2} \cdot S_i^{(p)}\right] + j \cdot \left[2 \cdot S_i^{(r)} + S_i^{(r)} \cdot S_i^{(d)} \cdot S_i^{(p)}\right]\right\}, \quad (2.33)$$

where $S_i^{(r)}$ is the restricted access signal.

The codes on a pilot channel are tiered (see Section 2.1.1.6). Primary code has a period of 4 ms on the data channel and 100 ms on the pilot channel. The primary code

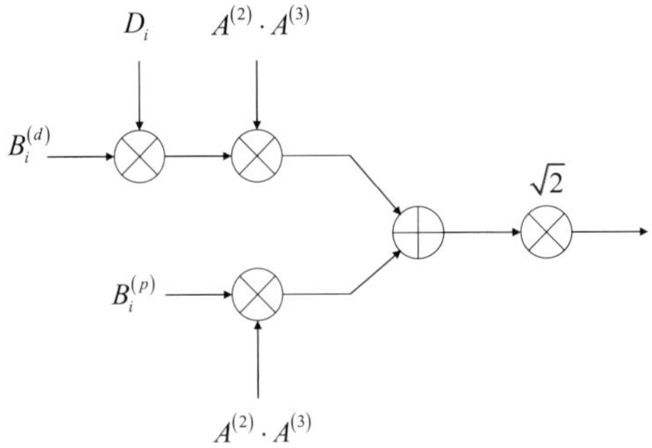

Figure 2.15 Galileo signal generator. After [5].

has length 4092 chips, and they are pseudorandom memory codes, i.e. optimized codes that need to be stored in memory. They are given in the Galileo ICD [5]. The secondary code is a fixed code sequence with a length of 25 chips. It is the same for all satellites. In hexadecimal notation it is defined as

$$B^{(2)} = 380AD90. \tag{2.34}$$

The value of the first chip of the secondary code controls the polarity of the first epoch of the primary code sequence by an XOR operation. The secondary code chips are always one chip of the secondary code per period of the primary code.

2.1.5 BeiDou signal

BeiDou transmits an L1 signal on the central frequency 1561.098 MHz:

$$S_i = S_i^{(r)} \cdot S_i^{(d)}, \tag{2.35}$$

$$S_i^{(d)} = A \cdot B_i \cdot B_i^{(2)} \cdot D_i. \tag{2.36}$$

The signals are CDMA, with spread code chipping rate 2.046 Mps.

The secondary code $B_i^{(2)}$ is a 20 bit Neuman–Hoffman code and has 1 ms duration. The secondary code sequence is as follows:

$$B_i^{(2)} = \{0, 0, 0, 0, 0, 1, 0, 0, 1, 1, 1, 0, 1, 0, 1, 0, 0, 1, 1, 1, 0\}. \tag{2.37}$$

(GPS uses Neuman–Hoffman code on an L5 safety of life (SoL) signal.)

The civil signal $S_i^{(d)}$ is combined with the restricted signal $S_i^{(r)}$ using QPSK modulation. The spread code is a truncated Gold code: one last chip is removed. The generation polynomials are as follows:

$$\begin{aligned} G1(X) &= 1 + X + X^7 + X^8 + X^9 + X^{10} + X^{11}, \\ G2(X) &= 1 + X + X^2 + X^3 + X^4 + X^5 + X^8 + X^9 + X^{11}, \end{aligned} \tag{2.38}$$

with initial states

$$G1_0 = G2_0 = \{0, 1, 0, 1, 0, 1, 0, 1, 0, 1, 0\}. \tag{2.39}$$

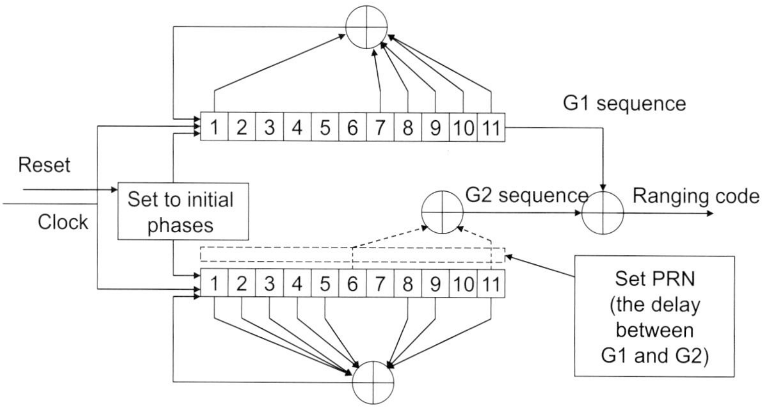

Figure 2.16 BeiDou signal generator. After [6].

The delay is set by choosing specific tap outputs and combining them with an exclusive OR operation in a way similar to GPS (see Figure 2.16). The range codes are different from those for GPS even for the same taps, because the polynomials and their initial states are different.

2.2 GNSS signal propagation error models

2.2.1 Effects of signal propagation through the atmosphere on GNSS

The specifics of the propagation of RF signals through the atmosphere on GNSS frequencies are important in two respects.

First, it is important to consider the effects of the errors caused by signal propagation through the atmosphere on the positioning accuracy. This is important for all positioning and navigation applications, from mobile positioning to extremely precise geodetic applications.

Second, for many geophysical applications a large amount of information about Earth's atmosphere can be derived from these signals. This aspect is essential for many very important geophysical applications.

In this book we consider only the first aspect, the effects of signal propagation on positioning accuracy. We refer readers who are interested in the geophysical applications, which include research on the atmosphere, the ionosphere, earthquake studies, and more, to Petrovski and Tsujii [1].

For our purposes of error consideration, the atmosphere can be divided into two parts: the *troposphere* and the *ionosphere* (see Figure 2.17).

The troposphere is part of the Earth's atmosphere up to 16 km altitude. It contains approximately 80% of the atmosphere's mass and 99% of its water vapor and aerosols. The word troposphere came from the Greek *tropos*, which can be translated as "change."

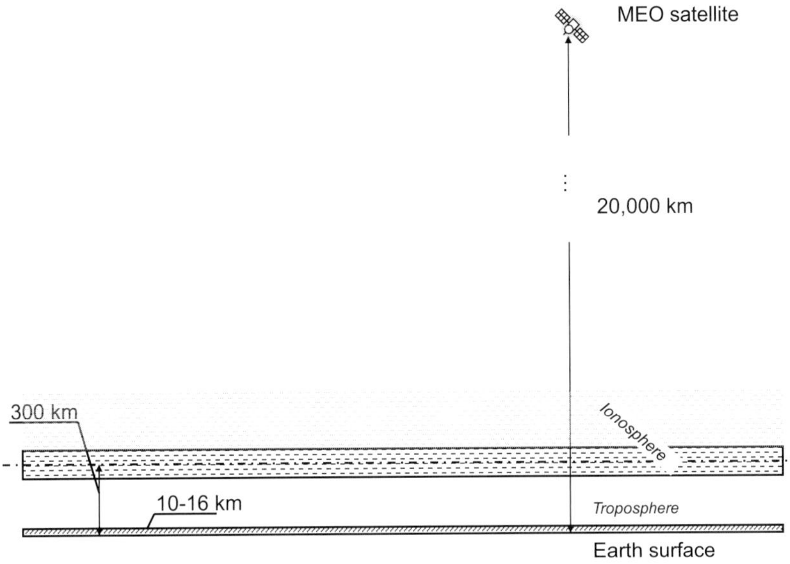

Figure 2.17 Troposphere and ionosphere.

The ionosphere, located between 300 km and 500 km above the Earth's surface, is the upper part of the Earth's atmosphere, containing a layer of ionized electrons.

Signal propagation through the atmosphere causes code measurements to show a delay, and so the measured distance to the satellite appears longer than it is. The amount of delay is different for each satellite and depends on the length of the signal path in the atmosphere. If a satellite has a low elevation angle, its signal travels for a longer time in the atmosphere and therefore it is delayed more than for satellites at a higher elevation. Therefore, in the time before good tropospheric models had been developed, it was customary to use only satellites with an elevation angle more than 15° for positioning. The minimum elevation angle, below which satellites are not accepted for positioning, is called the elevation mask angle or just the *mask angle*. The mask angle criterion is applied in order to eliminate low-elevation satellites from positioning tasks because of the associated large errors.

Regarding carrier phase measurements, the effect is different for different parts of the atmosphere. Carrier measurements indicate delay due to propagation in the troposphere and advance due to propagation in the ionosphere. This difference in the code phase measurements is very important. In particular, it puts limits on how long carrier measurements can be applied to smooth code measurements without reset.

Another important aspect of GNSS frequencies is that the code delay in the ionosphere is equal in magnitude to the carrier advance; the difference is only in the sign of the error. The tropospheric delays are the same for code and carrier. For both tropospheric and ionospheric delays, the amount of delay depends on conditions in the atmosphere and on path length. Therefore, models for atmospheric delay can be presented as a product of delay model and mapping function:

$$d = M(z, \alpha) \cdot d_z, \tag{2.40}$$

where $M(z)$ is the mapping function, d_z is the zenith delay model, z is the zenith angle, and α is the azimuth angle. The zenith angle is given by

$$z = \frac{\pi}{2} - \beta, \tag{2.41}$$

where β is the satellite elevation angle, an angle between the line-of-sight (LOS) from the receiver to the satellite and its projection on the horizontal plane.

The delay model d_z takes into account atmospheric conditions and calculates the delay if the satellite is in zenith. The mapping function M maps this delay into the line-of-sight to the satellite, resulting in the slant delay d.

2.2.2 Algorithms for tropospheric delay calculation

Tropospheric models describe the slant delay in the troposphere as a product of a zenith-angle-dependent mapping function and the zenith delay. The physical delay model is broken down into hydrostatic and wet delay models as follows:

$$d_T = M(z)(d_H + d_W), \tag{2.42}$$

where $M(z)$ is the mapping function, d_H is the zenith hydrostatic delay (in meters), d_W is the zenith wet delay (in meters), and z is the zenith angle (in radians).

The models that are generally in use differ in the way the zenith delays and mapping functions are calculated. Sometimes, different mapping functions are used for the dry and wet parts of the tropospheric delay. The dry, or hydrostatic, component is responsible for 90% of tropospheric delay and can be well modeled. It demonstrates slow temporal variations on the order of one centimeter every six hours. The wet component has a much higher variability, and it is hard to model due to its dependence on highly variable water vapor pressure. The wet component can achieve up to 40 cm. The accuracy of the hydrostatic zenith delay is at the millimeter level, whereas the wet delay is calculated within an error of a few centimeters. Therefore, if the atmospheric pressure measurement is available, the hydrostatic delay is fixed and the wet delay is estimated. Moreover, the wet delay is sometimes divided into two terms, which are dependent on the azimuth of satellite, in order to take into account the asymmetry of the meteorological conditions.

2.2.2.1 Black and Eisner model

A reasonable approximation of the mapping function for tropospheric delay for navigation applications is given by the inverse cosine of the zenith angle, i.e.

$$M(z) = \frac{1}{\cos z.} \tag{2.43}$$

There are many more precise alternatives suitable for geodetic applications.

For navigation applications, another mapping function is given by Black and Eisner:

$$M(z) = \frac{1.001}{\sqrt{0.002001 + \sin^2\beta}}, \tag{2.44}$$

where β is the satellite elevation angle.

2.2.2.2 Saastamoinen tropospheric delay model

The Saastamoinen model [16] implicitly contains both components and mapping functions:

$$d_T = \frac{0.002277}{\cos z}\left[p_H + \left(\frac{1255}{T} + 0.05\right)p_W - \tan^2 z\right]. \tag{2.45}$$

Input values for temperature T, atmospheric pressure p_H, and humidity H are derived from a standard atmosphere model as functions of receiver height, h:

$$p_H = p_{H_0}(1 - 0.0000226(h - h_0))^{5.225}[\text{millibar}], \tag{2.46}$$

$$T = T_0 - 0.0065(h - h_0)[\text{Celsius}], \tag{2.47}$$

$$H = H_0 \cdot e^{-0.0006396(h-h_0)}[\%]. \tag{2.48}$$

The partial pressure of water vapor can be obtained from humidity as follows:

$$p_W = H \cdot e^{(-37.2465+0.213166T-0.000256908T^2)}. \tag{2.49}$$

Temperature must be transformed from the Celsius to the Kelvin scale using

$$T[\text{kelvin}] = T[\text{Celsius}] + 273.16. \tag{2.50}$$

The reference values are given by the standard atmosphere model for altitude $h = 0$ as follows:

$$\begin{aligned} p_H &= 1013.25 \text{ mbar}, \\ T &= 18^\circ \text{ C}, \\ H &= 50\%. \end{aligned} \tag{2.51}$$

2.2.2.3 Niell mapping function

More precise mapping functions are given by the Niell model [17]. The mapping functions are given separately for the dry and for the wet components. The dry hydrostatic mapping function depends on the latitude, the height above sea level, and the day of the year. The wet mapping function depends on receiver latitude.

Mapping functions can be simplified as a function of elevation angle for angles near 90° as a $1/\sin\beta$ approximation (2.43). For lower elevation angles, the non-uniform and finite-width spherical shell model requires a more complicated function. The Niell model uses a three-term continued fraction as a mapping function:

Table 2.3. Coefficients for hydrostatic and wet mapping functions

	Latitude				
Coefficient	15°	30°	45°	60°	75°
			Hydrostatic mapping function		
a_{avg}	1.2769934e-3	1.2683230e-3	1.2465397e-3	1.2196049e-3	1.2045996e-3
b_{avg}	2.9153695e-3	2.9152299e-3	2.9288445e-3	2.9022565e-3	2.9024912e-3
c_{avg}	62.610505e-3	62.837393e-3	63.721774e-3	63.824265e-3	64.258455e-3
a_{amp}	0.0	1.2709626e-5	2.6523662e-5	3.4000452e-5	4.1202191e-5
b_{amp}	0.0	2.1414979e-5	3.0160779e-5	7.2562722e-5	11.723375e-5
c_{amp}	0.0	9.0128400e-5	4.3497037e-5	84.795348e-5	170.37206e-5
			Wet mapping function		
a	5.8021897e-4	5.6794847e-4	5.8118019e-4	5.9727542e-4	6.1641693e-4
b	1.4275268e-3	1.5138625e-3	1.4572752e-3	1.5007428e-3	1.7599082e-3
c	4.3472961e-2	4.6729510e-2	4.3908931e-2	4.4626982e-2	5.4736038e-2

$$M(\beta) = \frac{\dfrac{1}{1+\dfrac{a}{1+\frac{b}{1+c}}}}{\dfrac{1}{\sin(\beta)+\dfrac{a}{\sin(\beta)+\frac{b}{\sin(\beta)+c}}}}, \tag{2.52}$$

where β is the satellite elevation angle. The coefficients for the hydrostatic and wet mapping functions are given in Table 2.3 [17]. For the hydrostatic mapping function the coefficients are functions of latitude and can be calculated as follows:

$$a(\lambda_i, t) = a_{avg}(\lambda_i) + a_{amp}(\lambda_i) \cdot \cos\left[2\pi \frac{\Delta t - \Delta t_0}{365.25}\right], \tag{2.53}$$

where λ_i is the tabular latitude, Δt is the time elapsed since January 0.0 in UT (universal time) days, and Δt_0 is the phase, equal to 28 days. The parameters for the specific latitude are calculated by linear interpolation.

If the receiver altitude is not zero, height corrections can be applied,

$$\begin{aligned} \Delta M(\beta) &= \frac{dM(\beta)}{dh} H, \\ \frac{dM(\beta)}{dh} &= \frac{1}{\sin(\beta)} - f(\beta, a_{ht}, b_{ht}, c_{ht}), \end{aligned} \tag{2.54}$$

where H is the height of the site above sea level and $f(\beta, a_{ht}, b_{ht}, c_{ht})$ is the three-term continued fraction similar to (2.52) with the coefficients $a_{ht} = 2.53\text{e-}5$, $b_{ht} = 5.49\text{e-}3$, $c_{ht} = 1.14\text{e-}3$ determined by a least squares fit to the height corrections at the nine elevation angles.

2.2.3 Algorithms for ionospheric delay calculation

Solar radiation and cosmic rays ionize the upper parts of the Earth's atmosphere and create free electrons and positively charged ions. Ions create atmospheric electricity, which manifests itself through thunderstorms and lightning. The amount of solar radiation that reaches the Earth's atmosphere depends on the time of day and the time of year. The amount of solar radiation (which is correlated with the number of sunspots and changes over an 11 yr period) also affects the level of ionization. A GNSS signal exhibits code delay and phase advance when moving through the ionosphere. This means that when a receiver calculates the range to a satellite from code and carrier phase measurements, it should account for these variations in signal propagation time. The code phase measurements from each satellite can be expressed as follows:

$$\rho_i = r_i + d_I^i + \delta t_r \cdot c, i = 1, \ldots, n, \tag{2.55}$$

where r_i is the distance to the ith satellite, δt_r is a receiver clock error, and d_I^i is the code delay due to the ionosphere.

The carrier is delayed in the troposphere according to the same model as for the code. However, in the ionosphere the carrier experiences an advance, not a delay. The magnitude of the carrier advance is precisely equal to the magnitude of the code delay. The code delay and phase advance of the GNSS signal occur in the ionosphere because the ionosphere is a dispersive medium, in which refractive index depends on signal frequency. For the detailed explanation of this phenomenon, see [1]. The corrected (for ionospheric delay carrier phase) measurements, calculated from a number of carrier waves, can be expressed similarly as follows:

$$\phi_i = \lambda_j N_i + \varphi_i - d_I^i + \delta t_r \cdot c, \tag{2.56}$$

where λ_j is a wavelength on the L_j frequency, φ_i is a carrier phase measurement, N_i is the number of whole waves between a receiver and the ith satellite, and d_I^i takes a minus sign because the carrier *advances* due to the ionosphere.

The parameters for the ionospheric model are transmitted in the navigation message and allow compensation for ionospheric error to up to 70%.

2.2.3.1 Single-layer ionosphere model

Figure 2.18 shows a single-layer model that assumes that all free electrons are concentrated in a shell of infinitesimal thickness. The height of this idealized layer is usually set to the expected height of the maximum electron density.

This model can represent a delay as a function of nominal distance to the satellite and an additional component which is a function of electron density. The delay depends on the total number of electrons integrated along the LOS. It is convenient to introduce the slant total electron content (TEC), which is defined as the electron density integrated along the LOS path, i.e.

$$TEC = \int N_e(s)ds. \tag{2.57}$$

It is expressed in TEC units (TECU), where 1 TECU is equal to 10^{16} electrons in a cylindrical volume with a 1 m^2 cross-section aligned along the LOS. The delay in the ionosphere can be expressed as follows:

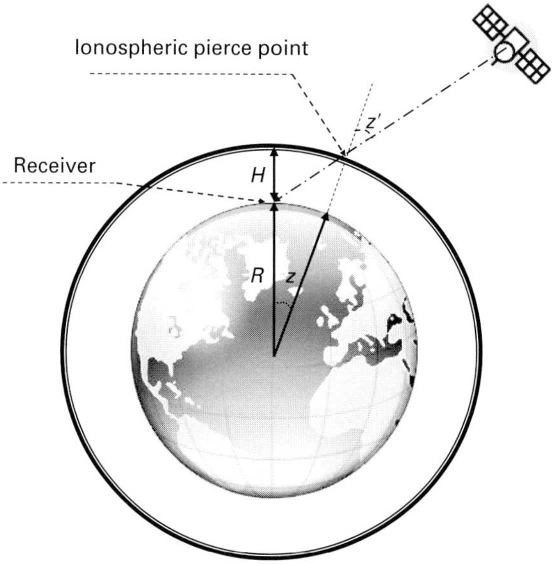

Figure 2.18 Single-layer ionosphere model.

$$d_I = \frac{K_x}{2} \cdot TEC \cdot \left(\frac{1}{f}\right)^2, \tag{2.58}$$

where

$$K_x \approx 80.6 \cdot 10^{16} \left[\frac{\mathrm{m}}{\mathrm{TECU} \cdot \mathrm{s}^2}\right]. \tag{2.59}$$

From equation (2.58), we can calculate a delay caused by the TEC for all GNSS frequencies. For example, for GPS L1=1.57542 GHz we get approximately 0.162 m/TECU.

The slant TEC is calculated along the LOS from a satellite to a receiver and is therefore a function of satellite and receiver antenna position. As it is unique to each user, the slant TEC cannot be used for mapping the ionosphere. Therefore it is necessary to introduce a vertical TEC (VTEC), which is the TEC along the local vertical. The maps of VTEC can be supplied to users, and each user should recalculate VTEC to get the LOS slant TEC for each satellite. Ionospheric maps are provided by, for example, the IGS (the International GNSS Service, formerly the International GPS Service) through the Internet in IONEX (ionosphere exchange format).

A reduced TEC map, broadcast by GPS satellites, is considered in this section. In order to recalculate VTEC to obtain the slant TEC, one uses the single-layer model (see Figure 2.18). In this model we assume that all electrons given by VTEC are concentrated in the single layer located at the specific altitude. An altitude of 350 km is chosen for the GPS broadcast model. CODE uses an altitude of 450 km for ionosphere analysis.

A mapping function needs to be constructed in order to transform VTEC to slant TEC values. These mapping functions are also required in order to construct VTEC maps from ground network observations. We have

$$TEC = M_{TEC}(z) \cdot VTEC. \tag{2.60}$$

The point at which the LOS penetrates the layer is called a *pierce point*. The satellite zenith distance at the pierce point can be expressed as follows (see [1] for details):

$$\sin z' = \frac{R}{R+H} \sin z, \tag{2.61}$$

where R is the radius of the Earth, H is the layer altitude, and z is the satellite zenith angle at the receiver location.

From geometrical considerations, the single-layer mapping function can be defined as follows:

$$M_{TEC}(z) = \frac{1}{\cos z'} = \frac{1}{\sqrt{1 - \sin^2 z'}}. \tag{2.62}$$

2.2.3.2 Ionospheric error compensation in GPS and BeiDou receivers

GPS and BeiDou signals use the Klobuchar ionospheric model. The Klobuchar model is a single-layer model, which takes only the first few members from a TEC representation in a harmonic series and thus defines a TEC map as a cosine-shaped bulge rotating synchronously with the Sun.

The mapping function for a GPS broadcast ionosphere model [18] implements a single layer at 350 km altitude and is expressed as

$$M_{TEC}(z) = 1 + 2\left(\frac{z+6}{96}\right)^3, \tag{2.63}$$

where z is the satellite zenith distance from the receiver (in degrees).

GPS satellites broadcast eight parameters for the single-layer model. An example of these parameters is shown in the GNSS software receiver panel in Figure 2.19 for GPS in July 2013.

The broadcast model should be used by a single-frequency user to correct for ionospheric delay using the algorithm described in the ICD. It is estimated that this model corrects for more than 50% of the single-user root-mean-square (RMS) error due to the ionosphere.

We present the algorithm according to Klobuchar [18]. We encourage the reader to refer to the ICD algorithms for receiver implementation. Ionospheric correction for the single user is calculated as follows. The obliquity factor F, which is similar to the mapping function that we considered for the tropospheric delay, is calculated as a function of satellite elevation β:

$$F(\beta) = 1 + 16 \cdot \left(0.53 - \frac{\beta}{\pi}\right)^3, \tag{2.64}$$

where the obliquity factor is used to calculate a slant delay from the vertical delay, and in this sense it is sometimes called the *slant factor*.

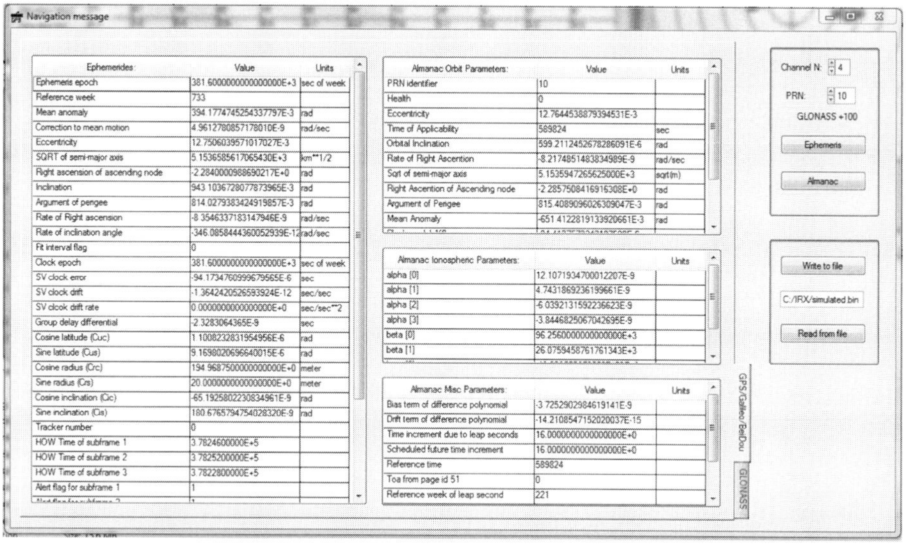

Figure 2.19 ARAMIS™ software receiver navigation message panel with almanac parameters for the Klobuchar model, July 2013.

Then a value x basically defines whether a signal ray goes through the ionospheric bulge:

$$x = \frac{2\pi(t - 50400)}{\sum_{n=0}^{3} b_n \phi_m^n}, \tag{2.65}$$

where ϕ_m is the user's geomagnetic latitude, t is the GPS time of week in seconds, and b_n, $n = 0,\ldots, 3$, are four coefficients transmitted by a satellite. The b_n parameters[1] define the size and shape of the bulge. The ionospheric delay is then calculated. If $|x| > 1.57$, the ray misses the bulge and goes through the area with uniform minimum TEC, and the delay is given by

$$d_I = c_0 F \cdot 5 \cdot 10^{-9} [\mathrm{m}]. \tag{2.66}$$

If $|x| \leq 1.57$, the signal ray passes through the bulge and an extra delay is added to the minimum, so we have

$$d_I = c_0 F \left[5 \cdot 10^{-9} + \left(\sum_{n=0}^{3} a_n \phi_m^n \right) \left(1 - \frac{x^2}{2} + \frac{x^4}{24} \right) \right] [\mathrm{m}], \tag{2.67}$$

where a_n, $n = 0,\ldots,3$, are another four coefficients transmitted by the satellite. The a_n parameters define the shape of the bulge and how much extra delay is added to the minimum.

[1] We denote the transmitted ionospheric parameters here as a_i, b_i instead of α_i, β_i as in the ICD in order not to confuse them with the satellite azimuth and elevation angles, previously denoted α, β.

2.2.3.3 Ionospheric error compensation in GLONASS receivers

A GLONASS navigation message does not provide a single-frequency user with information on the ionospheric parameters. GLONASS receivers can compensate for ionospheric errors using either (2.87) or the following analytical model [19]:

$$d_I = \frac{d_I^0}{\sqrt{1 - \left(\frac{R_E \cos\beta}{R_E + h_{LAYER}}\right)^2}}, \tag{2.68}$$

where d_I^0 is the minimum delay due to the ionosphere, R_E is the radius of the Earth, β is the satellite elevation angle, and h_{LAYER} is the single-layer altitude, which should correspond to the electron density profile maximum.

The minimum delay can be calculated, for example based on the Klobuchar model from (2.66). The slant factor should be adjusted correspondingly.

Most GLONASS receivers also support GPS. In single-frequency GPS/GLONASS receivers, ionospheric delay can be compensated for using the GPS broadcast Klobuchar model for GLONASS satellites. The correction algorithms should be adjusted for GLONASS frequencies. Additionally, GLONASS provides civil signals on two frequencies, which allows for direct measurements and compensation for the ionospheric delay.

2.2.3.4 Ionospheric error compensation in Galileo receivers

The previously considered ionospheric models are single-layer models. These models do not need to know a vertical electron density profile, and therefore they can be implemented using only ground-based observations, such as GNSS global networks. Recent advances in ionosphere mapping using occultation techniques and low Earth orbit (LEO) satellites (see [1] for details) allow us to obtain information about a vertical electron density profile and introduce corresponding multi-layer models.

The Galileo system adopts the NeQuick model, which is a multi-layer model, sometimes called a profiler. It was introduced first in [20] and then continuously developed, in particular for implementation in Galileo [21]. Graphical representations of a single-layer model, such as Klobuchar or the IGS global model, and a multi-layer model, such as NeQuick, are given in Figure 2.20.

NeQuick is expected to account for about 70% of the ionospheric error. The model uses empirical data collected from ionograms (obtained by probing the ionosphere with various frequencies). Based on these ionograms, NeQuick restores the electron density profile. The NeQuick model implementation in Fortran is available in source code free of charge from ITU-R (the radiocommunication sector of the International Telecommunication Union (ITU)). The inputs into the NeQuick model comprise the solar flux (F10.7) and data bases related to ionospheric maps and geomagnetic field information. These data also should be downloaded to a receiver periodically, once a month.

The Galileo broadcast ionospheric model parameters include coefficients which are required to compute the effective ionization level, Az. An effective ionization level is used in Galileo receivers instead of the solar flux and is computed as follows:

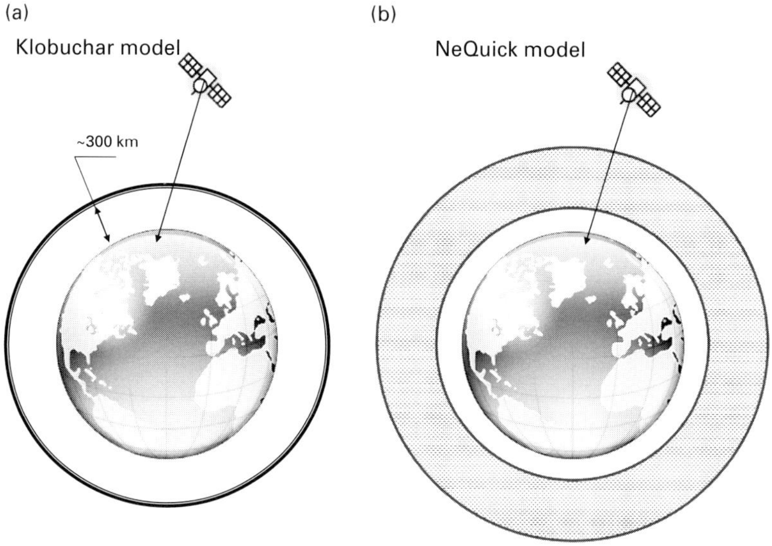

Figure 2.20 (a) Single-layer Klobuchar and (b) multi-layer NeQuick ionospheric models (not to scale). After [1].

$$Az = a_{i0} + a_{i1}\mu + a_{i2}\mu^2, \tag{2.69}$$

where μ is the modified dip latitude (MODIP), which is expressed as follows $\tan\mu = i_M/\sqrt{\cos\lambda}$. Here i_M is true magnetic inclination (dip) and λ is the latitude. The information on true magnetic inclination should be available in the receiver from a data base, which should be updated once every 5 years. The receiver also should download ITU-R long-term reference ionospheric data (FoF2) once per month. The NeQuick model then allows the calculation of the electron concentration at any point in the ionosphere. One can then integrate the electron density along the LOS between receiver and satellite. The resulting slant TEC can be directly transformed into the ionospheric error using (2.58). The obliquity factor F is calculated using (2.64).

For implementation in Galileo, the three parameters are calculated using a global reference station network and are broadcast to a user on the following day. Galileo transmits only three parameters in comparison with eight coefficients of the Klobuchar model, implemented in GPS and BeiDou. In addition to these parameters the Galileo navigation message has five disturbance flags. The Earth is divided into five regions by latitude. Galileo transmits disturbance flags, one for each latitude belt, which indicates whether the broadcast parameters can or cannot be used for that region.

2.2.3.5 Ionospheric error corrections from GEO/HEO satellites

GEO and HEO satellites, which are designated either as space-based augmentation systems (SBAS), such as WAAS, MSAS, EGNOS, or SDCM (Luch), or as part of GNSS, such as BeiDou use a similar grid-based ionospheric correction model; see [1] for details.

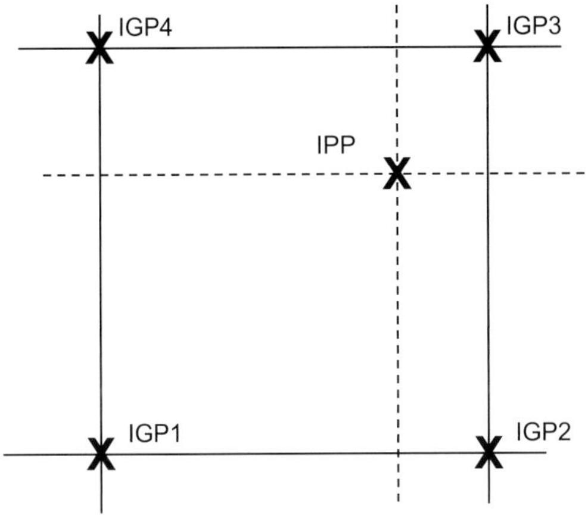

Figure 2.21 Application of a grid model.

The ionospheric delay is provided for an ionospheric grid point (IGP). The ionospheric delay correction in a receiver is computed at the ionospheric pierce point (IPP), which is the intersection of the signal path and the ionosphere layer with maximum electron density. The ionospheric delay at the IPP is computed using the interpolation method using the delay estimates of four surrounding IGPs, as shown in Figure 2.21.

The vertical ionospheric delay at the IPP is computed as follows:

$$d_{IPP}^{V} = (1-x)\cdot(1-y)\cdot d_{IGP1}^{V} + x(1-y)\cdot d_{IGP2}^{V} + x\cdot y\cdot d_{IGP3}^{V} + (1-x)\cdot y\cdot d_{IGP4}^{V}, \quad (2.70)$$

where

$$x = \frac{\lambda - \lambda_1}{\lambda_2 - \lambda_1}, \quad (2.71)$$

$$y = \frac{\phi - \phi_1}{\phi_2 - \phi_1}. \quad (2.72)$$

Then the slant ionospheric delay error and variance at the user are computed by multiplying the obliquity factor F as follows:

$$d_I = F \cdot I_{IPP}^{V}. \quad (2.73)$$

2.2.4 Ionospheric error compensation in multi-frequency GNSS receivers

It is intended that all GNSS will transmit civil signals on the second frequency band. Figure 2.22 shows the GNSS multi-frequency spectrum allocation. Most importantly, providing signals on more than two frequencies allows compensation for ionospheric error. GPS transmits civil signals on three frequencies; the GPS civil signal on L2 is called L2C, and GPS satellites also transmit a SoL signal on L5. An additional third frequency may not play a significant role for mobile applications, so we consider only L2C here. The L2C signal is already transmitted from some GPS satellites. In addition, L1C, L2C, and L5 signals are currently transmitted by the QZSS satellite in test mode.

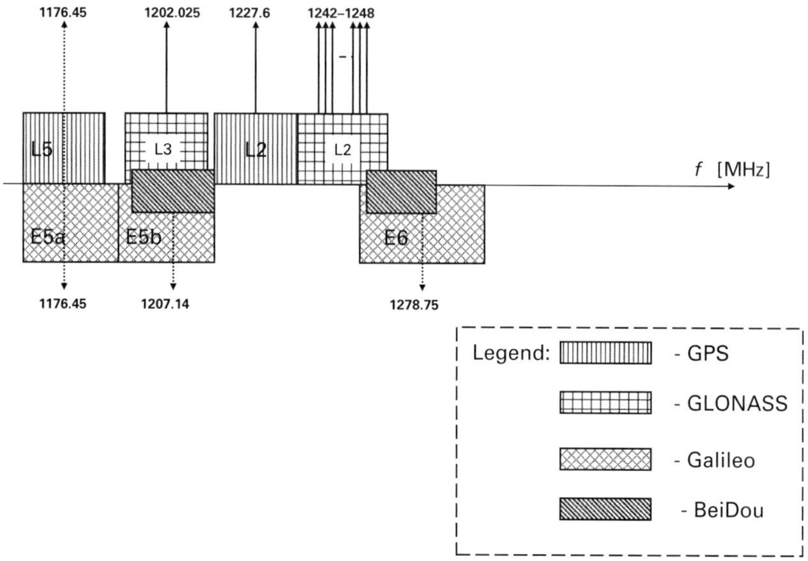

Figure 2.22 GNSS multi-frequency spectrum allocation.

GPS signals are generated on top of a common satellite clock with frequency 10.23 MHz. GPS frequencies used for civil signals are as follows:

$$\begin{aligned} \mathrm{L1} &= 1575.42 = 154 \times 10.23\ \mathrm{MHz}, \\ \mathrm{L2} &= 1227.6 = 120 \times 10.23\ \mathrm{MHz}, \\ \mathrm{L5} &= 1176.45 = 115 \times 10.23\ \mathrm{MHz}. \end{aligned} \tag{2.74}$$

GLONASS transmits civil signals on L1, L2, and L3, and has started to transmit civil CDMA signal on L3 in test mode. The signal uses a Kasami sequence. For L2, the GLONASS frequencies are defined as follows:

$$L2_k = L2_0 + k\Delta L_2, \tag{2.75}$$

where $L2_0 = 1246$ MHz, $\Delta L_2 = 4375$ kHz.

Depending on the signal design, dual-frequency users may also benefit from receiving a complete navigation message twice as quickly. For example, Galileo SoL E1-B and E5b-I users can receive even and odd pages at the same time from these two signals. Similar navigation data are transmitted on both frequencies, but odd pages are transmitted on one frequency at the same time as even pages on another frequency.

Let us consider ionospheric error compensation in detail. The ionospheric model delay depends on the frequency (2.58). This allows us to estimate ionospheric delay rather than calculate it if a receiver allows code measurements on two frequencies. A dual-frequency receiver is expected to be standard for mobile receivers in the future. All GNSS are to supply users with civil signals on two frequencies. However, the benefits of using a dual-frequency receiver may not always overcome its shortcomings related to its price, weight, and power consumption.

The ionospheric delay compensation can be achieved as follows. The corrected code phase observables may be calculated as [2]

$$P = \frac{P_j - \gamma_{i,j} \cdot P_i}{1 - \gamma_{i,j}}, \tag{2.76}$$

where P denotes the code phase measurements corrected for ionospheric delay, and P_i denotes the code phase measurements on frequency i. For L1 and L2C signal users,

$$\gamma_{1,2} = \left(\frac{f_1}{f_2}\right)^2 = \left(\frac{1575.42}{1227.6}\right)^2. \tag{2.77}$$

A similar algorithm should be applied to other civil dual-frequency signals. The algorithms for various combinations of L1-C/A, L2C, and L5-I/L5-Q users are similar to that in (2.76), and also include broadcast inter-signal correction (ISC) terms and group delay T_{GD} for each satellite:

$$P = \frac{(P_j - \gamma_{i,j} \cdot P_i) + c_0(ISC_j - \gamma_{i,j} \cdot ISC_i)}{1 - \gamma_{i,j}} - c_0 T_{GD}. \tag{2.78}$$

Correspondingly, for L1 and L5 signal users,

$$\gamma_{1,5} = \left(\frac{f_1}{f_5}\right)^2 = \left(\frac{1575.42}{1176.45}\right)^2, \tag{2.79}$$

and, for L2 and L5 signal users,

$$\gamma_{2,5} = \left(\frac{f_2}{f_5}\right)^2 = \left(\frac{1227.6}{1176.45}\right)^2, \tag{2.80}$$

where $\Delta P = P_i - P_j$ is the difference in code phase measurements on two frequencies. This difference includes errors resulting from unaccounted for differences in code phase measurements on both frequencies. This error includes inter-frequency hardware biases on satellite and receiver, noise, ionospheric errors of higher order, and so on.

For GLONASS, $f_1/f_2 = 9/7$, and correspondingly the ionospheric correction for a GLONASS L1 user can be described as follows:

$$d_I^{f_1} = \Delta P \left(1 - \frac{f_1^2}{f_2^2}\right)^{-1}, \tag{2.81}$$

$$d_I^{f_1} = 1.531 \cdot \Delta P, \tag{2.82}$$

where $\Delta P = P_i - P_j$.

Ionosphere-free observables cannot, however, remove higher-term components. The ray bending for two frequencies is different [1], and this also introduces an error which depends on satellite elevation.

Ionospheric corrections supplied by a server may provide even better accuracy than do dual-frequency measurements, if the server is located less than 10 km from a rover.

Let us consider ionosphere-free observables in such a way that ionospheric delay will be completely removed from a new linear combination as follows:

$$L_3 = \kappa_{1,3}L_1 + \kappa_{2,3}L_2, \tag{2.83}$$

where the coefficients are defined as

$$\begin{aligned} \kappa_{1,3} &= \frac{f_1^2}{f_1^2 - f_2^2}, \\ \kappa_{2,3} &= \frac{f_2^2}{f_1^2 - f_2^2}. \end{aligned} \tag{2.84}$$

For GPS L1 and L2, the coefficients are given by

$$\begin{aligned} \kappa_{1,3} &\approx 2.546, \\ \kappa_{2,3} &\approx -1.546. \end{aligned} \tag{2.85}$$

The main disadvantage of using ionosphere-free observables is that although they remove most of the ionospheric errors, the noise due to these observables is approximately three times higher than noise from L1 or L2 signals [22]. Assuming $\sigma(L_1) \approx \sigma(L_2)$, the noise in ionosphere-free observables can be expressed as follows:

$$\sigma(L_3) = \sqrt{\kappa_{1,3}^2\sigma^2(L_1) + \kappa_{2,3}^2\sigma^2(L_2)} = \sqrt{\kappa_{1,3}^2 + \kappa_{2,3}^2}\sigma(L_1) \approx 3\sigma(L_1). \tag{2.86}$$

This shows that noise in ionosphere-free observables is three times higher than L1 noise.

2.3 GNSS data

2.3.1 GPS and BeiDou navigation messages

All essential information related to satellite ephemerides, satellite clocks, and message time marks is encoded in binary messages, denoted D_i in equation (2.1).

GPS and BeiDou navigation messages also contain ionospheric parameters for the Klobuchar model to compensate for variable signal delay in the ionosphere.

Both GPS and BeiDou MEO satellites transmit L1 civil signal navigation messages with a rate of 50 bps. The complete navigation message for GPS is transmitted within 12.5 min, and that for BeiDou within 12 min. Figure 2.23 shows the structure of GPS L1 C/A and BeiDou MEO NAV messages.

Both navigation messages consist of 24 frames of 30 s duration. The most important and basic element of the navigation message is a subframe, which for both systems is 300 bit and lasts for 6 s. For both systems the immediate orbital information relating to the transmitting satellite is encoded in subframes 1, 2, and 3. The non-immediate information, such as the almanac data for the system, is encoded in subframes 4 and 5.

The subframe is important, because it has a time mark, which is necessary in order to construct pseudorange measurements. A navigation message subframe

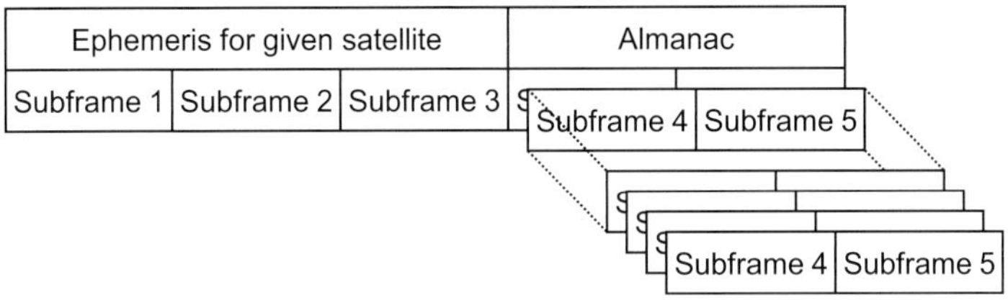

Figure 2.23 GPS L1 C/A and BeiDou MEO NAV data structures.

starts with the sequence called a preamble, which is used to identify the beginning of the subframes.

The preamble in GPS and BeiDou uses Barker code [23], whose autocorrelation function allows for the best alignment. However, the preamble is very short and may occur by accident many times in the body of the message. Therefore, the decoding algorithm must take special care to confirm that the sequence is indeed a preamble. This can be done, for example, by checking the sequential preambles, because, as soon as the algorithm has assumed that this is the preamble, the position of the next preamble is known to it. Alternatively, the preamble can be checked by decoding, for example, the date and time, which should be known with some certainty by the receiver.

The GPS preamble is an 8 bit modified Barker code given by

$$P = \{1, 0, 0, 0, 1, 0, 1, 1\}. \tag{2.87}$$

The BeiDou preamble is an 11 bit Barker code as follows:

$$P = \{1, 1, 1, 0, 0, 0, 1, 0, 0, 1, 0\}. \tag{2.88}$$

A Barker code is a very short (up to 13 bits) binary sequence such that the off-peak autocorrelation coefficients are minimal and either 0 or -1 for a sequence constructed of -1 and 1 values.

Any part of the GPS L1 navigation message with a length of 36 s contains enough information for a receiver to make a positioning. Therefore, a receiver can in general make a positioning with a GPS L1 C/A signal within 36 s after it is switched on. BeiDou orbital parameters are given in a format similar to the GPS navigation message, and use similar algorithms to derive satellite position.

2.3.2 Galileo navigation message

A Galileo navigation message page consists of two page parts (even and odd) transmitted sequentially over the same frequency. Each page part has 120 bits of information. The E1-B is also part of the SoL dual-frequency service, so it also transmits alert pages with another 120 bits on each page. The other part of the signal is an E5b-I signal, which is transmitted on L5.

This design provides a redundancy and additional error check mechanism. So if BeiDou now includes SBAS (GEO and HEO) satellites into their GNSS, transferring to them some of the GNSS functions, the new GPS and Galileo signals transfer some of

Table 2.4. GNSS civil signals on L1 at the time of writing

Parameter	GPS	GLONASS	Galileo	BeiDou
Signal	L1 CA	L1 SP	E1-B	B1 Data
Type	CDMA	FDMA[a]	CDMA	CDMA
Code	Gold code	M-code	memory code	balanced Gold code
Secondary code	no	no	no	yes
Code length	1023	511	4092	2046
Code sequence interval (ms)	1	1	4	2
NAV message size (bit)	37 500	7500	180 000/ 45 000[b]	36 000
NAV message period (min)	12.5	2.5	12/6[c]	12
NAV data bit rate (bps)	50	50	250	50/250[d]
Frame length (s)	30	30	30[e]	30
Subframe length (s)	6	2	2[e]	6
Time mark interval (s)	6	2[f]	2	6
Time frame	UTC(USNO)	UTC(SU)	TAI	BDT

[a] More precisely FDMA on top of CDMA.
[b] With redundancy and alert pages/without redundancy and alert pages.
[c] For single-frequency/dual-frequency user.
[d] For MEO/GEO(HEO).
[e] Galileo frame, subframe and page in this table corresponds to GPS/BeiDou superframe, frame, and subframe accordingly.
[f] Does not contain explicit time information.

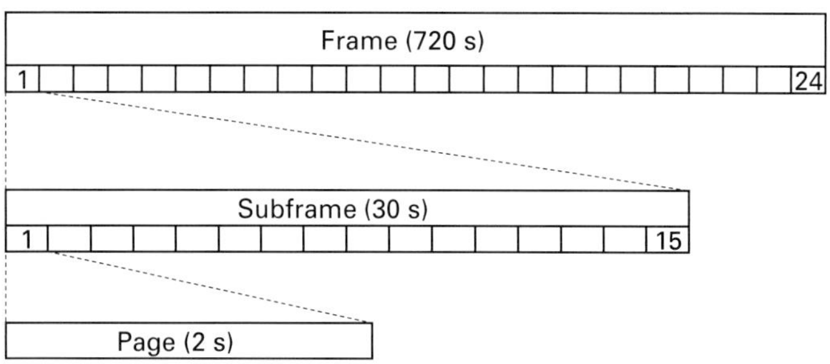

Figure 2.24 Galileo navigation message structure.

the SBAS functions, which in particular include high navigation data bit rate and integrity messages, to GNSS satellites.

A Galileo NAV message for an E1-B signal is shown in Figure 2.24. It has a 250 bps rate and is of 720 s duration. It consists of 24 30 s subframes; each subframe consists of 15 2 s pages. The Galileo page is similar to the GPS/BeiDou superframe and contains 500 bits.

The Galileo frame, subframe, and page are compared to the GPS/BeiDou superframe, frame, and subframe, respectively, in Table 2.4.

Each page has a preamble. The Galileo preamble is called the NAV synchronization pattern. It is a 13 bit Barker sequence truncated to 10 bits:

$$P = \{0, 1, 0, 1, 1, 0, 0, 0, 0, 0\}. \tag{2.89}$$

The orbital and clock parameters are given as Keplerian elements in a similar format to those for GPS and BeiDou. The algorithms for calculating satellite position are also similar, and are as given in Chapter 1. The ionospheric model is different from that of GPS and BeiDou, which use the Klobuchar model. The details of the ionospheric model are given in Section 2.2.3.4.

Galileo employs a convolutional encoding scheme for navigation data on all signals. This method is implemented using a Viterbi decoder. The same method is used for the GPS L2C signal. Each Galileo message uses interleaving. The interleaving on the transmitter side performs permutations of the bits of the navigation message. A simple block interleaving implements four M-row by N-column memories, two in the transmitter and two in the receiver. In the transmitter the encoder reads memory by row and writes to another memory by column. When the memory is full, the data are transmitted row by row. At the receiver the process is reversed and the correct sequence is restored. In this way the burst of errors caused, for example, by interference is distributed instead of affecting a number of neighboring bits. More complicated convolution interleaving is realized by reading driven by a shift register. For an E1-B navigation message, the block interleaver size (in symbols) is 240 and the block interleaver dimensions are 30 columns $\times$ 8 rows. A deinterleaving process on the receiver side restores the original order of bits. If parts of navigation message have been corrupted at certain localized points, for example as a result of multipath, obstructions, or interference, then the burst of corrupted bits is dispersed over a number of words. Pieces of data, corrupted by a small number of errors, can be restored through the redundancy provided in implemented error-correction algorithms.

2.3.3 Algorithm for constructing GPS/BeiDou/Galileo pseudorange measurements

2.3.3.1 GPS time mark

For the GNSS time scale, see Section 1.2. A GPS navigation message contains data necessary to convert GPS time to UTC that allow the conversion to be within 90 ns (1σ) accuracy. For many mobile applications, however, one may convert GPS time directly to calendar time.

The time mark is embedded in each subframe of the navigation message as a *Z-Count*. This gives the time of the week for each subframe, which corresponds to the time of transmission. The time of week (TOW) count occupies 19 least-significant bits of the Z-Count. The most-significant bits of the Z-Count give the current GPS week number.

A Z-Count epoch corresponds to 1.5 s. Therefore, the TOW count has a range of 403 199 epochs to account for 604 800 seconds in a week. The counter is reset to zero at the end of each week. Its zero state is defined as the epoch at the start of the present week at midnight on Saturday.

2.3.3.2 BeiDou time mark

For BeiDou, the time-related algorithm is the same as in the preceding section for GPS. The time mark, however, is different, in this case implemented with a seconds of week

(SOW) 20 bit count. The SOW occupies bits 19–26 and 31–42 in each subframe, and gives the number of seconds within a week since Saturday midnight in BeiDou Navigation Satellite System time (BDT). SOW identifies the time of transmission of the leading edge of the preamble first bit. The 13 bit week number gives the number of weeks since January 1, 2006 BDT.

2.3.3.3 Galileo time mark

The Galileo time mark is encoded in a 32 bit number which includes the week number and the TOW, which is synchronized to each satellite's version of Galileo system time (GST). It gives the SOW in a range up to 604 799 s and is reset to zero at the end of each week. GST starts at midnight on Saturday, August 21. The epoch time is given relative to the leading edge of the first chip of the first code sequence of the first page symbol.

The 12 bit week number provides a count from the origin of GST over a 4096 week range, which is about 78 years. Then the counter is reset to zero.

2.3.3.4 Pseudorange construction algorithm

Code phases are measured in the baseband processor first by acquisition, and then, if the receiver supports tracking, by tracking loops. Not all receivers necessarily have tracking loops. A snapshot receiver[2] may be able to make a positioning using only the acquisition. This is especially important for indoor positioning or for dynamical positioning in an environment where the satellite signal can be obstructed and corrupted by multipath. Code phase measurements and baseband processors are considered in detail in the following chapters. For now, we assume that code phase measurements have been made. Code phase measurements are ambiguous because code sequence is repeated; for a GPS C/A signal, this occurs every millisecond, and 1 ms is equivalent to about 300 km. For the purposes of positioning, we need to restore code measurements to relate to the full distance to the transmitting satellite.

As we have seen in Section 2.3.3.1, a GPS navigation message transmits a time mark, a Z-Count embedded into the navigation message every 6 s with its value incremented each 1.5 s. The Z-Count allows the restoration of the distance to a satellite from the code measurements in the form of a pseudorange.

The result of this restoration process is the pseudorange to a satellite; it does not yield the actual range because restored unambiguous measurements are still corrupted, usually with very large receiver clock error. The algorithm of pseudorange reconstruction should account for the following sequential tasks.

(1) Algorithm finds current epoch in the receiver, checks Z-Count, and verifies the preamble.
(2) Algorithm aligns receiver internal time scale called the fundamental time frame (FTF) with one satellite signal.
(3) In the final step all channels should be aligned with FTF.

[2] We define a snapshot receiver as a receiver which can make a positioning using only short chunks (snapshots) of GNSS signal with a duration of a few milliseconds.

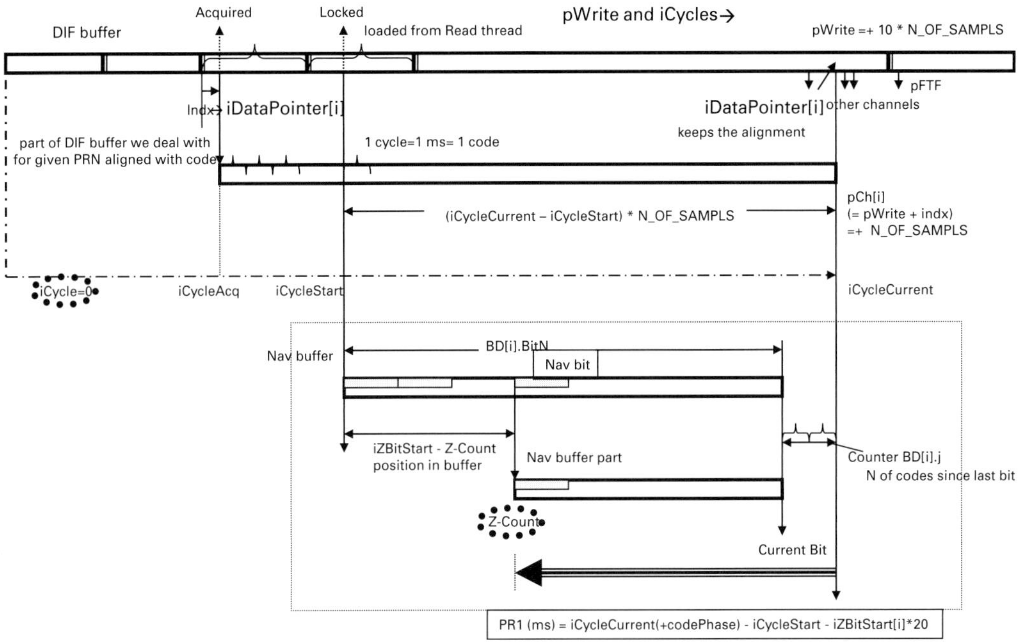

Figure 2.25 Constructing GPS code phase measurements.

(4) Pseudoranges can be constructed from the number of code cycles from the Z-Count by various ways, as depicted in Figure 2.25, for example as follows:

$$P = Z + \Delta P + M_C - M_0 - 20 \cdot B \quad [\text{ms}], \tag{2.90}$$

where P is the pseudorange, Z denotes the Z-Count time, ΔP is the code phase, B is the number of bits accumulated before the Z-Count, M_C is the current code period, and M_0 is the first code period at the epoch of the tracking loop lock. It is important to note that the FTF is not fixed to the receiver front-end clock through a pointer, because FTF moves with continuous time whereas the receiver front-end clock resolution is limited by samples, which are integers.

The same algorithm can be implemented for BeiDou and Galileo without modification.

2.3.4 GLONASS navigation message contents and structure

A GLONASS navigation message consists of repeated superframes with durations of 2.5 min (see Figure 2.26). Each superframe consists of 5 frames each with a duration of 30 s. Each frame consists of 15 subframes,[3] and each subframe has a duration of 2 s. All essential data for the transmitting satellite are transmitted within each frame.

[3] These elements are called strings in the ICD. We use the term subframes for academic purposes for consistency.

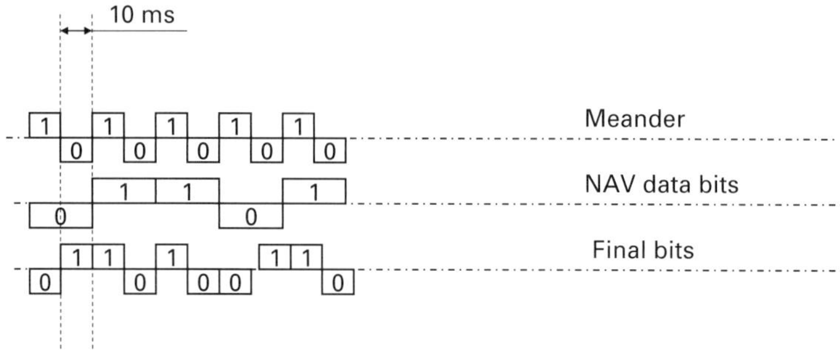

Figure 2.26 GLONASS navigation message structure.

We consider the GLONASS message in more detail because there is substantially less literature available than for GPS. Here we concentrate mainly on the part of navigation message that is essential for positioning. Immediate information is given by the following parameters.

Satellite slot number, n.

Start epoch of frame within a day, expressed in hours within a day, minutes within an hour, and half-minutes within a minute in seconds (t_{hour}, t_{min}, t_{sec}).

Health flag B_n. The flag takes value 0 if the satellite is healthy or 1 if it is not.

Parameter t_b, an index of time interval within 24 hr. The length of the interval is defined by flag P_1 in accordance with Table 2.5. Ephemerides are referenced to the middle point of this interval.

Flag P_1 indicates the size of interval t_b according to Table 2.5. No navigation data uploads should occur within the interval of applicability.

Flag P_2 indicates when t_b takes an odd value.

Flag P_3 indicates the number of satellites for which the current frame contains almanac data. The value 1 indicates five and the value 0 indicates four satellite almanacs in this frame.

Flag P_4 indicates whether a new set of ephemerides for a given satellite has been uploaded (1) or not (0).

Flag P denotes satellite operation mode in relation to the generation of parameters specifying relations with GPS and GLONASS time scales (τ_{GPS} and τ_C, respectively). Flag P is in use for GLONASS-M satellites only.

Health flag, l_n. The flag takes a value 0 if the satellite is healthy or 1 if it is not.

Parameter E_n is the ephemeris age in days. The parameter can take values from 0 to 31.

Parameter γ_n is the deviation (predicted by the carrier frequency) from the nominal value for epoch t_b caused by relativistic effects:

$$\gamma_n(t_b) = \frac{\hat{f}_n(t_b) - f_n}{f_n}, \tag{2.91}$$

where $\hat{f}_n(t_b)$ is the estimated carrier frequency value accounting for gravitational forces and relativity effects, and f_n is the nominal frequency for the transmitting satellite.

Table 2.5. Parameter value for designating interval of ephemeris validity

P_1 value	Length of ephemeris validity interval, t_b (min)
0	0
1	30
2	45
3	60

Table 2.6. Accuracy indicator value and corresponding accuracy

F_T	Measurement accuracy (m)
0	1
1	2
2	2.5
3	4
4	5
5	7
6	10
7	12

Parameter N_T is the current day number within a four-year interval, starting from January 1 of the previous leap year.

Parameter F_T is an accuracy indicator. It can take values from 0 to 14 (see Table 2.6). This parameter allows code observables to be weighted accordingly in positioning algorithms. This parameter is similar to GPS user range accuracy (URA).

Parameter M indicates the transmitting satellite type. The value 0 denotes GLONASS, and value 1 denotes the GLONASS-M satellite.

Orbital parameters are given by tabular values of coordinates, velocities, and acceleration along orbital axes X, Y, Z in the PZ-90.02 coordinate frame at epoch t_b.

Parameter τ_n is the shift between GLONASS time and the onboard clock.

GPS ephemerides are provided as a set of Keplerian osculating elements, and GLONASS ephemerides are provided in tabular format. Each navigation message also provides the user with a time mark to pin down the satellite position to a specific moment of time. A GPS L1 C/A or BeiDou MEO user receives a time mark within 6 s of the navigation message, which is the length of one frame. A GLONASS user receives a time mark within 2 s of the navigation message.

For GLONASS, a receiver needs to receive 15 subframes (strings), i.e. one frame. Each subframe takes 2 s. It takes 15 times 2 s plus an additional 2 s to ensure receipt of the first four subframes in the frame. Altogether, this accounts for 32 s. For GPS, it is necessary to receive five subframes. To ensure the reception of five whole subframes, each with 6 s length, the receiver needs 5 times 6 s plus an additional 6, which accounts for 36 s.

The preamble in GPS is located at the beginning of each subframe. The preamble in GLONASS is called the time mark, and it is located at the end of each subframe (string).

For GLONASS pseudorange measurements are constructed using time taken from the start of the day. This time is calculated based on the time mark provided in a subframe and a code sequence count, which calculates time in milliseconds since the time mark.

In order to obtain the contents of the navigation message and the time mark, we have to look at how it is encoded.

2.3.5 Subframe of a GLONASS navigation message

2.3.5.1 Algorithm for reading GLONASS subframe

(1) Collect navigation message.

(2) Remove meander from the recorded navigation message by applying to each bit an XOR operation with the previous bit.

(3) Find the time mark, which performs the same function as a preamble, with the difference that it is located at the end of the subframe rather than at the beginning (see Figure 2.27).

The time mark is repeated every 10 s and is denoted by the following sequence:

$$TM[30] = \left\{ \begin{array}{l} 1,1,1,1,1,0,0,0,1,1,0,1,1,1,0, \\ 1,0,1,0,0,0,0,1,0,0,1,0,1,1,0 \end{array} \right\}. \tag{2.92}$$

(4) Apply bitwise XOR ($\wedge$) operation on the message string and its copy, 1 bit shifted to the right. The string is of length 85 bits. The total subframe is 100 bits in length with a 15 bit time mark, which is excluded from further operations. The resulting string contains navigation data in binary format.

(5) Check parity of eight Hamming code bits.

(6) Read specified bits from the string and convert binary format to decimal.

(7) Scale the numbers according to the scale values.

Let us look now at the subframe contents in detail. Subframes 1 to 5 contain immediate information or ephemerides. Subframes 6 to 15 contain orbit parameters

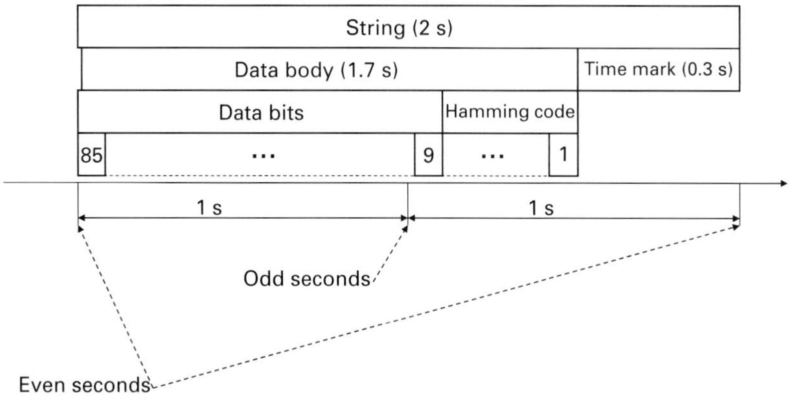

Figure 2.27 GLONASS subframe structure. After [4].

and clocks for all the satellites in a constellation. Satellite position is then calculated using algorithms given in Chapter 1.

2.3.5.2 Subframes containing immediate information

2.3.5.2.1 Subframe 1

Subframe 1 contains the following data.

(1) P_1 flag (bits 8 and 9).[4]

(2) Epoch of frame start within a day, in hours (t_{hour}) (bits 10–14).

Epoch of frame start within an hour, in minutes ($t_{\min}$) (bits 15–20).
Epoch of frame start within a minute, in seconds (t_{sec}) (bit 21). Note that t_{sec} can take two values, either 0 or 30 s. The epoch (in seconds) since the start of the day is then restored as follows:

$$t_{FRAME} = t_{hour} \cdot 60 \cdot 60 + t_{\min} \cdot 60 + t_{\text{sec}}. \tag{2.93}$$

(3) Coordinate (bits 52–77), velocity (bits 23–45), and acceleration (bits 47–50) along axis *X*, and their signs (bits 51, 22, and 46, respectively). A 0 value indicates a "+" sign and a 1 indicates a "−" sign. The corresponding values are calculated using signs and scale factors as follows:

$$X = X' \cdot sign(X) \cdot 2^{-11}, \tag{2.94}$$

$$\dot{X} = \dot{X}' \cdot sign(\dot{X}) \cdot 2^{-20}, \tag{2.95}$$

$$\ddot{X} = \ddot{X}' \cdot sign(\ddot{X}) \cdot 2^{-30}. \tag{2.96}$$

2.3.5.2.2 Subframe 2

Subframe 2 contains the following data.

(1) Flags B_n (bit 6) and P_2 (bit 9).

(2) An index of time interval within 24 hours, T_b (bits 10–16); scale factor is 15.

(3) Coordinate (bits 52–77), velocity (bits 23–45), and acceleration (bits 47–50) along axis *Y* and their signs (bits 51, 22, and 46, respectively). A 0 value indicates a "+" sign, and a 1 indicates a "−" sign. The corresponding values are calculated in a similar way to the *X* values.

2.3.5.2.3 Subframe 3

Subframe 3 contains the following data.

(1) Flag P_3 (bit 6).

(2) Carrier frequency deviation from predicted γ_n. Value in bits 8–17 and sign in bit 7.

[4] The bit number is given as 1-based.

(3) Coordinate (bits 52–77), velocity (bits 23–45), and acceleration (bits 47–50) along axis *Z* and their signs (bits 51, 22, and 46, respectively). A 0 value indicates a "+" sign, and a 1 indicates a "−" sign. The corresponding values are calculated in a similar way to the *X* values.

2.3.5.2.4 *Subframe 4*

Subframe 4 contains the following data.

(1) The scaled value of the onboard clock shift $\tau_n^{'}$ (bits 7–27) and its sign (bit 6). The value of the shift is restored by

$$\tau_n = \tau_n^{'} \cdot sign(\tau_n) \cdot 2^{-30}. \tag{2.97}$$

(2) Flag P_4 (bit 52).
(3) Accuracy flag F_T (bits 53–56).
(4) Day number N_T (bits 60–70).
(5) Satellite number *n* (bits 71–75).
(6) Satellite type parameter M (bits 76–77).

2.3.5.2.5 *Subframe 5*

Subframe 5 contains the following essential data.

(1) Health flag l_n, presented by bit 77. This health flag will indicate if a problem has occurred no later than within 10 s.
(2) Time scale parameter τ_C, the time offset between GLONASS system time and UTC (SU). Its resolution has been increased to 2^{-31} s. GLONASS time is connected to UTC, so unlike for GPS it is not a continuous time. Therefore, the leap second difference between time scales should be accounted for in the receiver algorithm.
(3) Parameter τ_{GPS} (see footnote 4, p. 81), the time offset between GPS and GLONASS time scales.

The values of parameters τ_C and τ_{GPS} are given by bits 18–48 and 56–76, respectively. The parameter signs are given by bits 17 and 55, respectively. The parameters are constructed as follows:

$$\tau_C = \tau_C{'} \cdot sign(\tau_C) \cdot 2^{-27}, \tag{2.98}$$

$$\tau_{GPS} = \tau_{GPS}{'} \cdot sign(\tau_{GPS}) \cdot 2^{-30}. \tag{2.99}$$

Figure 2.28 shows an example of a GLONASS ephemeris message decoded by the adaptive receiver applied for monitoring of ionospheric scintillation (ARAMIS™).

2.4 What's in a sat's name?

At first sight, it would seem that a GNSS user would never be interested in the details concerning the satellites themselves. However, one may be interested in the name of the specific space vehicle for a few reasons. Figure 2.29 shows an artist's representation of the Galileo satellite. All GNSS satellites look similar, featuring the same distinguishing

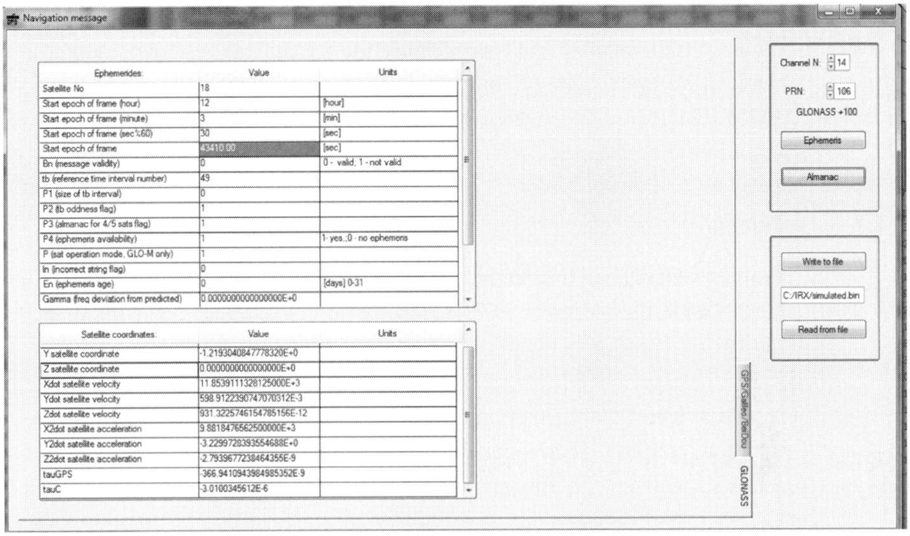

Figure 2.28 ARAMIS™ receiver panel with decoded GLONASS ephemerides.

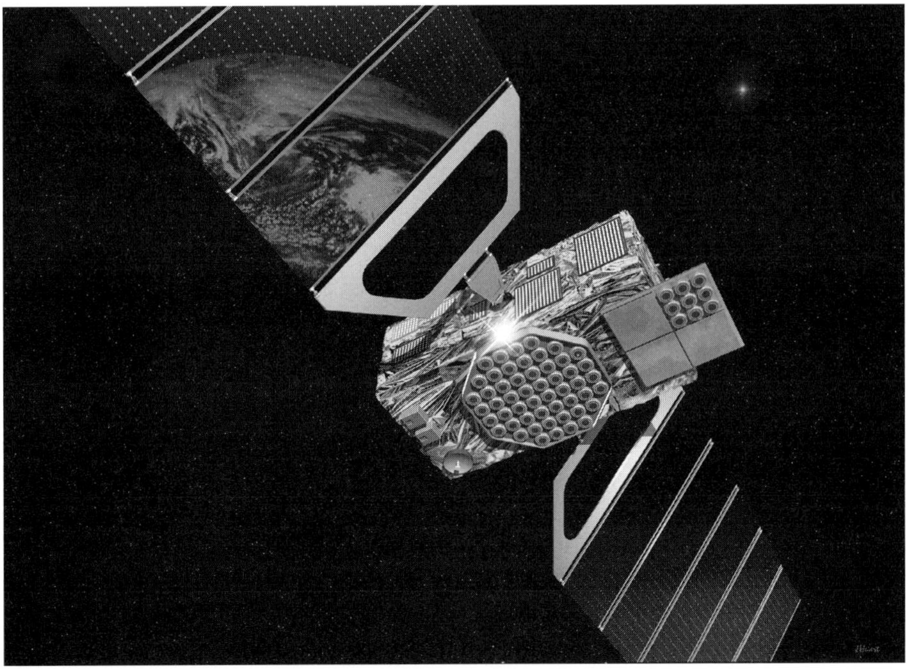

Figure 2.29 Galileo satellite (artist representation). Courtesy of ESA: J. Huart.

design elements. The most important feature is a transmitting antenna array looking down on the Earth. The space vehicle must maintain its orientation in relation to the Earth at all times. The second most common element is the solar battery panels. The space vehicle body is covered with a material that allows the undesirable effects of solar radiation to be reduced.

2.4.1 Models

Probably the least important reason for mobile applications is that, for GPS and GLONASS, the available signals depend on the satellite model. This does not concern the basic L1 signal, which is always available.

In the past, the GLONASS constellation went through a period when only a few satellites were available. Potential investors in the GLONASS user equipment have therefore sought reassurance. One element that might provide this is the GLONASS-K satellite, which is an entirely new concept based on a non-pressurized platform. The GLONASS-K satellite's estimated service life has been increased to 10–12 yr, and a third civilian L-range frequency has been added. The real life span has been proved to be much longer than that guaranteed for GLONASS. GLONASS-K is a small-sized spacecraft that is considerably lighter than previous models, which makes it less costly to put into orbit, and a wider range of launch vehicles could be used. The mass of a GLONASS-K satellite has been decreased to 700 kg, compared with 1415 kg for GLONASS. After the complete constellation is deployed, it will require one Soyuz launch per year to maintain the constellation in full.

A further reassurance for an investor in GNSS equipment is that the GPS III generation of satellites from Lockheed Martin will no longer have SA capability.

2.4.2 Signals

Table 2.7 shows GPS and GLONASS satellites currently in service or planned for launch and the signals that are available from them. GPS satellites Block IIR-M, IIF, and subsequent blocks will have two additional PRN ranging codes: the L2 civil-moderate (L2 CM) code and the L2 civil-long (L2 CL) code.

An L1 civil (L1C) signal will be available from all GPS satellites, excep Blocks II/IIA, IIR/IIR-M, and IIF. The L1C signal will be available with the first Block III launch, scheduled for 2014. Starting from Block IIF, satellites may transmit an L5 signal.

GLONASS is beginning to implement CDMA signals on the single frequency on L3. The first GLONASS satellite to transmit CDMA on L3 was GLONASS-K in 2011. The GLONASS L3 CDMA signal increases GLONASS's accuracy due to a chip rate of 10.23 megachips/s (similar to that for GPS). Because the chip length for the new signal is halved, the accuracy of the code phase measurements may double.

2.4.3 Geometry

Another important implication resulting from the satellite type is that of the space vehicle geometry, which affects the forces experienced from the Sun's radiation

Table 2.7. GPS and GLONASS satellites currently in service

System	Satellite	Launches	In orbit	In plan	Civil signals
GPS	Block IIA	1990–1997	9	-	L1 C/A
	Block IIR	1997–2004	12	-	L1 C/A
	Block IIR-M	2005–2009	7	-	L1 C/A, L2C
	Block IIF	2010–2013	3	10	L1 C/A, L2C, L5 SoL[a]
	Block IIIA	2014–	0	12	L1 C/A, L1C, L2C, L5 SoL
	Block IIIB	–	0	N/A	
	Block IIIC	–	0	N/A	
GLONASS	GLONASS-M	2003–2016	28 (including 3 spare)		L1 SP, L2SP, L3 CDMA[b]
	GLONASS-K1	2011, 2013	1		L1 SP, L2SP, L3 CDMA
	GLONASS-K2	2015–2024	0		L1 SP, L2SP, L1 CDMA, L2 CDMA, L3 CDMA
	GLONASS-KM	–	0		

[a] L5 is transmitted from some satellites in demo mode.
[b] GLONASS-M transmits L3 CDMA, except GLONASS-M launched before 2014.

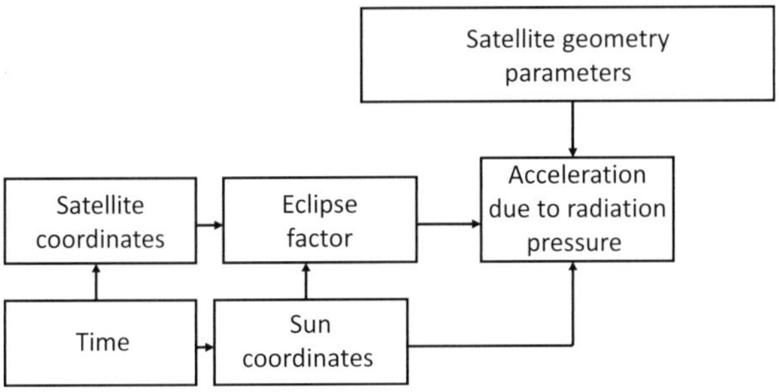

Figure 2.30 Sun radiation pressure implementation model.

pressure. This is important for high-precision geodetic applications, orbit determination tasks, and may also be useful for offline AGNSS, when high-accuracy models are necessary to propagate ephemerides inside the receiver over a prolonged time. An application of the Sun radiation pressure model to calculate the forces affecting a satellite is given in Figure 2.30. The satellite geometry model is based on satellite coordinates in relation to the Sun. Satellite shape, size, and material also affect the forces caused by solar radiation.

2.4.4 Clock

Clock parameters are important for instant positioning when predicted ephemerides can be used. One can predict satellite ephemerides accurately over a very long time.

Predicted ephemerides can be used in a receiver instead of the broadcast ephemerides. This would allow the receiver to make a first positioning much sooner. For

Table 2.8. GLONASS satellites[a]

Name	GLONASS	GLONASS-M	GLONASS-K1	GLONASS-K2
Lifetime (yr)	3	7	10	10
Years of production	1982–2005	2003–2016	2011–2013	2014–
Clock stability	$5{\cdot}10^{-13}$	$1{\cdot}10^{-13}$	$5{\cdot}10^{-14}$ to $1{\cdot}10^{-13}$	$1{\cdot}10^{-14}$ to $5{\cdot}10^{-14}$
Civil signals	L1 SP	L1 SP, L2 SP, L3 CDMA	L1 SP, L2 SP, L3 CDMA	L1 SP, L2 SP, L1, L2, L3 CDMA
Status	out of service	in service	in service	design stage

[a] After [24].

example, for a GPS L1 C/A signal the time would reduce to 6 s from 36 s, and for GLONASS it would be 2 s instead of 32 s.

However, satellite clock errors are indistinguishable from orbit errors. It is also much more difficult to predict satellite clock errors.

Clocks are different for different satellites. Table 2.8 shows, for example, the GLONASS satellite clock parameters for various satellite models. The highest precision is expected from Galileo onboard clocks, which have an onboard hydrogen maser, whereas other satellites use rubidium (GPS, BeiDou) or cesium (GPS, GLONASS) clocks [24],[25]. The drift value, the stability, and even the drift model may be different, because different types of atomic clock are used. Often, an assembly of clocks onboard a satellite is used instead of a single precise clock. This affects the structure and parameters of the clock drift models and consequently the algorithms for clock prediction. Such algorithms should include a receiver autonomous integrity monitoring (RAIM)-type algorithm in order to exclude most erroneous satellites by use of a voting method. A weighting algorithm can also be used, which accounts for clock accuracy and reliability, if the dependence on these models of satellite vehicle type is known.

References

[1] I. Petrovski and T. Tsujii, *Digital Satellite Navigation and Geophysics: A Practical Guide with GNSS Signal Simulator and Receiver Laboratory*. Cambridge: Cambridge University Press, 2012.

[2] Navstar GPS Space Segment/Navigation User Segment Interfaces, GPS Interface Specification IS-GPS-200, Rev F. Global Positioning Systems Directorate, Sept. 2011.

[3] Navstar GPS Space Segment/Navigation User Segment L1C Interface, GPS Interface Specification IS-GPS-800, Rev B. Global Positioning Systems Directorate, Sept. 2011.

[4] Global Navigation Satellite System GLONASS, Interface Control Document, Navigational radiosignal in bands L1, L2, Edition 5.1. Russian Institute of Space Device Engineering, Moscow 2008.

[5] European GNSS (Galileo) Open Service Signal In Space Interface Control Document (OS SIS ICD), Issue 1.4. European Union and European Space Agency, Sept. 2010.

[6] BeiDou Navigation Satellite System Signal In Space Interface Control Document. Open Service Signal B1I (Version 1.0). China Satellite Navigation Office, Dec. 2012.

[7] J. J. Spilker, GPS signal structure and theoretical performance, in *Global Positioning System: Theory and Applications, Vol. I*, B. W. Parkinson and J. J. Spilker, Eds. Washington, D.C.: American Institute of Aeronautics and Astronautics Inc., 1996.

[8] J. M. Lipton and K. P. Dabke, Spread spectrum communications based on chaotic systems, *Int. J. Bifurcation Chaos*, vol. 6, pp. 2361–2374, 1996.

[9] E. Lorenz, Deterministic nonperiodic flow, *J. Atmos. Sci.*, vol. 20, no. 2, pp. 130–141, 1963.

[10] R. Shaw, Strange attractor, chaotic behavior and information flow, *Z. Naturforsch A*, vol. 36, pp. 80–112, 1981.

[11] S. Dadras and H. Momeni, A novel three-dimensional autonomous chaotic system generating two-, three- and four-scroll attractors. *Phys. Lett. A*, vol. 373, pp. 3637–3642, 2009.

[12] J. Betz, Binary offset carrier modulations for radionavigation, *J. Inst. Navigat.*, vol. 48, pp. 227–246, 2001.

[13] Interface Specification for QZSS, Draft 1.2. Japanese Aerospace Exploration Agency (JAXA), Mar. 15, 2010.

[14] R. Gold, Optimal binary sequences for spread spectrum multiplexing, *IEEE Trans. Inform. Theory*, vol. 13, pp. 619–621, 1967.

[15] Y. Urlichich, V. Subbotin, G. Stupak, V. Dvorkin, A. Povalyaev, and S. Karutin, GLONASS. Developing strategies for the future, *GPS World*, pp. 42–49, Apr. 2011.

[16] J. Saastamoinen, Contribution to the theory of atmospheric refraction, *Bull. Geod.* pp. 105–106, 1972.

[17] A. E. Niell, Global mapping functions for the atmosphere delay at radio wavelengths, *J. Geophys. Res.*, vol. 101, no. B2, pp. 3227–3246, Feb. 10, 1996.

[18] J. A. Klobuchar, Ionospheric time-delay algorithm for single-frequency GPS users, *IEEE Trans. Aerosp. Electron. Syst.*, vol. 23, pp. 325–331, 1987.

[19] A. Perov and V. Harisov, Eds., *GLONASS: Design and Operations Concepts*, 4th edn. Moscow: Radiotechnica, 2010 (in Russian).

[20] G. Di Giovanni and S. M. Radicella, An analytical model of the electron density profile in the ionosphere, Adv. *Space Res.*, vol. 10, no. 11, pp. 27–30, 1990.

[21] B. Arbesser-Rastburg and N. Jakowski, Effects on satellite navigation, in *Space Weather – physics and effects*, V. Bothmer and I. Daglis, Eds. Berlin: Springer, pp. 383–402, 2007.

[22] S. Schaer, *Mapping and Predicting the Earth's Ionosphere Using the Global Positioning System*, Geodätisch-geophysikalische Arbeiten in der Schweiz, Vol. 59. Zürich: Schweizerische Geodätische Kommission, Institut für Geodäsie und Photogrammetrie, Eidg. Technische Hochschule, 1999.

[23] P. Borwein and M. Mossinghoff, Barker sequences and flat polynomials, in *Number Theory and Polynomials*, J. McKee and C. Smyth, Eds., LMS Lecture Notes 352. Cambridge: Cambridge University Press, pp. 71–88, 2008.

[24] S. Averin, GLONASS system: present day and prospective status and performance, Presented at *European Navigation Conf. GNSS-2006*, Manchester, May 7–10, 2006.

[25] Mallette, L., White, J., and Rochat, P., Space qualified frequency sources (clocks) for current and future GNSS applications, Presented at *PLANS2010 Conf.*, Indian Wells, CA, May 6, 2010.

3 Standalone positioning with GNSS

The Global Navigation Satellite System comprises a constellation of satellites and a ground segment that determines the satellite coordinates. The main idea behind GNSS is to measure distances between satellites and users located on the surface of the Earth or in the lower atmosphere.

Satellites move around the Earth along precisely known orbits, so we can define a satellite's position precisely at any given moment. GNSS satellites transmit signals on frequencies between 1 and 2 GHz. These signals can be received by using a GNSS receiver. When the receiver acquires the satellite signal, the user can estimate the time required for the signal to propagate to the user from the satellite. By referring to information about satellite orbits and measured distances to the satellites, a user can calculate receiver position as an intersection of four spheres in the four-dimensional space-time continuum. If the receiver time could be perfectly synchronized with the system or satellite time frame, only three satellites would be required to determine receiver position in three-dimensional space (Figure 3.1).

It is important to note that the coordinates that we calculate are the coordinates of the antenna phase center. This may be important for some applications, for example when the antenna is located at some distance from the receiver. In landslide monitoring systems [1], the receiver is located about 3 km from the antenna, using a fiber-optic cable. This system is intended for use in hazardous areas; the receiver has been located at a safe distance and connected to a set of remotely located antennas. The signal from each antenna is processed by the same receiver, and thus the position of each antenna can be calculated using these data. The time taken by the signal to travel through the lengthy cable is simply added to the receiver clock error, and then compensated for in the final calculation.

3.1 Application of pseudorange observables

3.1.1 Code phase measurements

In this chapter we consider only conventional positioning, which uses time marks for example from navigation messages. This means that code phase measurements are transformed into unambiguous pseudorange observables, either using information from a navigation message or via external assist information. It also means that the code phase measurements are aligned with each other.

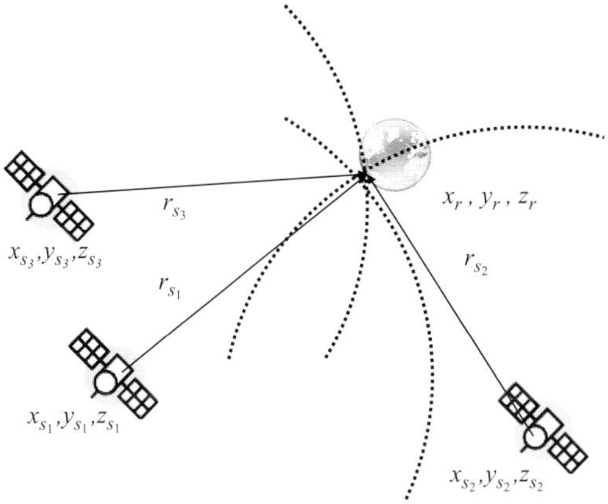

Figure 3.1 Positioning with GNSS.

Code phase measurements are ambiguous outside the code length. The code length for a GPS L1 C/A signal is approximately 300 km. A receiver can measure code phase within one code sequence (in the case of a GPS L1 C/A signal, for example, this is 300 km), but without reading navigation data it cannot find out how many whole code sequences are on their way from the satellite. We define the pseudoranges ρ_{s_i} as the sum of ambiguous code phase measurements $\Delta\rho_{s_i}$ and the time mark Z_i from the signal navigation message or assist data:

$$\rho_{s_i} = \Delta\rho_{s_i} + Z_i. \tag{3.1}$$

This representation of pseudoranges becomes useful when we consider positioning without decoding the navigation message. The time mark from a navigation message allows us to decode Z_i and restore pseudorange measurements. We distinguish in this book between pseudorange and code phase observables (see the List of definitions).

Chapters 8 and 9 describe positioning without a time mark, using instead various assist information. After the time mark has been decoded from the navigation message, the receiver can output pseudoranges.

The code and phase measurements contain the range information contaminated by errors due to the satellite clock error, signal propagation, and other factors. These errors can be corrected either outside of the receiver in the case of post-processing, or in the receiver navigation processor using real-time corrections. The raw measurements from the receiver normally have only pseudorange ambiguities resolved, with all other errors uncompensated. This allows more flexibility in the processing engine, especially external to the receiver. A navigation processing engine external to the receiver may have access to more precise information on the GNSS error budget, and may also be more powerful; therefore the achievable positioning accuracy will not be limited by the receiver internal resources, in terms of both information and computational power. Besides, the error compensation is carried out in the receiver navigation processor, and, as we show later, nowadays the navigation process may not necessarily be part of the receiver.

The unambiguous pseudorange measurements are contaminated with errors and therefore can be expressed with those errors shown explicitly as follows:

$$P_i = Z_i + \Delta\rho_{s_i} = r_i + c(\delta t_s + \delta t_r) + T_i + I_i + O_i + M_i + \delta P_i + \cdots \quad i = 1, \ldots, n, \tag{3.2}$$

where P_i is the pseudorange observable from satellite i, Z_i is the unambiguous time measurement from the navigation message, $\Delta\rho_{s_i}$ is the code phase measurement (which is ambiguous beyond the code length), r_i is the distance to the satellite, δt_s and δt_r are satellite and receiver clock errors, respectively, T_i, I_i, O_i, M_i are tropospheric, ionospheric, orbital, and multipath errors, respectively, and δP_i is the code phase noise.

The left-hand side of equation (3.2) shows code phase measurements and time mark information. The right-hand side shows that, apart from the distance to the satellite r_i, the measurements contain various errors.

3.1.2 Carrier phase measurements

Similar to equation (3.2) for the code phase measurements, the carrier phase measurements can be expressed as follows:

$$\Phi_i = \lambda N_i + \Delta\varphi_i, \tag{3.3}$$

where Φ_i is the carrier phase observable from satellite i, $\Delta\varphi_i$ is the carrier phase measurement (which is ambiguous outside the carrier wavelength), and N_i is the carrier phase ambiguity number.

The carrier phase measurements are those measurements inside the carrier wavelength. In order to restore these measurements to attain the distance to the satellite, one needs in addition to estimate the number of whole carrier wavelengths on the signal path from the satellite to the antenna.

With main errors explicitly specified, the phase measurements can be expressed as follows:

$$\Phi_i = \lambda N_i^{(0)} + \Delta\varphi_i = r_i + c(\delta t_s + \delta t_r) + T_i + I_i + O_i + M_i^{(2)} + \delta\Phi_i + \cdots, \tag{3.4}$$

where as before δt_s, δt_r are satellite and receiver clock errors, respectively, $T_i + I_i + O_i + M_i^{(2)}$ are tropospheric, ionospheric, orbital, and multipath errors, and $\delta\Phi_i$ is carrier noise.

The ambiguity number here becomes the initial estimate $N_i^{(0)}$, which often cannot be used as an ambiguity number in the algorithms prior to compensation for errors on the right-hand side of equation (3.4). In comparison, the time mark on the right-hand side of equation (3.2) is used in the construction of pseudorange observables prior to error compensation in positioning algorithms.

Equations (3.2) and (3.4) are almost equivalent, except that the carrier noise and carrier multipath are much smaller than the code noise and code multipath, which makes the carrier measurements very valuable:

$$\delta\Phi_i \ll \delta P_i, \tag{3.5}$$

$$M_i^{(2)} \ll M_i. \tag{3.6}$$

The problem with using the carrier phase measurements is in their ambiguity outside the carrier wavelength. The ambiguity number N_i can be restored to the true value only as a result of processing measurements from more than one receiver; these algorithms are discussed in Chapter 4. For standalone positioning, the carrier measurements can be used for smoothing code measurements, which we discuss in the following sections.

3.1.3 Pseudorange equations

For standalone mobile positioning, all errors, besides the receiver clock error δt_r, are compensated for explicitly. The receiver clock error can be expressed as follows:

$$\delta t_r = t_r^{\langle s\rangle} - t_r^{\langle r\rangle}, \tag{3.7}$$

where $t_r^{\langle s\rangle}$ denotes an epoch of signal reception in the GNSS time scale and $t_r^{\langle r\rangle}$ denotes an epoch of signal reception in the receiver time scale.

Note that a time mark gives us an approximate estimation of satellite transmission time $t_s^{\langle r\rangle}$. This allows us to calculate satellite position. The algorithm may take a few iterations, because the satellite is moving at a very high velocity (for GPS, it is about 4 km/s). Once the receiver knows the satellite position from the navigation message, and measures pseudoranges to a few satellites, we have a system of equations.

Therefore, pseudorange measurements are considered as the sum of the distance to the satellite and the receiver clock error. Equation (3.1) can now be expressed via satellite distance as

$$\rho_{s_i} = r_i + c\cdot\delta t_r, \tag{3.8}$$

i.e.

$$\rho_{s_i} = \sqrt{(x_r(t_r) - x_{s_i}(t_{s_i}))^2 + (y_r(t_r) - y_{s_i}(t_{s_i}))^2 + (z_r(t_r) - z_{s_i}(t_{s_i}))^2} + \delta t_r\cdot c, \tag{3.9}$$

where $x_{s_i}, y_{s_i}, z_{s_i}$ are the coordinates of the ith satellite at the epoch of signal transmission t_{s_i}, $x_r(t_r), y_r(t_r), z_r(t_r)$ are unknown receiver coordinates at the epoch of signal reception t_r, and δt_r is receiver clock error.

Equations (3.9) are the main pseudorange equations. In addition, these equations show that the distance to the satellite cannot be measured exactly because of large receiver clock error δt_r.

Equations (3.9) can be written in matrix form as follows:

$$\overline{Z} = [A(\overline{X})], \tag{3.10}$$

where $\overline{X}$ is a state vector,

$$\overline{X} = \begin{bmatrix} x_r \\ y_r \\ z_r \\ \delta t_r \\ N_{s_1} \\ N_{s_2} \\ \dots \\ N_{s_n} \end{bmatrix}, \tag{3.11}$$

and $\overline{Z}$ is an observation vector,

$$\overline{Z} = \begin{bmatrix} P_{s_1} \\ P_{s_2} \\ \dots \\ P_{s_n} \\ \varphi_{s_1} \\ \varphi_{s_2} \\ \dots \\ \varphi_{s_n} \end{bmatrix}. \tag{3.12}$$

For standalone positioning, in this chapter we consider only code phase measurements, and thus the state vector and observation vector are simplified as follows:

$$\overline{X} = \begin{bmatrix} x_r \\ y_r \\ z_r \\ \delta t_r \end{bmatrix}, \tag{3.13}$$

$$\overline{Z} = \begin{bmatrix} P_{s_1} \\ P_{s_2} \\ \dots \\ P_{s_n} \end{bmatrix}. \tag{3.14}$$

In order to solve a positioning task, we need to find a fourth-order vector $\overline{X}$ using a vector of pseudorange measurements $\overline{Z}$ of nth order. In order for the system of equations (3.10) to have a solution, the condition $n \geq 4$ must be satisfied. The system of equations (3.10), which we need to solve, is non-linear. The system can be linearized due to the fact that satellites are located very far away from a user.

The linearized equations may be written in matrix form as follows:

$$\overline{Z} - \overline{Z}_0 = [H] \times \Delta\overline{X}, \tag{3.15}$$

where matrix $[H]$ is called a measurement matrix,

$$[H] = \begin{bmatrix} \frac{\partial P_{s_1}}{\partial x} & \frac{\partial P_{s_1}}{\partial y} & \frac{\partial P_{s_1}}{\partial z} & 1 \\ \frac{\partial P_{s_2}}{\partial x} & \frac{\partial P_{s_2}}{\partial y} & \frac{\partial P_{s_2}}{\partial z} & 1 \\ \cdots & \cdots & \cdots & \cdots \\ \frac{\partial P_{s_n}}{\partial x} & \frac{\partial P_{s_n}}{\partial y} & \frac{\partial P_{s_n}}{\partial z} & 1 \end{bmatrix}, \tag{3.16}$$

$$[H] = \begin{bmatrix} e_{1x} & e_{1y} & e_{1z} & 1 \\ e_{2x} & e_{2y} & e_{2z} & 1 \\ \cdots & \cdots & \cdots & \cdots \\ e_{nx} & e_{ny} & e_{nz} & 1 \end{bmatrix}, \tag{3.17}$$

where the partial derivatives vector is, in fact, the unit direction vector from an approximate position $x_{r_0}, y_{r_0}, z_{r_0}$ to a satellite as follows:

$$\overline{e}_i = \left[\frac{\partial}{\partial x} P_{S_i}, \frac{\partial}{\partial y} P_{S_i}, \frac{\partial}{\partial z} P_{S_i} \right], \tag{3.18}$$

$$\Delta \overline{X} = \begin{bmatrix} x_r - x_{r_0} \\ y_r - y_{r_0} \\ z_r - z_{r_0} \\ \delta t_r \end{bmatrix}, \tag{3.19}$$

and $\overline{Z}_0$ is a vector of measurements predicted for point $x_{r_0}, y_{r_0}, z_{r_0}$.

For standalone positioning, we linearize equations relative to an a-priori estimate of the receiver position, which can come from an analytical solution, although the numerical solution implemented with modern numerical algorithms should converge in most cases, even with the initial position chosen arbitrarily.

We can find a solution of this system as long as its determinant is not zero. In other words, if the number of linear independent equations (equal to the number of satellites with good geometry) is more than the number of unknowns such as the antenna coordinates and receiver clock error.

3.1.4 Satellite coordinates

In order to complete our calculation of the satellite equations, we need to introduce the estimated satellite positions. The satellite coordinates are estimated by the ground segment and uploaded to the satellites as part of the navigation message. In order to receive all the necessary information about the satellite orbits, the receiver must receive a significant part of the navigation message. For the GPS L1 C/A signal, it is one complete frame. (It may in fact include parts of two different frames, but it would still work because the necessary orbit information is repeated in each frame.) The length of one frame is 30 s for a GPS L1 signal. The information includes the orbital parameters

```
     2.10           N: GPS NAV DATA                         RINEX VERSION / TYPE
teqc  2002Mar14     GSI, JAPAN          20090522 04:44:18UTCPGM / RUN BY / DATE
Linux 2.0.36|Pentium II|gcc -static|Linux|486/DX+           COMMENT
teqc  2002Mar14     GSI, JAPAN          20090520 03:15:31UTCCOMMENT
     2              NAVIGATION DATA                         COMMENT
DAT2RIN 3.53        GSI, JAPAN          20MAY09 10:03:55    COMMENT
                                                            COMMENT
    7.4510D-09  2.2350D-08 -5.9600D-08 -1.1920D-07          ION ALPHA
    8.6020D+04  8.1920D+04 -1.3110D+05 -5.2430D+05          ION BETA
    7.450580596920D-09 2.753353101070D-14   503808     1532 DELTA-UTC: A0,A1,T,W
    15                                                      LEAP SECONDS
                                                            END OF HEADER
 2 09  5 20  0  0  0.0 1.544686965640E-04 5.684341886080E-13 0.000000000000E+00
    6.900000000000E+01 8.003125000000E+01 4.917704909960E-09-1.452650735020E-02
    4.025176167490E-06 9.066797443670E-03 9.216368198390E-06 5.153657329560E+03
    2.592000000000E+05 2.216547727580E-07-2.839270790000E-01-6.891787052150E-08
    9.416867444840E-01 1.944375000000E+02 2.816546211150E+00-8.417136321270E-09
    1.071473187950E-10 1.000000000000E+00 1.532000000000E+03 0.000000000000E+00
    2.400000000000E+00 0.000000000000E+00-1.722946763040E-08 6.900000000000E+01
    2.520180000000E+05
 2 09  5 20  2  0  0.0 1.544733531770E-04 5.684341886080E-13 0.000000000000E+00
    8.300000000000E+01 8.228125000000E+01 4.865917002660E-09 1.035608460950E+00
    4.287809133530E-06 9.067845880050E-03 8.540228009220E-06 5.153659574510E+03
    2.664000000000E+05 2.589076757430E-07-2.839869050340E-01 1.769512891770E-07
    9.416879791870E-01 2.023437500000E+02 2.816603926200E+00-8.258558281680E-09
    3.821587837850E-11 1.000000000000E+00 1.532000000000E+03 0.000000000000E+00
    2.400000000000E+00 0.000000000000E+00-1.722946763040E-08 8.300000000000E+01
    2.592180000000E+05
```

Figure 3.2 RINEX, ver. 2 navigation file format.

of the transmitting satellite. An example of navigation parameters in a broadcast navigation message given in a RINEX navigation file format is shown in Figure 3.2. RINEX (Receiver Independent Exchange Format) [2] was originally developed at the Astronomical Institute of the University of Berne for raw GPS data, including code and carrier phase, Doppler, signal-to-noise ratio, and satellite ephemerides. The format was developed to facilitate exchange of GPS data during the European GPS campaign EUREF 89, which involved more than 60 GPS receivers from four different manufacturers. The version 3 RINEX format files are divided into the following types:

(1) observation data file,
(2) navigation data file,
(3) meteorological data file.

At the time of the writing, the latest RINEX version (3.01) supported GLONASS, Galileo, BeiDou, and SBAS, as well as GPS on all frequencies.

Ephemeris data from GNSS navigation messages are usually available from high-end receivers.

The broadcast message provides the orbit parameters in real time, but what about the IGS or other similar services? In fact, there is very little difference. The IGS mostly provides precise ephemerides for post-processing in geodetic applications. However, it also provides predicted (over one day) ephemerides that are of similar or even greater accuracy than broadcast ephemerides, although they lack the integrity of the latter data. However, one needs to remember that broadcast ephemerides are in fact also predicted values. In general, it is possible to predict ephemerides for MEO satellites with very high accuracy over much longer periods (more than a week). Unfortunately, satellite clocks are less predictable, and errors in their predicted values may significantly degrade positioning accuracy. For instant positioning, a user would need to receive the orbit and clock information from an outside source in order not to have to wait for the reception of the navigation message frame. Such information can come from sources such as the IGS.

The outside information can be delivered in three main ways.

(1) It can be received by data link in real time. This type of system works for cellular phone applications.
(2) It can be preloaded in advance using predicted ephemerides. Predicted ephemerides can be downloaded for free from the IGS for a limited prediction length of one day. The predicted ephemerides can also be calculated for a longer prediction interval using either proprietary reference station network data or data from IGS, available freely.
(3) It can be calculated within a receiver using previous orbit information from broadcast or predicted ephemerides and geodetic data related to the Earth's rotation parameters.

In addition, the receiver can record snapshots of the RF signal. The recorded signal can then be used later in the positioning calculation, when orbit information becomes available. Such applications include photo tags, bird tracking, and others.

Finally, when the range measurements are available from the satellite signal from three or more satellites, and orbits for those satellites are available from the satellite or other sources, one can compute the position of the receiver.

3.1.5 Minimum number of satellites for positioning

The number of unknowns in the system of equations should be less than or equal to the number of equations, which is defined by the number of available satellites. If the lines-of-sight (LOS) to different satellites are collinear, these satellites provide the same information. Such equations are linear dependent and can be reduced to one equation. The quality of satellite geometry in general can also be referred to in terms of dilution of precision, or DOP, which will be discussed later in this chapter.

Because the number of measurements should be more than or equal to the number of unknowns, four satellites would be required to locate a position in three-dimensional space. The fourth satellite is required to calculate the receiver clock error. Satellite clocks are very precise, but receiver clocks are much cheaper and, if not compensated for, would render the whole system unusable.

Four satellites only are required for three-dimensional positioning only if all satellites belong to the same GNSS. If one or more belongs to different GNSS, then an additional measurement is required for each additional system in order to estimate the time difference between the system time frames.

The time difference between system time frames should also be provided within the navigation message of the corresponding system. Although this time-shift estimate is of limited accuracy, it should be sufficiently accurate for mobile applications. In this case, if a navigation message can be received, the additional satellite is not required. If positioning is carried out without a navigation message, this information can come from assist data or be kept in the receiver from the previous positioning.

When there are not enough satellites due perhaps to an obstructed environment, the number of positioning satellites required can be reduced by introducing an assumed receiver altitude, and thus removing one unknown from the equations. This altitude constraint can also come with assist data or from a digital map.

In principle, if the receiver clock is stable enough, the introduction of applying a proper clock model and estimation of its parameters at the previous epoch when enough satellites were available would allow positioning with just two satellites in view for a short period of time.

Furthermore, in this case, if the receiver is in a device integrated with a digital map, the number of required satellites during the movement along a street can be reduced to just one.

If a user has limited visibility, with not enough satellites or satellites with poor geometry, which means that satellites are visibly close to each other on the sky from the antenna position, we can add constraints to these equations. An example of an additional constraint could be height, which could be assumed to be fixed to the previously estimated value. In this case, the receiver should be searching for coordinates in a local horizontal frame because the vertical coordinate in the horizontal frame is fixed and can therefore be removed. Then the state vector order is reduced, and the minimum required number of satellites is also reduced by one.

The equations in the system must be linearly independent. This means that they should not be derived from other equations that are already in the system by multiplication and addition. If there are linearly dependent equations, they would make the system determinant equal or close to zero, which would make the system poorly defined and the solution unobtainable. In the real physical world this situation will arise when vectors pointing to the satellites from the antenna position are close to being collinear, or the LOS between the antenna and satellites are close to being parallel to each other (see Figure 3.3); i.e. if one equation can be derived from another by a sequence of linear operations, then this equation does not count. In a real situation, when two satellites are on the same LOS as the user, the second satellite does not provide any additional information. This is an extreme case. It is, however, clear from this example that the information from a satellite will be more valuable if the angles between the satellite LOS are larger. The system determinant $\|D\|$ in this case can be a good indicator of the quality of the geometry. These quality criteria are formalized in the following.

Code phase measurements or pseudoranges, together with optional carrier phase measurements, Doppler measurements, and navigation data from a receiver, are called *raw* measurements. The final processed measurements, if such processing is carried out within the receiver, include positioning information and optionally some additional information about satellites that can be called *final* measurements. There are some well-established standards that a receiver should meet in order to output these measurements. They usually appear in NMEA format, and the raw measurements are usually given in a RINEX O format. NMEA is a combined electrical and data specification for communication between marine electronic devices, including GPS receivers. It is

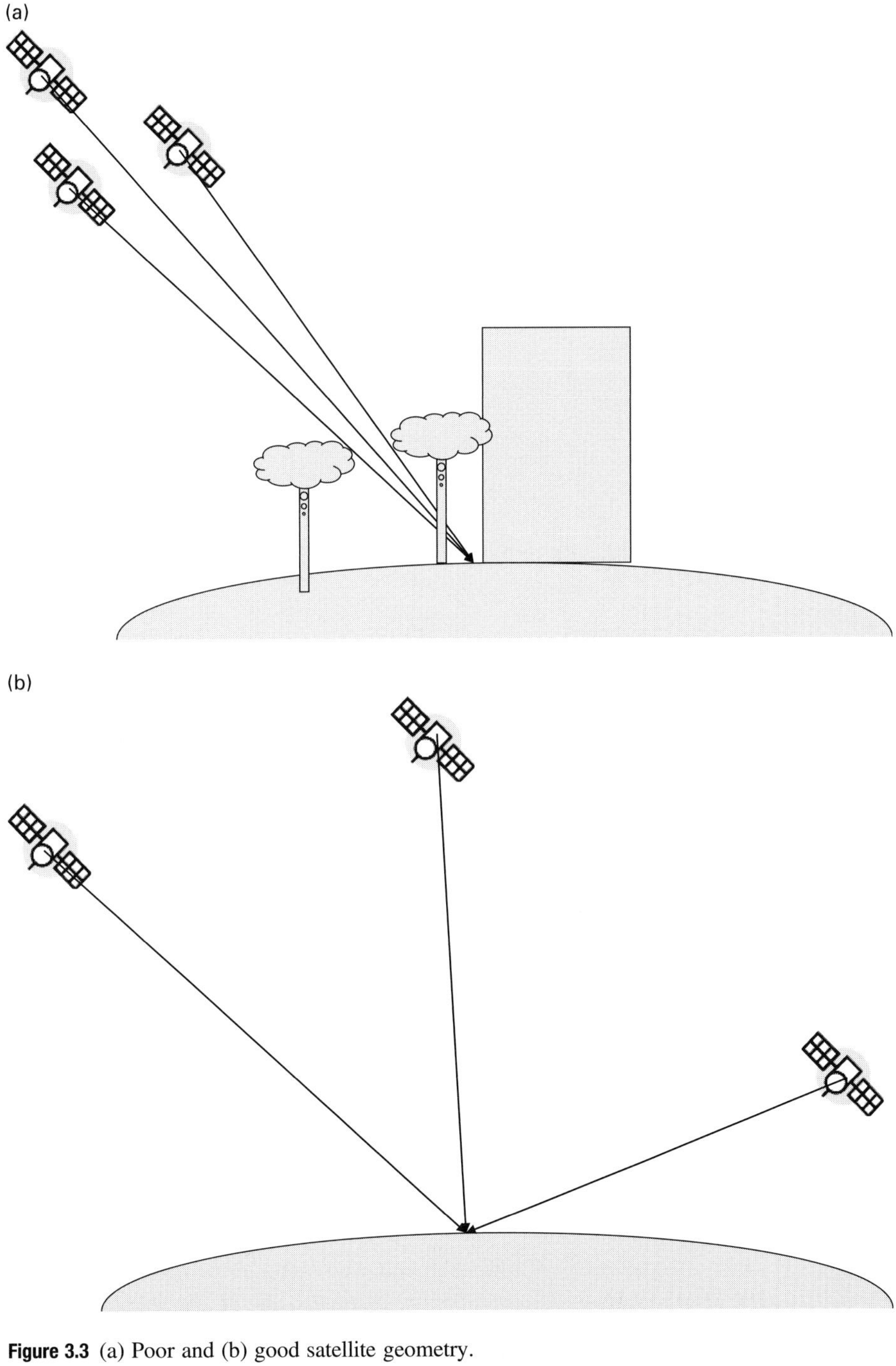

Figure 3.3 (a) Poor and (b) good satellite geometry.

Table 3.1 NMEA content example

Position	Content	Example	Explanation
1	header	$GPGGA	GP is for GPS receivers, GGA – NMEA sentence type
2	UTC time in hours, minutes, and seconds	090507	time of fix is 5 min 7 s after 9 am (GMT)
3	latitude	3540.468	35° 40.4689' (decimal min, not s)
4	N or S	N	north
5	longitude	13931.896	139° 31.8960'
6	E or W	E	east
7	GPS quality indicator (0 = invalid; 1 = standalone; 2 = Diff. fix)	1	single positioning fix
8	number of satellites in use	8	not all satellites in view
9	HDOP	1.42	horizontal dilution of position
10	antenna altitude	87.672	above or below geoid (mean sea level)
11	antenna height unit	M	meters
12	geoidal separation	29.1	difference between WGS-84 ellipsoid and geoid
13	units of geoidal separation	M	meters
14	DGPS update age	10	time in seconds since last update from reference station
15	reference station ID number	0515	
16	checksum	*5B	always begins with *

defined by the US-based National Marine Electronics Association. Mobile GNSS receivers usually have output positioning information in NMEA format. In particular, a GGA sentence for displaying GNSS fix data must be used. The content of an NMEA GGA output message is given in Table 3.1. The other important NMEA sentences include GLL for geographic position and latitude/longitude data, GSA for the GPS DOP and active satellites list, and GSV for the GPS satellites in view list.

3.2 Navigation solution algorithms

3.2.1 Least-squares estimation (LSE) solution

We need to solve the system of equations (3.15), which can be rewritten as follows without loss of generalization:

$$\overline{Z} = [H] \times \Delta\overline{X}, \tag{3.20}$$

where $\overline{Z}$ is the vector of measurements, consisting of the differences between predicted and measured observables, $\Delta\overline{X}$ is an unknown state vector which represents an error in positioning, and $[H]$ is the measurement matrix.

In real life, equations (3.20) incorporate measurement errors and therefore should be written as follows:

$$\overline{Z} = [H] \times \Delta\overline{X} + \vec{\zeta}, \tag{3.21}$$

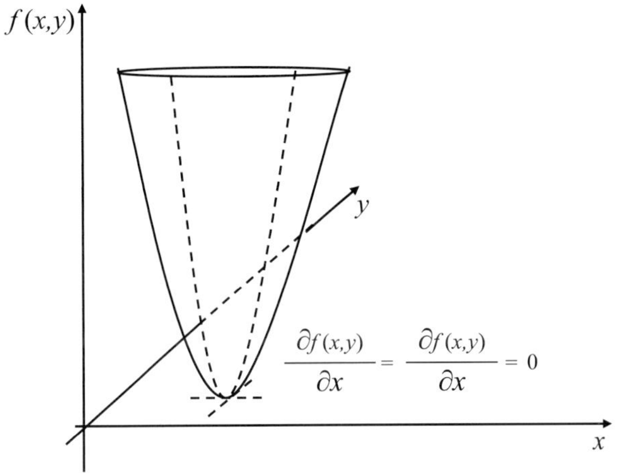

Figure 3.4 Downhill search algorithms find minimum of function when its partial derivatives become zero.

where $\overline{\xi}$ is a vector of measurement errors. Then we can look for the solution $\Delta\hat{\overline{X}}$ of the system (3.21) which minimizes the sum of squared components $\sum_{i=1}^{n} \zeta_i^2$ of the residual vector:

$$\overline{\zeta} = [H] \times \Delta\hat{\overline{X}} - \overline{Z}. \tag{3.22}$$

The minimum value of this criterion is achieved when its partial derivatives are zero. Figure 3.4 shows an illustrative example of a similar criterion for a more simple function. Equations for this in matrix form can be written as follows:

$$2 \cdot [H]^T \left| [H] \times \Delta\hat{\overline{X}} - \overline{Z} \right| = 0, \tag{3.23}$$

where $[H]^T$ is the transpose matrix of $[H]$.

In turn, this leads to what are called *normal* equations, which we can also derive formally by left-multiplication of both sides of equation (3.20) by matrix $[H]^T$, i.e.

$$[H]^T \cdot [H] \cdot \Delta\hat{\overline{X}} = [H]^T \cdot \overline{Z}. \tag{3.24}$$

The unique solution of this system,

$$\Delta\hat{\overline{X}} = ([H]^T \times [H])^{-1} \times [H]^T \times \overline{Z}, \tag{3.25}$$

exists if the Gramian matrix

$$[G] = [H]^T \times [H] \tag{3.26}$$

is invertible (i.e. if its rank is equal to or higher than n). If the column vectors of the measurement matrix are linearly dependent, the Gramian matrix becomes singular, i.e. its determinant is equal to zero. The number of linearly independent vectors must be equal to or larger than n. The matrix columns are linearly dependent if we can obtain one column vector from another through linear operations such as addition, subtraction, and multiplication by a constant. In a practical sense, the linear dependency means that the corresponding measurements carry no new information. This can happen, as we discussed in Section 3.1.5, if LOS vectors from a receiver to satellites are collinear or close to collinear.

For practical applications, we can assign a weight to each observable. For example, we can put a low weight on observables from low-elevation satellites because those observables are more corrupted with uncompensated atmospheric errors. Then the weighted least-squares estimation (LSE) solution can be written as follows:

$$\Delta\hat{\overline{X}} = ([H]^T \times [W] \times [H])^{-1} \times [H]^T \times [W] \times \overline{Z}. \tag{3.27}$$

The solution can usually be found within a few iterations. (The number of iterations also depends on how far the a-priori position is from the true position.) Note that one cannot obtain a general solution of (3.20) by taking the inverse of the measurement matrix because usually it is not a square matrix. The number of measurements usually exceeds the number of unknowns. When there is extraneous information in these measurements, we would like to use it in an optimum way to minimize errors. This LSE operation is sometimes called *pseudoinverse*.

In order to make a numerical solution more robust to round-off errors, and more economical in terms of memory size and calculation speed, the matrices in equation (3.15) are transformed to various products of factors. These transformations are denoted as *decomposition* or *factorization*. We consider here only *QR* decomposition, or factorization. Gauss first solved the system of equations (3.15) by elimination using back-substitution.

If matrix $[Q]$ is orthogonal, then, by definition,

$$[Q]^T[Q] = [I], \tag{3.28}$$

where $[I]$ is the identity matrix.

Matrix $[H]$ can always be expressed as the product of an orthogonal matrix $[Q]$ and an upper triangular matrix $[R]$ in the following way:

$$[H] = [Q] \times \begin{bmatrix} R_{11} & R_{12} & \cdots & R_{n1} \\ 0 & R_{22} & \cdots & R_{n2} \\ \vdots & \vdots & \ddots & \vdots \\ 0 & 0 & 0 & R_{nn} \end{bmatrix}. \tag{3.29}$$

Then, the initial equation can be rewritten as follows:

$$[Q][R] \times \Delta\hat{\overline{X}} = \overline{Z}. \tag{3.30}$$

Using (3.28), equation (3.30) can be rewritten as

$$[R] \times \Delta\hat{\overline{X}} = [Q]^T \times \overline{Z}, \tag{3.31}$$

and then solved by back-substitution. Back-substitution means that we find $\Delta\hat{X}_n$ directly from the last row of (3.31) because $[R]$ is a triangular matrix. Then we put our known $\Delta\hat{X}_n$ into the next row and find $\Delta\hat{X}_{n-1}$, and so on.

The algorithms for calculating QR factorization and back-substitution are well developed and are included in most mathematical software libraries.

3.2.2 Analytical solution

The system of equations (3.15) may also be solved analytically. There are a few analytical methods that we could consider. The first analytical method developed for GNSS equations was the Bancroft method [3].

We will use the Lorentz inner product, which is defined as

$$\langle \vec{g}, \vec{h} \rangle \triangleq \vec{g}^{\,T} [M] \, \vec{h}, \tag{3.32}$$

where $\vec{g}, \vec{h} \in R^4$, and

$$[M] = \begin{bmatrix} \|I\|_{3\times 3} & 0 \\ 0 & -1 \end{bmatrix}. \tag{3.33}$$

The pseudorange equations can be rewritten via squaring and grouping to yield

$$\frac{1}{2}\left\langle \begin{bmatrix} \vec{r}_i \\ \rho_i \end{bmatrix}, \begin{bmatrix} \vec{r}_i \\ \rho_i \end{bmatrix} \right\rangle - \left\langle \begin{bmatrix} \vec{r}_i \\ \rho_i \end{bmatrix}, \begin{bmatrix} \vec{r} \\ b \end{bmatrix} \right\rangle + \frac{1}{2}\left\langle \begin{bmatrix} \vec{r} \\ b \end{bmatrix}, \begin{bmatrix} \vec{r} \\ b \end{bmatrix} \right\rangle. \tag{3.34}$$

Here, $\vec{r}_i$ is the radius vector of the ith satellite, $\vec{r}$ is a radius vector defining user position, and $b = c\,dt$. If we have four such equations for four satellites, we can write them in matrix form:

$$\vec{\alpha} - [B][M]\begin{bmatrix} \vec{r} \\ b \end{bmatrix} + \Lambda\,\vec{\tau} = 0, \tag{3.35}$$

$$\Lambda = \frac{1}{2}\left\langle \begin{bmatrix} \vec{r} \\ b \end{bmatrix}, \begin{bmatrix} \vec{r} \\ b \end{bmatrix} \right\rangle, \tag{3.36}$$

where

$$\vec{\tau} = \begin{bmatrix} 1 \\ 1 \\ 1 \\ 1 \end{bmatrix}$$

and $\vec{\alpha}$ is a 4×1 vector with each component described as

$$\alpha_i = \frac{1}{2}\left\langle \begin{bmatrix} \vec{r}_i \\ \rho_i \end{bmatrix}, \begin{bmatrix} \vec{r}_i \\ \rho_i \end{bmatrix} \right\rangle.$$

Solving (3.35) for vector $\begin{bmatrix} \vec{r} \\ b \end{bmatrix}$ yields

$$\begin{bmatrix} \vec{r} \\ b \end{bmatrix} = [M][B]^{-1}(\Lambda\,\vec{\tau} + \vec{\alpha}). \tag{3.37}$$

However, Λ is a function of $\begin{bmatrix} \vec{r} \\ b \end{bmatrix}$, so finally we can rewrite (3.35) as

$$\left\langle [B]^{-1}\ \vec{\tau}, [B]^{-1}\ \vec{\tau}\ \right\rangle \Lambda^2 + 2\left[\left\langle [B]^{-1}\ \vec{\tau}\ , [B]^{-1}\ \vec{\alpha}\ \right\rangle - 1\right]\Lambda + \left\langle [B]^{-1}\ \vec{\alpha}, [B]^{-1}\ \vec{\alpha}\ \right\rangle = 0, \tag{3.38}$$

which is a quadratic equation in Λ.

Equation (3.38) describes the case for four measurements. If we have more than four satellites, it can be extended as follows by multiplying equation (3.35) by $[B]^T$:

$$\left\langle [C]\ \vec{\tau}, [C]\ \vec{\tau}\ \right\rangle \Lambda^2 + 2\left[\left\langle [C]\ \vec{\tau}, [C]\ \vec{\alpha}\ \right\rangle - 1\right]\Lambda + \left\langle [C]\ \vec{\alpha}, [C]\ \vec{\alpha}\ \right\rangle = 0, \tag{3.39}$$

where

$$[C] = \left([B]^T[B]\right)^{-1}[B]^T. \tag{3.40}$$

This equation gives us two solutions for Λ. We can choose the correct one either by applying some constraints, such as it should be on the surface of the Earth, or by finding the solution for a different set of measurements which leads to only one consistent solution for Λ.

3.2.3 Kalman-filter solution

For mobile applications, when a rover receiver is moving, it may be useful to introduce rover velocities as additional variables into the state vector, which we need to estimate, as

$$\overline{X} = \begin{bmatrix} x_r \\ y_r \\ z_r \\ V_x \\ V_y \\ V_z \\ \delta t_r \end{bmatrix}. \tag{3.41}$$

In this case, we also can add velocity measurements into the measurement vector:

$$\overline{Z} = \begin{bmatrix} P_{S_1} \\ P_{S_2} \\ \cdots \\ P_{S_n} \\ \dot{P}_{S_1} \\ \dot{P}_{S_2} \\ \cdots \\ \dot{P}_{S_n} \end{bmatrix}. \tag{3.42}$$

Now we can also introduce additional constraints because we have additional information about the system model. Therefore, the system equations which describe the system dynamics

$$\dot{\vec{X}} = [A] \times \vec{X} + \vec{\xi}, \tag{3.43}$$

can be added to the measurement equations (3.20), which can be rewritten as

$$\overline{Z} = [H] \times \overline{X} + \vec{\zeta}, \tag{3.44}$$

where $\vec{\xi}$ is the system noise with covariance matrix $[Q]$, and

$$[A] = \begin{bmatrix} 0 & 0 & 0 & 1 & 0 & 0 & 0 \\ 0 & 0 & 0 & 0 & 1 & 0 & 0 \\ 0 & 0 & 0 & 0 & 0 & 1 & 0 \\ 0 & 0 & 0 & 0 & 0 & 0 & 0 \\ 0 & 0 & 0 & 0 & 0 & 0 & 0 \\ 0 & 0 & 0 & 0 & 0 & 0 & 0 \\ 0 & 0 & 0 & 0 & 0 & 0 & 0 \end{bmatrix}.$$

The measurement noise $\vec{\zeta}$ covariance matrix is denoted by $[R]$. Assuming that system and measurement noise can be described as white noise, the optimal recursive solution for such a system is given by a Kalman filter, which was introduced by Rudolf Kalman in 1960 [4]. The filter was implemented by NASA for Apollo and later Space Shuttle trajectory estimation. Since then, various implementations and variations of Kalman filters have played major roles in almost all real-time control and estimation systems.

The main limitation of the Kalman filter concerns the conditions that the system and measurement noise should satisfy. The system and measurement noise are assumed to be uncorrelated white noises with known auto-covariance functions $[Q]$ and $[R]$.

We present a general description of a Kalman filter due to its importance. We will use a continuous-time representation of the Kalman filter to retain the notation used in previous material. The dynamics of the variable estimates is given by the following equation:

$$\frac{d}{dt}\hat{\vec{X}} = [A] \times \hat{\vec{X}} + [K] \times \vec{v},$$

where $\vec{v}$ describes an innovation process presenting the prediction based on the system dynamics $[H] \times \vec{X}$ close to measurements $\vec{Z}$:

$$\vec{v} = \vec{Z} - [H] \times \hat{\vec{X}}.$$

The system model provides the predicted behavior of the state variables. The predicted state of these variables is calculated by the dynamic model using the variable values from the previous epoch. The measured values of the same variable are calculated using measurements at the current epoch. The measured and calculated values of the variables are combined in an optimum manner, by using information about the system and measurement noise encapsulated in the Kalman gain matrix $[K]$:

$$[K] = [P] \times [H]^T \times [R]^{-1},$$

where $[P]$ is the error covariance matrix. The propagation of the covariance matrix is described by the Riccati equation, which is named after an Italian mathematician, Count Jacopo Francesco Riccati (1676–1754):

$$\frac{d}{dt}[P] = [A] \times [P] + [P] \times [A]^T + [Q] - [K] \times [R] \times [K]^T.$$

Practically any technical system can be formalized into a model and measurements. In general, a Kalman filter can be used for any such system, as long as we have enough information about the system and measurement noises to reduce them to Gaussian white noise processes. For GNSS, a Kalman filter is usually implemented to combine information from GNSS with other sensors. It can also be used in receiver tracking loops to combine, in the optimum way, measurements and our knowledge of the system represented by the system model.

3.2.4 Brute-force solution

This method allows the solution of the non-linear equation (3.10) directly without linearization. Residuals can be found for a grid of assumed positions $\overline{X}_i$ as follows:

$$\vec{v}_i = \overline{Z} - [A(\overline{X}_i)]. \tag{3.45}$$

The solution found as a minimum grid value can be further improved by a standard minimization method. The problem with this solution is that the density of the grid should be chosen in such a way that it will exclude the possibility of finding a wrong solution, which can be a local minimum. The correct position should be at an absolute, global minimum of the function $A(\overline{X})$. Such a high-density grid may require too much computer power which, depending on initial conditions for the search and constraints, would make real-time implementation impossible. Why do we mention this method? Because the brute-force method is still used, in particular for instant positioning (i) when there are good initial conditions for the search and additional constraints and (ii) for post-processing.

3.3 Multi-system positioning

3.3.1 Generalized equations

In this section, we consider how the positioning task changes if we use two or more GNSS for positioning. A GLONASS navigation message contains parameters that allow one to calculate the difference between GLONASS and GPS time, $\Delta\tau$. The parameters allow the GPS/GLONASS receiver to connect time system frames with an accuracy of 30 ns, which roughly translates to about 9 m error. A much more accurate way is to estimate the time difference between two time frames by introducing it as an extra unknown into equation (1.19),

$$\overline{Z} = [A(\overline{X})], \tag{3.46}$$

with the state vector $\overline{X}$ now defined by

$$\overline{X} = \begin{bmatrix} x_r \\ y_r \\ z_r \\ \delta t_r \\ \Delta\tau_{GPS/GLONASS} \end{bmatrix}. \tag{3.47}$$

This allows us to find $\Delta\tau$ with the same level of accuracy (which is normally much better) as that for other coordinates. GPS time can then be translated to UTC or calendar time directly.

The downside of this method is that it requires an extra satellite to provide an extra measurement for the extra unknown. Therefore, using all four systems (GPS, GLONASS, Galileo, and BeiDou) for positioning would require a minimum of seven satellites instead of four, i.e.

$$\overline{X} = \begin{bmatrix} x_r \\ y_r \\ z_r \\ \delta t_r \\ \Delta\tau_{GPS/GLONASS} \\ \Delta\tau_{GPS/Galileo} \\ \Delta\tau_{GPS/BeiDou} \end{bmatrix}. \tag{3.48}$$

3.3.2 Time-shift calculation using navigation message data

GPS to Galileo time-shift parameters for Galileo signal users are transmitted in the Galileo navigation message. GPS to BeiDou time-shift parameters for BeiDou signal users are transmitted in the BeiDou navigation message. The parameters are the time-scale offset A_0 and its rate of change A_1 at the reference epoch t_{ref}, specified as a week and time of week.

The new GPS navigation messages CNAV data for an L2C signal and CNAV-2 data for an L1C signal also contain similar parameters for time offset between GPS and GLONASS, GPS and Galileo, and GPS and other GNSS [5],[6].

In general, the time offset is calculated as follows:

$$\Delta t_{GPS/GNSS_i} = t_{GNSS_i} - t_{GPS} = A_0 + A_1(t - t_{ref}), \tag{3.49}$$

where $GNSS_i$ stands for GLONASS, Galileo, or BeiDou.

The GLONASS-M satellite transmits the time offset between GPS and GLONASS, τ_{GPS}, which is kept within 30 ns. The time offset is then calculated as follows:

$$\Delta t_{GPS/GLONASS} = t_{GPS} - t_{GLONASS} = \Delta T + \tau_{GPS}, \tag{3.50}$$

where ΔT is the integer part of the time offset and is defined using the GPS navigation message, i.e. it is the difference between the GPS and GLONASS time scales in UTC leap seconds.

If a receiver uses broadcast corrections, the number of unknowns is reduced and it is not necessary to solve for these offsets in the system of equations (3.46).

3.4 Error budget for GNSS observables

3.4.1 Error budget contents

GNSS errors can be described in the form of an *error budget*. The error budget consists of the following errors (see Table 3.2).

Table 3.2. GNSS error budget

User range error	Code phase noise	Orbits	Satellite clock	Tropospheric[a]	Ionospheric[a]	Multipath
Value [m]	<1	~1	~1	2–25	1.5–15	0–10

[a] Before compensation.

(1) Receiver-related errors, including noise and hardware biases.
(2) Satellite-related errors, including satellite clock error, satellite orbit errors, satellite transmitter errors, including noise and biases.

The following two categories of errors have been considered in detail in Chapter 2.

(3) Propagation errors in dispersive medium due to ionosphere.
(4) Propagation errors in non-dispersive medium due to troposphere.

The following two categories will be considered next.

(5) DOP arising from geometrical properties of the satellite constellation.
(6) Multipath.

3.4.2 Geometrical factors

We now consider how the satellite constellation geometry affects position error. To do that, we define a coordinate vector $\Delta\overline{X}$, which is the difference between a satellite's assumed and estimated position, in a local horizontal coordinate system (instead of the Cartesian ECEF frame), with east, north, up, and time components as follows:

$$\Delta\overline{X} = \begin{bmatrix} \Delta E \\ \Delta N \\ \Delta H \\ \delta t_r \end{bmatrix}. \tag{3.51}$$

Then, the components of the measurement matrix $[H]$ are unit direction vectors from the a-priori position to a satellite:

$$[H] = \begin{bmatrix} e_{1x} & e_{1y} & e_{1z} & 1 \\ e_{2x} & e_{2y} & e_{2z} & 1 \\ \cdots & \cdots & \cdots & \cdots \\ e_{nx} & e_{ny} & e_{nz} & 1 \end{bmatrix}. \tag{3.52}$$

Measurements are made with errors, $\delta\overline{Z}$. Such errors in measurements propagate to the state vector estimate via equation (3.25). Therefore, we can write an equation for positioning error caused by measurement errors as follows:

$$\delta\Delta\hat{\overline{X}} = ([H]^T \times [H])^{-1} \times [H]^T \times \delta\overline{Z} + \overline{\xi}. \tag{3.53}$$

The errors $\delta\hat{\overline{X}}$ are of stochastic nature, so we are interested in their statistics, rather than specific instances. The first moment, which is the mathematical expectation, can be expressed as follows:

$$\mathrm{E}(\delta\Delta\hat{\overline{X}}) = ([H]^T \times [H])^{-1} \times [H]^T \times \mathrm{E}(\delta\overline{Z}) + E(\overline{\xi}). \tag{3.54}$$

The second moment for the positioning error, which is a covariance, can be written as

$$\begin{aligned} \mathrm{cov}(\delta\Delta\hat{\overline{X}}) &= \mathrm{E}(\delta\Delta\hat{\overline{X}} \cdot \delta\Delta\hat{\overline{X}}^T) \\ &= ([H]^T \times [H])^{-1} \times [H]^T \times \mathrm{cov}(\delta\overline{Z}) \times [H] \times ([H]^T \times [H])^{-1}. \end{aligned} \tag{3.55}$$

Equation (3.54) describes the error propagation through geometry, which is called the dilution of precision (DOP). If we make the assumption that all components of the measurement covariance matrix are uncorrelated and have equal variances, then we have

$$\mathrm{cov}(\delta\overline{Z}) = \sigma^2 \cdot [I], \tag{3.56}$$

where $[I]$ is the identity matrix. Then (3.55) becomes

$$\mathrm{cov}(\delta\Delta\hat{\overline{X}}) = \sigma^2 \cdot ([H]^T \times [H])^{-1} \triangleq \sigma^2 \cdot [D]. \tag{3.57}$$

The diagonal elements of the covariance matrix $\mathrm{cov}(\delta\Delta\hat{\overline{X}})$ define the DOP. The vertical DOP (VDOP) for height error variation, horizontal DOP (HDOP) for horizontal positioning error variation, positional DOP (PDOP), time DOP (TDOP), and geometric DOP (GDOP) are defined as follows:

$$\begin{aligned} d_{VDOP} &= \sqrt{d_{33}} \\ d_{HDOP} &= \sqrt{d_{11} + d_{22}} \\ d_{PDOP} &= \sqrt{d_{11} + d_{22} + d_{33}} \\ d_{TDOP} &= \sqrt{d_{44}} \\ d_{GDOP} &= \sqrt{d_{11} + d_{22} + d_{33} + d_{44}}, \end{aligned} \tag{3.58}$$

where d_{ii} is an ith diagonal element of matrix $[D]$.

The specific DOP values are subsets of GDOP, which looks at the whole system of equations, while other criteria consider the specific subset, defined by unknowns of interest.

The accuracy of range measurements is limited by various errors, including errors due to (i) signal propagation in the atmosphere, (ii) the receiver, and (iii) the antennas and their surroundings. Therefore, an error in positioning results from the errors in the range measurements multiplied by the geometry factor.

This means that errors in pseudoranges, such as those caused by atmospheric errors, propagate to create larger positioning errors. These positioning errors could be very large when poor satellite geometry exists. In this sense, it is extremely advantageous to have more satellites. Figure 3.5 shows the DOP for GPS and the GPS+GLONASS+Galileo constellations, calculated for a specified interval of time.

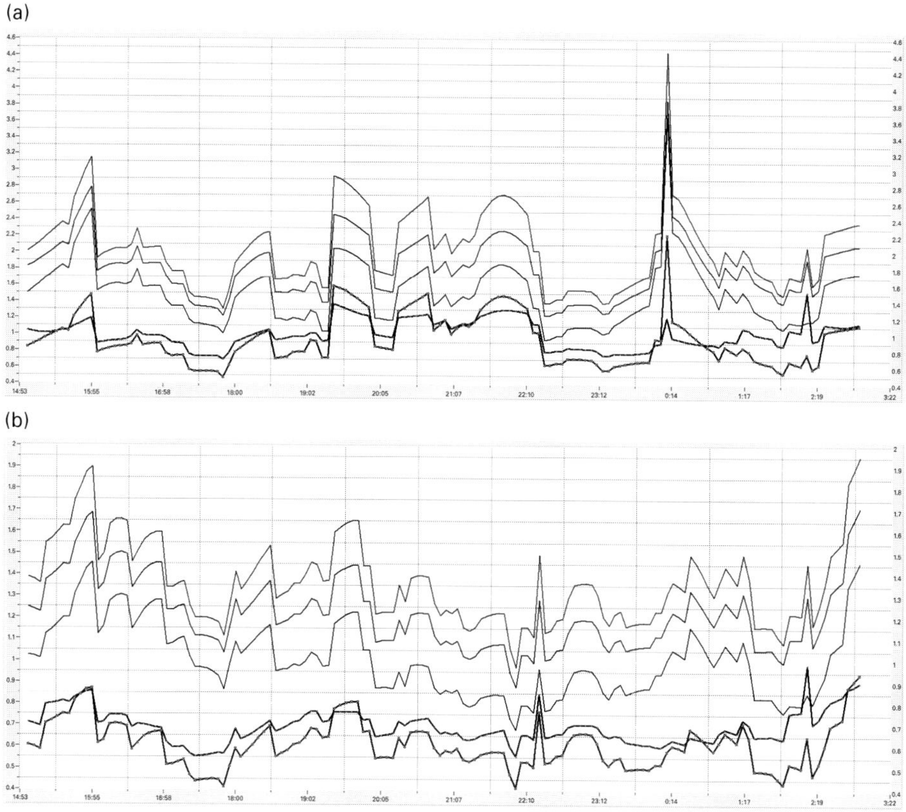

Figure 3.5 DOP with (a) GPS and (b) GPS + GLONASS + Galileo constellation.

3.4.3 Multipath

The difference between multipath and most of the other GNSS errors is that multipath affects the baseband processor. A multipath error is created by a signal coming to an antenna along different paths, which result from signal reflections on various surfaces.

A GNSS signal may experience multipath as it comes not only directly from a satellite, but also after being reflected by various surfaces. We can describe a GPS signal as follows (see Chapter 2):

$$A = A_0 \sin(\omega t + \varphi) \cdot B, \tag{3.59}$$

where B is a C/A code. We can omit the navigation message here because it is affected via the code only. The multipath signal then can be expressed in the following way:

$$A_M = k_M \cdot A_0 \cdot \sin(\omega t + \varphi + \varphi_M) \cdot B(\varphi_M), \tag{3.60}$$

where the signal is attenuated by k_M and delayed by φ_M. The total signal coming to a receiver then can be expressed as a sum of the signals:

$$A = A_D + \sum_{i=1}^{n} A_{M_i} = A_0 \left(\sum_{i=0}^{n} \left[k_{M_i} \cdot \sin\left(\omega t + \varphi + \varphi_{M_i}\right) \cdot D(\varphi_{M_i}) \right] \right), \quad k_{M_0} = 1 \tag{3.61}$$

where A_D is a direct signal. The phase delay is calculated from the delay as follows:

$$\varphi_{M_i} = 2\pi f \tau + \varphi_{R_i}, \tag{3.62}$$

where f is the carrier frequency, and φ_{R_i} is the phase shift caused by the reflector and the antenna phase pattern.

The way multipath error is simulated in a GNSS simulator in comparison to other errors provides us with an insight about the nature of multipath error. This approach is considered in detail in [7].

In a receiver baseband processor, instead of one signal, correlators will encounter two or more signals with the same spread code. The shapes of the correlation plots for two or more signals will be different, and a receiver will try to follow the envelope of the signals rather than a direct signal.

The reflected signal is not always a parasitic signal. In the case of indoor positioning, reflected signals are sometimes the only ones available. Therefore, indoor positioning is often realized as snapshot positioning using chunks of code, sometimes reflected from various surfaces.

It is also possible that a receiver acquires only a reflected signal or that a reflected signal is stronger than the direct signal and $k_M > 1$. Such situations can happen, for example, in urban canyons, when the direct signal is attenuated by vegetation.

References

[1] I. Petrovski, M. Ishii, H. Torimoto, *et al.* LAMOS-BOHSAI: landslide monitoring system based on high-speed sequential analysis for inclination, in *ION-GPS 2000*, Salt Lake City, UT, Sept. 2000, pp. 84–94.

[2] W. Gurtner and L. Estey, RINEX: the receiver independent exchange format, Astronomical Institute University of Bern, UNAVCO, Jun. 22, 2009. Available at http://igscb.jpl.nasa.gov/igscb/data/format/rinex301.pdf.

[3] S. Bancroft, An algebraic solution of the GPS equations, *IEEE Trans. Aerosp. Electron. Syst.*, vol. 21, pp. 56–59, 1985.

[4] R. Kalman, A new approach to linear filtering and prediction problems, *Trans. ASME (J. Basic Engineering)*, vol. 82 D, pp. 35–45, 1960.

[5] Navstar GPS Space Segment/Navigation User Segment Interfaces, GPS Interface Specification IS-GPS-200, Rev F. Global Positioning Systems Directorate, Sept. 2011.

[6] Navstar GPS Space Segment/Navigation User Segment L1C Interface, GPS Interface Specification IS-GPS-800, Rev B, Global Positioning Systems Directorate, Sept., 2011.

[7] I. Petrovski and T. Tsujii, *Digital Satellite Navigation and Geophysics: A Practical Guide with GNSS Signal Simulator and Receiver Laboratory*. Cambridge: Cambridge University Press, 2012.

4 Referenced positioning with GNSS

4.1 Requirements for code and carrier differential positioning

In Chapter 3 we discussed positioning with code phase measurements. If one wants to use carrier phase measurements, which are much more accurate, the task becomes more complicated. This is because there are no time marks on a carrier, and therefore the carrier measurements are ambiguous down to one carrier wave.

With main errors explicitly specified, code and carrier phase measurements are constructed as follows:

$$\Phi_i = \lambda N_i^{(0)} + \Delta\varphi_i = r_i + c(\delta t_s + \delta t_r) + T_i - I_i + O_i + M_i + \delta\,\Phi_i + \cdots, \qquad (4.1)$$

where Φ_i is the carrier phase observable from satellite i in meters, $\Delta\varphi_i$ is the carrier phase measurement, which is ambiguous outside of one wavelength, N_i is an ambiguity number, δt_s and δt_r are satellite and receiver clock errors, respectively, T_i, I_i, O_i are tropospheric, ionospheric, and orbital errors, respectively, M_i is the multipath error, and $\delta\Phi_i$ is the carrier noise.

The ambiguity number is shown in equation (4.1) as an initial estimate, $N_i^{(0)}$, which is different from the true number because of measurement errors. The errors on the right-hand side of equation (4.1) must be compensated for in advance of the ambiguity resolution process, which means that $N_i^{(0)}$ cannot be used as an ambiguity number in the algorithms prior to this.

If one sets up a reference receiver at a known position, most of the errors in equation (4.1) can be estimated. These estimates, called *corrections*, can be used in order to compensate for similar errors in a rover receiver. If such a receiver is in the vicinity of the reference receiver, then the errors in the reference receiver should be close enough to the errors in the mobile receiver in order to be compensated for and allow the ambiguity resolution methods to work. Some errors, however (for example, those related to the receiver hardware and antenna environment, in particular multipath), cannot be corrected for by using the reference receiver.

Different methods exist to resolve carrier ambiguity. These methods come from geodesy, where they were initially designed for geodetic tasks carried out in a post-processing mode. To be able to use carrier ambiguity resolution in real time for mobile applications requires the components discussed in the following.

(1) **Reference station**. The reference station must be located within a few kilometers of a mobile device at a known position. This reference station should be able to make carrier phase measurements at the same time as the user. When a reference station is used, the user receiver is called a rover receiver, or simply a rover.

(2) **Carrier phase measurements**. The rover receiver should be able to make carrier phase measurements. Not all receivers are capable of providing carrier phase measurements at the quality sufficient for carrier positioning.

(3) **Data link**. A data link is required in order to send measured raw data either from the rover to the reference station, which in this case may be called a control center, or from the reference station to the rover. Which way the raw data go depends on where the rover position should be calculated. The data sent by the reference station to the rover receiver are called corrections, and they are usually in RTCM format (RTCM refers to the Radio Technical Commission for Maritime Services; see Section 8.4).

The data link should fulfill some latency requirements, so the corrections data can be applied when they arrive at a rover. If the corrections are older than a few seconds, they cannot be used for carrier phase measurements.

(4) **Number of visible satellites**. More than four satellites should be available for a defined period of time. This time is required for ambiguity resolution, and it depends on the satellite geometry, the number of satellites, and the overall signal error budget.

(5) **RTK algorithms**. The real-time ambiguity resolution, a.k.a. RTK (real-time kinematic), algorithms must be implemented to process the raw data from the rover and the reference station.

A reference station can also be used to provide corrections for code-only positioning. It always provides corrections for code measurements anyway. These corrections allow compensation for all common errors between a rover and reference stations. In this case, a general term differential GPS (GNSS), or DGPS (DGNSS), is applied. If one uses carrier phase data and corrections, the term may be further specified as carrier differential GPS (GNSS), or CDGPS (CDGNSS).

Code phase and carrier phase measurements for DGPS can be constructed as follows:

$$P_i^k = r_i^k + c(\delta t_s + \delta t_r^k) + T_i^k + I_i^k + O_i + \delta P_i^k + \cdots, \tag{4.2}$$

$$\Phi_i^k = N_i^k + \frac{1}{\lambda}\left(\Delta\varphi_i^k + c(\delta t_s + \delta t_r^k) + T_i^k - I_i^k + O_i + \delta\,\Phi_i^k + \cdots\right), \tag{4.3}$$

where P_i^k denotes the pseudorange measurements of receiver k from satellite i in meters, Φ_i^k denotes the carrier phase measurements of receiver k from satellite i in cycles, $\Delta\varphi_i$ is an unambiguous carrier phase measurement within a wavelength, N_i^k is an ambiguity number, δt_s and δt_r^k are satellite and receiver clock errors, respectively, T_i^k, I_i^k are tropospheric and ionospheric errors, respectively, O_i is an error due to orbit degradation, δP_i^k, $\delta\Phi_i^k$ are code and carrier noise, respectively, and r_i is the distance to satellite i. The ionospheric error I_i^k has the same value, but opposite sign, for code and carrier phase observables.

4.2 Spatial correlations in error budget

For differential positioning, it is important that errors at the reference station and at the rover position are of about the same magnitude. However, if a rover is located far enough away from the reference station, the magnitude of these errors may be different. The speed of spatial decorrelation is different for each error.

4.2.1 Decorrelation of satellite orbital errors

The errors in equations (4.2) and (4.3) include those for the satellite clock. As we have discussed earlier, from the receiver's point of view, orbital errors per se are indistinguishable from satellite clock errors at a given epoch. They may decorrelate differently though. Orbital errors decorrelate because the error value depends on difference between the LOS to the real and assumed satellite positions (see Figure 4.1).

For orbit error, decorrelation has a relatively small effect on the position accuracy and can be estimated as follows [1]:

$$\Delta b = b\frac{\Delta O}{r}, \tag{4.4}$$

where b is the baseline (the vector between the rover and the reference station), ΔO is the satellite orbit error, r is the distance to the satellite, and Δb is the error in position (or baseline) estimation due to the orbit error.

Satellite clock and orbital errors are practically the same for a rover within 10 km of the reference station.

The satellite clock error decorrelation depends on the difference in signal propagation time. The difference between satellite clock error as seen at two distant receivers can be expressed as follows:

$$\delta t_s(t_r^{(2)}) - \delta t_s\left(t_r^{(1)}\right) \cong \delta t_s \frac{d(\delta t_s)}{dt} \cdot \left(t_r^{(2)} - t_r^{(1)}\right), \tag{4.5}$$

where $t_r^{(j)}$ is the signal reception time at the jth receiver. Since the satellite clock drift $\delta\dot{t}_s$ is normally smaller than 10^{-3} m/s, and the difference in transmission time is negligibly small, especially if the receivers are in close proximity, the satellite clock error can be considered the same for both receivers.

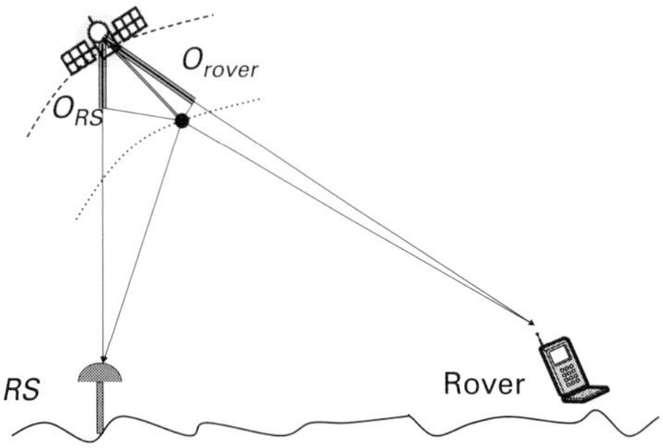

Figure 4.1 Decorrelation of satellite orbital errors.

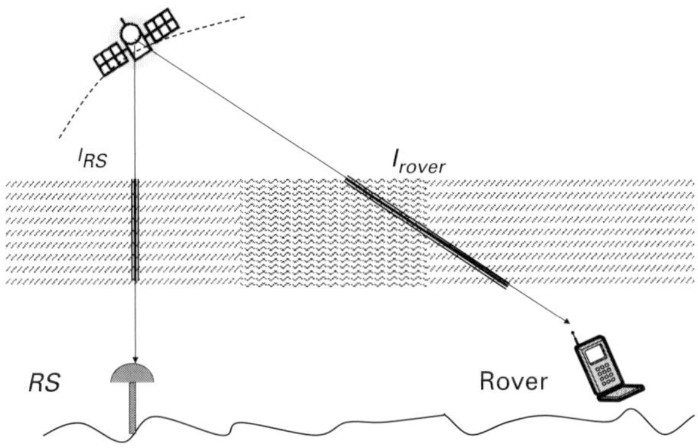

Figure 4.2 Decorrelation of ionospheric errors.

4.2.2 Decorrelation of tropospheric errors

Tropospheric models can be very accurate, but residual errors (those that remain after the model is applied) are poorly correlated. For tropospheric models, the distance does not play a significant role. Tropospheric delay errors are much less correlated, especially if there is a height difference between the two receivers.

4.2.3 Decorrelation of ionospheric errors

Ionospheric error depends on the length of the signal path in the ionosphere belt (see Figure 4.2) and on ionospheric conditions (the total electron content, TEC). Ionospheric errors are well correlated within approximately 10 km, but beyond that distance they may decorrelate quickly to the point where RTK methods are rendered ineffective. This means that the RTK algorithm can provide the incorrect position, although corrections remain valid within code-positioning accuracy.

The ionospheric error in fact is one of those errors that limit the RTK applicability to the vicinity of a reference station. One way of extending the area of applicability is to implement a network RTK method, and in particular its implementation known as a VRS (virtual reference station). This chapter looks at the network RTK method, and Chapter 8 discusses its implementation as a VRS in detail.

In Section 4.3, we consider how we can combine measurements from a rover and a reference receiver.

4.3 Observables

4.3.1 Single-difference observables

By looking at equations (4.2) and (4.3), we can assert that some errors may be eliminated by forming a difference between equations that describe the rover and reference station observables. Single-difference code phase and carrier phase

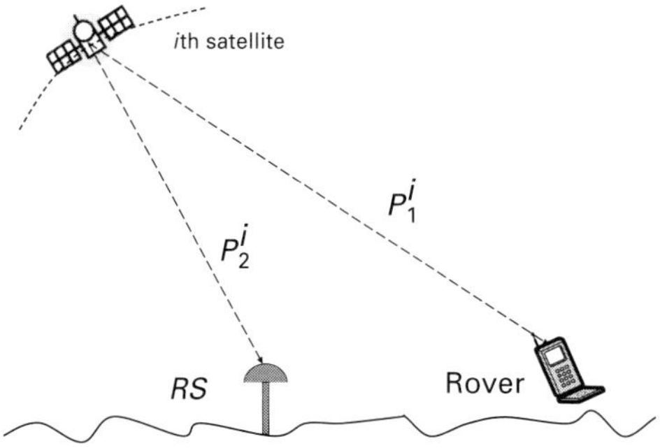

Figure 4.3 Single-difference observables.

observables for the ith satellite are defined as the difference between the measurements from the rover (receiver 1) and the reference station (receiver 2); see Figure 4.3:

$$\Delta P^i_{1,2} = P^i_1 - P^i_2, \tag{4.6}$$

$$\Delta \Phi^i_{1,2} = \Phi^i_1 - \Phi^i_2. \tag{4.7}$$

If the baseline length between two receivers is relatively short, most of the orbit related errors are canceled. If the signal propagation paths for both receivers are similar, the propagation errors are also canceled. The rule of thumb again is that ionospheric errors are well compensated if the reference receiver is within 10 km of a rover. The clock and orbit errors are well compensated for over even greater distances.

Forming a differencing equation from (4.3) results in the following:

$$\Delta \Phi^i_{1,2} = N_i^{(1)} - N_i^{(2)} + \frac{1}{\lambda}\Big(\Delta\varphi_i^{(1)} - \Delta\varphi_i^{(2)} + c(\delta t_r^{(1)} - \delta t_r^{(2)}) + \Delta T_i^{(1,2)} - \Delta I_i^{(1,2)} + \Delta O_i^{(1,2)} + \delta\Phi_i\Big). \tag{4.8}$$

Taking into account decorrelation properties of the error budget (Section 4.2), this can be rewritten as follows:

$$\Delta \Phi^i_{1,2} = \Delta N_i + \frac{1}{\lambda}\Big(\Delta\varphi_i^{(1)} - \Delta\varphi_i^{(2)} + c(\delta t_r^{(1)} - \delta t_r^{(2)}) + \delta\,\Phi_i\Big), \tag{4.9}$$

where ΔN is an unknown single-difference ambiguity number given by

$$\Delta N = N_i^{(1)} - N_i^{(2)}. \tag{4.10}$$

This value should be estimated along with other parameters, such as the receiver coordinates.

4.3.2 Double-difference observables

Double differences can be formed using measurements from two receivers (1 and 2) and two satellites (i and j). A double-difference observable is defined as the difference

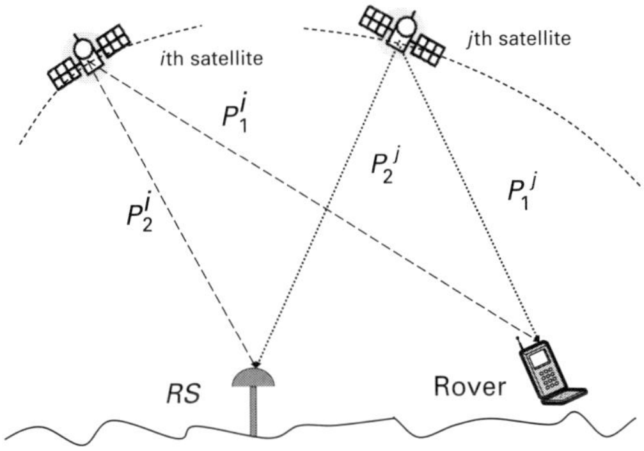

Figure 4.4 Double-difference observables.

between two single differences (Figure 4.4). Denoting a double-difference operator as $\nabla\Delta$, we can calculate the double differences as follows for code phase measurements:

$$\nabla\Delta P = P_i^{(1)} - P_j^{(2)} - P_i^{(1)} + P_j^{(2)}. \tag{4.11}$$

The double differences for the carrier phase are constructed as follows:

$$\nabla\Delta\Phi = \Phi_i^{(1)} - \Phi_j^{(2)} - \Phi_i^{(1)} + \Phi_j^{(2)}. \tag{4.12}$$

Consequently, double-difference carrier phase measurements can now be expressed as

$$\nabla\Delta\Phi = \frac{1}{\lambda}(\nabla\Delta\rho + \nabla\Delta T + \nabla\Delta I + \nabla\Delta O + \nabla\Delta\delta\Phi) + \nabla\Delta N, \tag{4.13}$$

where $\nabla\Delta\Phi$ denotes the carrier phase double differences, $\nabla\Delta N$ is the (unknown) double-difference ambiguity number, $\nabla\Delta\rho$ denotes the double-difference ranges, $\nabla\Delta T$ is the error due to the change in tropospheric conditions over the baseline, $\nabla\Delta I$ is the error due to the change in ionospheric conditions over the baseline, $\nabla\Delta O$ is the error due to orbit degradation, and $\nabla\Delta\delta\Phi$ is the error due to noise and multipath.

What is the difference between using single differences and double differences? As we can see by comparing equations for single- and double-difference observables, the equation for the double-difference observables contains far fewer errors. The receiver clock error is canceled (because the reception time of the signals from satellites i and j is the same, so the clock biases are also the same), and other errors become significantly smaller. However, this reduction in errors comes at a price. The dynamics of the observables has also been drastically reduced (see Table 4.1). This means that convergence of the related algorithms would take longer for double differences than for single differences. However, this mainly occurs in situations for which the solution is not strong enough, for example for orbit determination, especially for GEO and HEO satellites, rather than for positioning algorithms.

Double-difference equations are more noisy and make more demands on the quality of the carrier phase measurements.

Also, double differencing may have a different effect on the decorrelation of various errors. For example, the residual orbital error resulting from decorrelation

Table 4.1. Dynamics of difference observables

	Zero differenced	SD	DD
Satellite clock error	3.0 m	0	0
Receiver clock error	< ms	< ms	0
Geometry	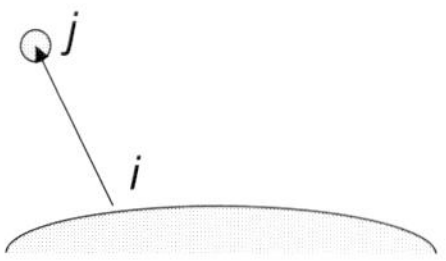	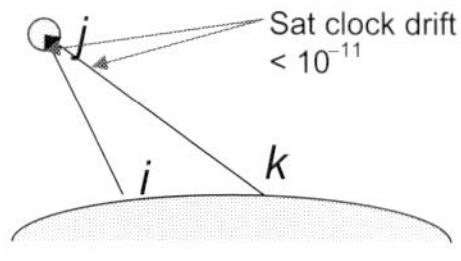	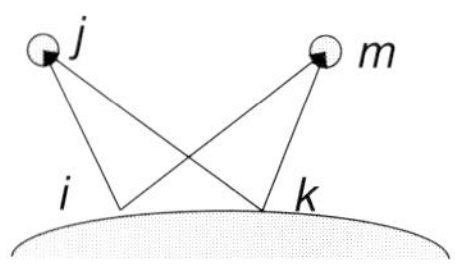
Observations and its dynamics	$\rho^j_i = r_{ji} + \delta t_i + \delta t^j + \Sigma\Delta_{ij}$ $\rho^j_i = r_{ji} + \Sigma\Delta_{ij}$ 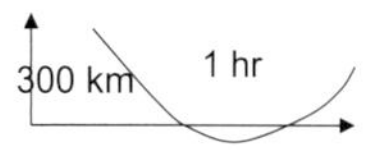	$SD^j_{ik} = r_{ji} - r_{jk} + \delta t_i - \delta t_k + \Sigma\Delta_{ijk}$ 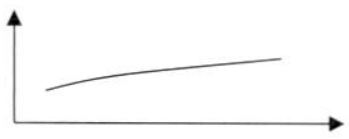	$DD^{jm}_{ik} = r_{ji} - r_{jk} - r_{mi} + r_{mk} + \Sigma\Delta_{ijmk}$ 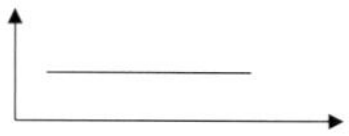

for the double-difference equation, $\nabla\Delta O$, can be estimated to be from 0.5 to 2 cm for a 50 km baseline for broadcast ephemerides, and even less than 5 mm for precise ephemerides.

4.3.3 GLONASS inter-frequency bias

Each GLONASS satellite transmits on a slightly different frequency. This leads to additional terms when considering GLONASS ambiguities due to inter-frequency bias. To account for this, double-difference equations can be formed as follows:

$$\nabla\Delta\Phi = \frac{1}{\lambda}(\nabla\Delta\rho + \nabla\Delta T + \nabla\Delta I + \nabla\Delta O + \nabla\Delta\delta\Phi) + \nabla\Delta N + \frac{\Delta\lambda}{\lambda}\Delta N_\lambda, \quad (4.14)$$

where ΔN_λ is an additional single-difference ambiguity due to inter-frequency bias, and $\Delta\lambda$ is the difference between the carrier wavelengths of a pair of satellites. The main problem with this is that the additional parameter compromises the integer character of double-difference ambiguities. This term only affects double-difference equations.

4.3.4 Triple-difference observables

We can go one step further and construct triple differences by taking the difference between sequential double-difference measurements (Figure 4.5). In this way, we can remove the ambiguity number $\nabla\Delta N$ from the state vector:

$$\Delta\nabla\Delta\Phi = \Delta\nabla\Delta\varphi - \Delta\nabla\Delta I + \Delta\nabla\Delta T + \Delta\nabla\Delta O + \Delta\nabla\Delta M + \Delta\nabla\Delta\delta\Phi. \quad (4.15)$$

When the observation rate is high and the ionosphere and troposphere are stable, the propagation delay errors may be negligible. However, the further loss in dynamics,

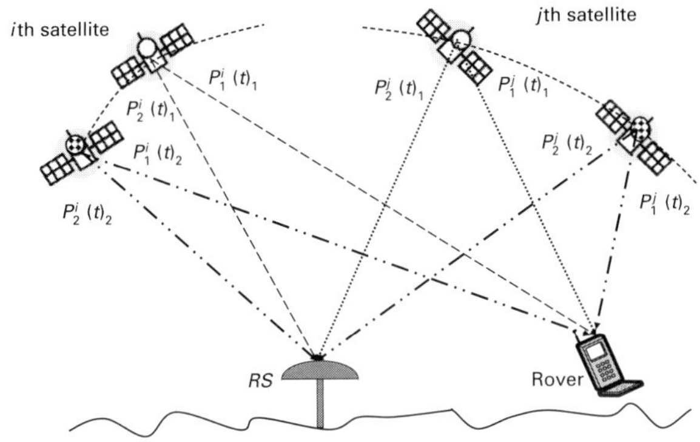

Figure 4.5 Triple-difference observables.

which we discussed in Section 4.3.2, practically renders these equations useless for positioning. However, they can serve as an indicator of cycle slip (a sudden jump that occurs when a carrier tracking loop misses a whole number of wavelengths in carrier phase measurements), because a cycle slip will appear as a jump in carrier phase measurements $\Delta\nabla\Delta\varphi$ which immediately shows up in low dynamics $\nabla\Delta\nabla\Phi$ as a spike.

If the initial position of a vehicle is known, the positioning using a triple-difference method can easily be performed as a dead reckoning. This may be useful, for example, for indoor positioning with pseudolites (see Chapter 8). However, positioning errors will accumulate, as always occurs for dead reckoning. In addition, a cycle slip will introduce an unrecoverable error.

4.3.5 Double-difference equations for multi-systems

The inter-frequency biases will also affect double-difference equations when satellites with signals on different frequencies are used to form double differences in the same system of equations. This is not the same situation as in (4.14). For a GLONASS SP signal, a term caused by the difference between the carrier wavelengths of a pair of satellites, $(\frac{\Delta\lambda}{\lambda}\Delta N_\lambda)$, should be added to each satellite pair, whereas for a multi-system such a term is only required for system pairs on different frequencies. For a GPS/GLONASS dual user, the term must account for the difference, i.e.

$$\Delta\lambda = k \cdot \Delta\lambda_{GLO} + \Delta\lambda_{GPS/GLO}, \tag{4.16}$$

where $\Delta\lambda_{GLO}$ is the difference in wavelength between two GLONASS adjusted frequencies, $k = -7,\ldots,0,\ldots,6$ (see Section 2.1.3), and $\Delta\lambda_{GPS/GLO}$ is the difference in wavelength between the GPS and GLONASS central frequencies.

For example, for $k = 0$, $\Delta\lambda_{GPS/GLO} = -0.01722$, and, for $k = 6$, $\Delta\lambda_{GPS/GLO} = -0.01932$. The difference seems not to be large, but it affects ambiguity-resolution algorithms due to the fact that this number accumulates over the large distances to a satellite. This effect also corrupts the integer nature of the ambiguities.

4.4 Real-time kinematic method

4.4.1 Code and carrier phase difference equations

We now consider differential positioning using one reference station. In this case, the linearization of GPS equations is performed in a manner similar to that considered in Chapter 3, but in relation to a known position of the reference receiver. Differential positioning can then be expressed as the task of finding a baseline vector which is the difference between the unknown position of the receiver and the known position of the reference station.

If the distance between the reference station and the rover is not large, then most of the errors in equations (4.2) and (4.3) are similar for the rover and reference receiver. Some errors in the error budget are more strongly correlated than others.

In matrix form, the linearized equations can be written as follows:

$$\Delta\overline{Z} = [H] \times \Delta\overline{X}, \tag{4.17}$$

where the state vector $\Delta\overline{X}$ now includes the baseline, the receiver clock error, and ambiguity numbers to become

$$\overline{X} = \begin{bmatrix} \Delta x_r \\ \Delta y_r \\ \Delta z_r \\ \Delta\delta t_r \\ \Delta N_{s_1} \\ \Delta N_{s_2} \\ \cdots \\ \Delta N_{s_n} \end{bmatrix}. \tag{4.18}$$

The baseline vector is the vector between the rover and the reference station. The position of the reference station is known, so the rover coordinates can be estimated. That is not true for the receiver clock error, which now can be estimated as $\Delta\delta t_r$ only in respect to the reference station clock error. This is because the receiver clock error is not stable enough to perform the absolute referenced estimate.

The observation vector $\Delta\overline{Z}$ is a vector of code and carrier phase single-difference observables:

$$\overline{Z} = \begin{bmatrix} \Delta P_{s_1} \\ \Delta P_{s_2} \\ \cdots \\ \Delta P_{s_n} \\ \Delta\Phi_{s_1} \\ \Delta\Phi_{s_2} \\ \cdots \\ \Delta\Phi_{s_n} \end{bmatrix}. \tag{4.19}$$

The measurement matrix $[H]$ for code phase measurements only is constructed from the unit direction vectors from an approximate position x_{r_0}, y_{r_0}, z_{r_0} to a satellite as follows:

$$[H] = \begin{bmatrix} e_{1x} & e_{1y} & e_{1z} & 1 \\ e_{2x} & e_{2y} & e_{2z} & 1 \\ \dots & \dots & \dots & \dots \\ e_{nx} & e_{ny} & e_{nz} & 1 \end{bmatrix}. \quad (4.20)$$

For carrier phase measurements, the extended measurement matrix looks similar, with the exception that it includes ambiguity numbers:

$$[H'] = \begin{bmatrix} e_{1x} & e_{1y} & e_{1z} & 1 & 1 & 0 & \dots & 0 \\ e_{2x} & e_{2y} & e_{2z} & 1 & 0 & 1 & \dots & 0 \\ \dots & \dots & \dots & \dots & \dots & \dots & \dots & \dots \\ e_{nx} & e_{ny} & e_{nz} & 1 & 0 & 0 & \dots & 1 \end{bmatrix}, \quad (4.21)$$

or

$$[H'] = [\,\|H\|\|I\|\,]. \quad (4.22)$$

The equations for the double-difference observables can be formed in a similar way. The state vector no longer contains the receiver clock errors, as they are canceled at the double differencing, i.e.

$$\overline{X} = \begin{bmatrix} x_r \\ y_r \\ z_r \\ \Delta\nabla N_{s_2} \\ \Delta\nabla N_{s_3} \\ \dots \\ \Delta\nabla N_{s_n} \end{bmatrix}. \quad (4.23)$$

The extended measurement matrix is as follows:

$$[H'] = \begin{bmatrix} e_{2x}-e_{1x} & e_{2y}-e_{1y} & e_{2z}-e_{1z} & 1 & 1 & 0 & \dots & 0 \\ e_{3x}-e_{1x} & e_{3y}-e_{1y} & e_{3z}-e_{1z} & 1 & 0 & 1 & \dots & 0 \\ \dots & \dots & \dots & \dots & \dots & \dots & \dots & \dots \\ e_{nx}-e_{1x} & e_{ny}-e_{1y} & e_{nz}-e_{1z} & 1 & 0 & 0 & \dots & 1 \end{bmatrix}, \quad (4.24)$$

where satellite 1 was chosen as a key satellite to form all the double differences. (It is best to use the satellite with the longest predicted visibility period and an elevation higher than 15° as the key satellite.)

The advantage now is that all the components of the state vector are constant for a static receiver, which means that we can accumulate measurements from different epochs. This would not be immediately possible for single differences because of the changing receiver clock error, which cannot be removed from the state vector as it defines the receiver coordinates along with three-dimensional space coordinates.

The approach via the construction of single- or double-difference equations, which we discussed in Section 4.3, and then solving for coordinates, has many advantages. We can use this approach for the code and phase differential positioning and also for the network RTK method.

A different, and more simple, approach would be to calculate range corrections using reference station measurements and their predicted values based on the reference station's known position. In the approach just described we do not need to use the position of the reference station until after the baseline is estimated. The range corrections can then be applied to the rover range measurements directly. The reference station then estimates its clock error and removes it from the corrections. In the case of double differences, we do not need to use the position of the reference station until after the baseline is estimated.

The corrections can be supplied through various data links, such as a TV broadcast signal [2], satellites (SBAS), and the Internet [3]. More details on the data link implementation are considered in Chapter 8.

4.4.2 RTK positioning algorithm

Ambiguity-resolution methods find the ambiguity number, which minimizes the following residuals for all satellites. For single-difference observables,

$$\sum_{i=1}^{n} \Delta v_i = \sum_{i=1}^{n} \left(\Delta r_i - \Delta\Phi_i - \lambda \cdot \Delta N_i - c(\delta t_r^{(1)} - \delta t_r^{(2)}) - \Delta\delta\Phi_i \right), \tag{4.25}$$

where Δr_i is the single difference of the distances to the ith satellite from the rover and the reference station, λ is the wavelength, and $\delta t_r^{(1)} - \delta t_r^{(2)}$ is the relative rover clock error.

All ambiguity-resolution algorithms are basically brute-force search methods. The search is looking for a set of Δr_i and ΔN_i values which minimizes the residuals. The Δr_i and ΔN_i values are functions of the estimated rover position, and the minimum of the residuals should give us the rover position estimate with an accuracy at the centimeter scale.

The persistent solution, i.e. the one that minimizes the residuals for all epochs, found using the integer assumption for the number of sequential epochs will yield a good candidate for the ambiguity numbers and rover position (Figure 4.7). Increasing the number of satellites allows one to decrease the time required for RTK positioning algorithms down to a few epochs.

The single-difference residual equation (4.25) is not very useful for RTK purposes because it contains a changing clock error. Therefore we must again consider double-difference residuals:

$$\sum_{i=1}^{n-1} \nabla\Delta v_i = \sum_{i=1}^{n-1} \nabla\Delta r_i - \nabla\Delta\Phi_i - \lambda \cdot \nabla\Delta N_i - \nabla\Delta\delta\Phi_i, \tag{4.26}$$

where i is the satellite number. The sum is for $n - 1$, because the double difference is taken using one satellite as a key satellite.

For the RTK algorithm, we search the residuals to be minimum ($\nabla\Delta v_i \rightarrow 0$), and we can rewrite the double-difference equations as follows:

$$\overrightarrow{\nabla\Delta y} = [A]\overrightarrow{\nabla\Delta N} + [B]\,\overrightarrow{\nabla\Delta r} + \vec{\zeta}\,, \tag{4.27}$$

where $\overrightarrow{\nabla\Delta y}$ is the vector of double-difference measurements, $\overrightarrow{\nabla\Delta N}$ is the vector of double-difference ambiguities, $\overrightarrow{\nabla\Delta r}$ is a baseline vector, and ζ is the noise. The corresponding variance–covariance matrices are $[Q_A]$, $[Q_B]$, $[Q_z]$, $[Q_{AB}]$.

These equations contain only constant $\nabla\Delta r_i$ and $\nabla\Delta N_i$ parameters for a static receiver and can be used across multiple epochs, thus accumulating a larger number of measurements while keeping the number of unknowns the same.

The carrier phase measurements for the following epochs become unambiguous as soon as we fix the ambiguity numbers to certain values. Therefore, the initial search grid can be established in either a coordinate or an ambiguities domain. If it is established in the coordinate domain, double-difference distances to the satellites are calculated based on the rover grid-point coordinates, and the ambiguities are calculated using fixed rover and satellite coordinates:

$$\nabla\Delta r|_j \Rightarrow \nabla\Delta N|_j, \tag{4.28}$$

where the $\nabla\Delta r_i|_j$ and $\nabla\Delta N_i|_j$ are double-difference distances to the satellites at grid point j.

The initial ambiguities can be estimated from the code phase measurements as follows:

$$\nabla\Delta N|_j \Rightarrow \nabla\Delta r|_j. \tag{4.29}$$

The minimum of the residuals over the number of epochs gives an estimate of the ambiguities and position, where constraints should also be applied to each residual so as not to exceed a certain threshold:

$$\begin{cases} \sum\limits_{i=0}^{i=n-1} \nabla\Delta\upsilon_i \Rightarrow \min, \\ \nabla\Delta\upsilon_i < \nabla\Delta\upsilon_{TRESHOLD}. \end{cases} \tag{4.30}$$

The search in the ambiguity domain has the advantage that the ambiguities are integers. Therefore the search creates additional constraints and limits the search grid to only a limited number of points. The constraints also may exist in some particular cases for (4.28), for example if a rover is on the roof of a building, which would fix the vertical coordinate, thus making the position two dimensional. The vertical coordinate exists in the position domain, but this constraint can also be transformed into the ambiguities domain.

The most widely used method, RTK LAMBDA, is described in [4] and [5]. The software (developed by the authors of RTK LAMBDA), along with the source code that implements the method, is available from Delft University of Technology upon request.

4.4.2.1 Float solution

At first, the ambiguity resolution finds an approximate float solution $(\nabla\Delta\tilde{N}, \nabla\Delta\tilde{r})$ for equation (4.27) without making any assumptions about the integer nature of the ambiguities. The float solution can therefore be based on unambiguous code measurements. After that, the carrier phase ambiguities are determined by choosing the best fit to the measurements from a number of candidates. Assuming that the receiver antenna is static (the movement of the rover antenna can also be accounted for), the ambiguity is resolved by making use of the change of satellite constellation with time. Figure 4.6 shows the changes of positions corresponding to each ambiguity candidate with the change of satellite constellation. Since the position solution corresponding to

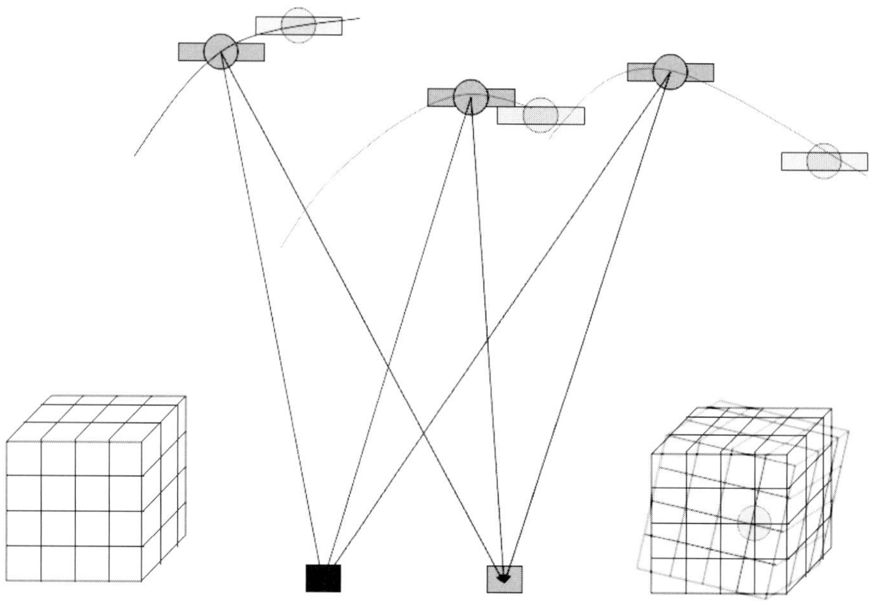

Figure 4.6 RTK method with a search cube in the ambiguity space.

the correct ambiguity does not change with time, the ambiguity will be resolved after a change in satellite constellation geometry.

4.4.2.2 Integer solution

At the next step, the algorithm finds an improved solution for ambiguities ($\nabla\Delta \widehat{N}$, $\nabla\Delta \widehat{b}$) assuming that they can be integers only. The simplest RTK method is to round off $\Delta\nabla\tilde{N}$ to the nearest integer, i.e.

$$N^k : \nabla\Delta \widehat{N} = int\,(\nabla\Delta\Phi - \nabla\Delta P). \tag{4.31}$$

This method does not take into account the correlations between ambiguities, as described by matrix $[Q_A]$.

The LAMBDA method [4],[5] developed by the Delft University of Technology takes into account the correlations between ambiguities. The ambiguity-resolution task is transformed from ambiguity k-dimensional space to a new n-dimensional space, in which the ambiguities are uncorrelated:

$$N^k \rightarrow Z^n. \tag{4.32}$$

The LAMBDA method transforms the problem from the ambiguity space, where geometrically the search space looks like a very long prolate n-dimensional ellipsoid, into an equivalent space, where the search space is transformed to a much less prolate shape, thus significantly decreasing the number of candidates. The prolate ellipsoid shape corresponds to a high correlation between the ambiguities, whereas in the transformed space the transformed ambiguities are almost uncorrelated.

The estimation can be performed using least-square estimation with the integer constraints. If Z^n is an n-dimensional space of integers, the required estimation reads as follows:

$$Z^n : a_{fix} = min(\tilde{N} - z)^T Q_A{}^{-1}(\tilde{N} - z). \quad (4.33)$$

To provide the solution for the total number of ambiguities, a brute-force method, for example, can be used.

4.4.2.3 Validation

At the next step, the algorithm validates the solution, i.e. it checks that the value found at the previous step is correct. This step is necessary because the previous step will always give a solution, but it can be the wrong one. Validation can be provided by a statistical test in either the measurement domain or the positioning domain. If only one ambiguity candidate set is retained, it is considered as the solution. If more than one candidate is retained, similar statistical tests should be performed for the next epoch.

4.4.3 Network RTK method

We consider a network RTK method following [6], [7], and [8], in which case we consider a number of reference stations rather than just one. Double-difference carrier phase measurements for the ith and jth receivers (reference station or rover) are expressed as follows:

$$\nabla\Delta\Phi = \frac{1}{\lambda}(\nabla\Delta\rho + \nabla\Delta T + \nabla\Delta I + \nabla\Delta O + \nabla\Delta U) + \nabla\Delta N, \quad (4.34)$$

where $\nabla\Delta\Phi$ denotes the carrier phase double differences, $\nabla\Delta N$ is the (unknown) double-difference ambiguity number, $\nabla\Delta\rho$ denotes the double-difference ranges, $\nabla\Delta T$ is the error due to the change in tropospheric conditions over the baseline, $\nabla\Delta I$ is the error due to the change in ionospheric conditions over the baseline, $\nabla\Delta O$ is the error due to orbit degradation, and $\nabla\Delta\delta\Phi$ is the error due to noise and multipath.

A network RTK system requires components as discussed in the following subsections.

4.4.3.1 Network of reference stations

A network of reference stations comprises a number of high-grade GNSS receivers with high-quality antennas that should be installed in a low-multipath environment with full sky view. The antenna positioning must be surveyed using geodetic software and external network data, for example the IGS or GSI (Geographical Survey Institute, in Japan). Nowadays, an estimation of receiver position to millimeter accuracy in post-processing mode is a routine task.

Rovers and reference stations should be able to see the same satellites. The role of thumb is that 100 km of distance on the surface of the Earth is equivalent to 1° change in elevation of an MEO satellite. In the case of a single reference station, the usual distance between it and a rover does not usually exceed 10 km, so the same constellation is available for both receivers. For an RTK network, however, the distance becomes much greater, and this issue becomes important to keep in mind.

The number of unknowns that we have to solve for in the case of a network is increased due to new ambiguities arising from the new stations. For single-difference equations, the state vector becomes

$$\overline{X} = \begin{bmatrix} \nabla\Delta N_{s_1}^{R_1} \\ \nabla\Delta N_{s_2}^{R_1} \\ \cdots \\ \nabla\Delta N_{s_n}^{R_1} \\ \cdots \\ \nabla\Delta N_{s_1}^{R_k} \\ \nabla\Delta N_{s_2}^{R_k} \\ \cdots \\ \nabla\Delta N_{s_n}^{R_k} \\ \cdots \end{bmatrix}, \quad (4.35)$$

where $\nabla\Delta N_{s_n}^{R_k}$ denotes the single-difference ambiguities for satellite n and reference station k. The difference between this state vector and that in (4.23) is that here the baseline $[x_{ij};y_{ij};z_{ij}]^T$ between any ith and jth reference station is excluded, because reference station coordinates are surveyed in advance.

The observation vector $\Delta\overline{Z}$ now includes observations from all reference station single-difference observables:

$$\overline{Z} = \begin{bmatrix} \nabla\Delta P_{s_1}^{R_1} \\ \cdots \\ \nabla\Delta P_{s_2}^{R_2} \\ \cdots \\ \nabla\Delta P_{s_n}^{R_n} \\ \nabla\Delta \Phi_{s_1}^{R_1} \\ \nabla\Delta \Phi_{s_2}^{R_2} \\ \cdots \\ \nabla\Delta \Phi_{s_n}^{R_k} \end{bmatrix}. \quad (4.36)$$

4.4.3.2 Control center

The control center is basically a computer equipped with specialized software. The control center calculates corrections for the key reference station at the secondary reference station position as follows:

$$\nabla\Delta\upsilon = \nabla\Delta\Phi - \frac{1}{\lambda}\nabla\Delta\rho - \nabla\Delta N, \quad (4.37)$$

where $\nabla\Delta\rho$ is the double difference of the true range to a satellite from two reference stations (the key reference station and the secondary reference station).

The control center can resolve ambiguities from a key reference station somewhere in the middle of network to all reference stations one by one. If reference station A is the key reference station, the control center needs to find ambiguities from reference station A to reference station B and reference station C. To find the correct double-difference ambiguities in real time is rather a complicated task because the control center should be able to recalculate ambiguities instantly if cycle slip on one of the reference stations or other problems occur. Therefore it is important to minimize this task without sacrificing accuracy. The control center may skip the step of resolving ambiguities between reference station B and reference station C. In that case, the network configuration will be a star configuration (see Figure 4.7(a)) instead of a closed-loop configuration (see Figure 4.7(b)). This strategy is advantageous for RTK networks because if an error had occurred on one baseline, it would not propagate to other baselines through the network constraints [6],[8].

Based on the corrections $\nabla\Delta\upsilon$ calculated for the reference station positions, the control center can recalculate them into the grid points (Figure 4.7(c)). Grid points

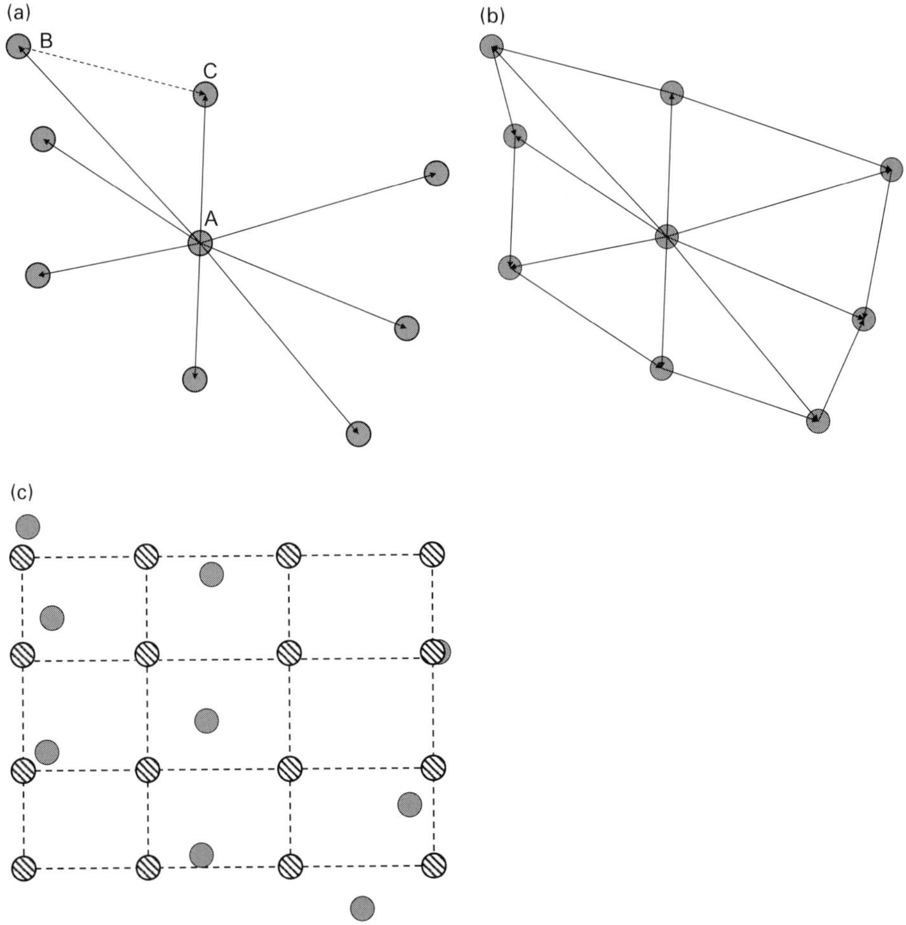

Figure 4.7 Resolving network ambiguities. (a) Star configuration; (b) closed-loop configuration; (c) grid points.

can be used if the link from the control center to a rover is unidirectional. If a rover can send its coordinates to the control center, the control center calculates the correction to the virtual reference station (VRS) with zero baseline to the user. For a unidirectional link, the user calculates the VRS data using the grid data.

The calculated corrections therefore consist of main errors in the receiver error budget:

$$\nabla\Delta\upsilon = \frac{1}{\lambda}(\nabla\Delta T + \nabla\Delta I + \nabla\Delta O + \nabla\Delta\delta\Phi). \tag{4.38}$$

Multipath errors should be minimized by carefully checking the antenna environment and using special types of antenna. The main error is due to the ionosphere, i.e. $\nabla\Delta I$. The ionosphere considered as a propagation medium for GNSS signal manifests itself in three ways. (i) Ambiguity resolution fails if the overall error exceeds approximately half the wavelength, and the ionospheric error is the biggest one in the error budget. (ii) The ionosphere interferes with the correction calculation because of TEC fluctuations between LOS to the grid points. (iii) Excess activity in the ionosphere causes phase scintillation in the GNSS signal, which results in cycle slip (see details on scintillation theory and its effects in [9]).

It is possible to decrease ionospheric error via real-time ionosphere mapping as a function of the station coordinates and satellite elevation [8]:

$$\nabla\Delta I = F(\nabla\Delta\upsilon) \rightarrow F(\beta, \varphi, \Lambda), \tag{4.39}$$

where φ is latitude, Λ is longitude, and β is satellite elevation. The mapping function is less accurate than a satellite-to-station estimate, but it has the advantage that it does not depend on a particular satellite and therefore $\nabla\Delta\upsilon$ can be quickly restored without resolving the ambiguity:

$$\nabla\Delta\upsilon \leftarrow F(\beta, \varphi, \Lambda). \tag{4.40}$$

The ionospheric mapping function can be estimated and then used to assist in ambiguity resolution, which in turn is used for ionospheric error estimation. The algorithm requires time for initialization when implemented in real time.

References

[1] G. Beutler, R. Weber, U. Hugentobler, M. Rothacher and A. Verdun, GPS satellite orbits, in *GPS for Geodesy*, 2nd edn., P. J. G. Teunissen and A. Kleusberg, Eds. Berlin: Springer-Verlag, 1998.

[2] I. Petrovski, K. Sasano, B. Townsend, M. Ishii, and H. Torimoto, TV broadcast utilization for code and carrier phase differential GPS, in *Proc. GNSS 2001*, Seville, May 2001.

[3] H. Hada, I. Petrovski, H. Sunahara, *et al.*, DGPS and RTK positioning using the Internet, *GPS Solutions*, vol. **4**, no. 1, pp. 34–44, 2000.

[4] P. J. G. Teunissen, GPS carrier phase ambiguity fixing concepts, in *GPS for Geodesy*, 2nd edn., P. J. G. Teunissen and A. Kleusberg, Eds. Berlin: Springer-Verlag, 1998.

[5] P. J. De Jonge and C. C. J. M. Tiberius, *The LAMBDA Method for Integer Ambiguity Estimation: Implementation Aspects*, Delft Geodetic Computing Centre LGR series, no. 12. Delft: University of Technology, 1996.
[6] B. Townsend, I. Petrovski, *et al.*, Concept for a regional satellite-based carrier-phase correction service, in *Proc. 16th Int. Tech. Meeting of the Satellite Division of The Institute of Navigation (ION GPS/GNSS 2003)*, Portland, OR, Sept. 9–12, 2003, pp. 2631–2636.
[7] G. Lachapelle, P. Alves, *et al.*, DGPS RTK using a reference network, in *Proc. 13th Int. Tech. Meeting of the Satellite Division of the Institute of Navigation (GPS 2000)*, Salt Lake City, UT, Sept. 20–22, 2000.
[8] I. Petrovski, E. Cannon, G. Lachapelle, *et al.*, The issues of practical implementation of the commercial RTK network service, in *Proc. 14th Int. Tech. Meeting of the Satellite Division of The Institute of Navigation (ION GPS 2001)*, Salt Lake City, UT, Sept. 11–14, 2001, pp. 2654–2664.
[9] I. Petrovski and T. Tsujii, *Digital Satellite Navigation and Geophysics: A Practical Guide with GNSS Signal Simulator and Receiver Laboratory*. Cambridge: Cambridge University Press, 2012.

Part II

From conventional to software GNSS receivers and back

5 Generic GNSS receivers

5.1 GNSS receiver overview

5.1.1 Digest of GNSS receiver operation

A GNSS receiver estimates position by measuring distances to a few satellites. When it measures the distance to the satellite, it in fact measures a signal propagation time from the satellite to the antenna. The satellite signal bears a time mark, which indicates when the signal was transmitted.

First, we look at the GPS L1 C/A signal. The signal can be considered to consist of three layers, namely the carrier layer, the code layer, and the navigation data layer (see Figure 5.1). These three layers have different time resolutions. Other signals may have additional layers. The GLONASS L1 SP signal, for example, has an additional meander sequence. Galileo, BeiDou, modernized GPS, and GLONASS have additional code layers. The length of the navigation message bit for the GPS L1 C/A signal is 20 ms. One can multiply this time interval by the speed of light to calculate the length of the navigation bit. For a GPS L1 C/A signal, the code sequence has 1023 bits, and this repeats itself each millisecond. For mobile devices, we are mostly interested in code positioning only. The 1 ms time is approximately equal to 300 km. The length of one code chip is therefore 300 km divided by 1023 code chips, i.e. about 300 m. In order to achieve positioning accuracy of a few meters, a receiver must be able to measure the code phase with an accuracy of one-hundredth of a code bit. A receiver can measure a code phase by aligning the incoming signal from a satellite with an internal code replica.

From the receiver's point of view, a signal can be disassembled even further. Almost all GNSS receivers implement digital signal processing, in which a receiver works with a signal after the signal has been down-converted and digitized in a receiver front end. Figure 5.2 depicts a functional diagram of a GNSS receiver. The sampling rate of digitization is defined by the receiver clock and must meet the requirements of the Nyquist criterion,

$$f_S \geq 2B, \tag{5.1}$$

where f_S is the minimum sampling frequency and B is the signal bandwidth.

For a snapshot receiver, which does not use tracking, the sampling rate basically defines the achievable accuracy of measurements. If a snapshot receiver uses only short

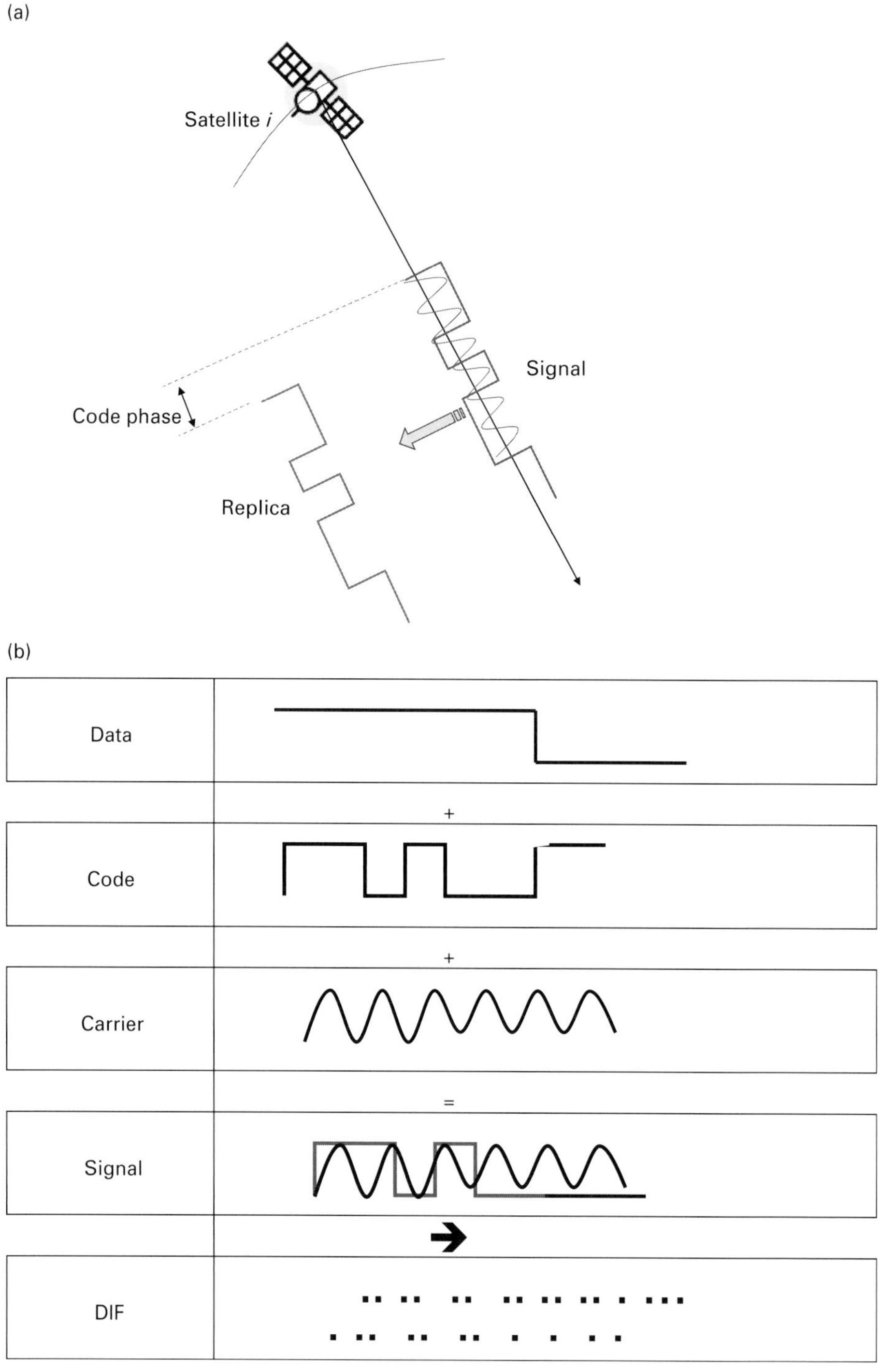

Figure 5.1 GPS L1 C/A signal in chips, waves, and samples.

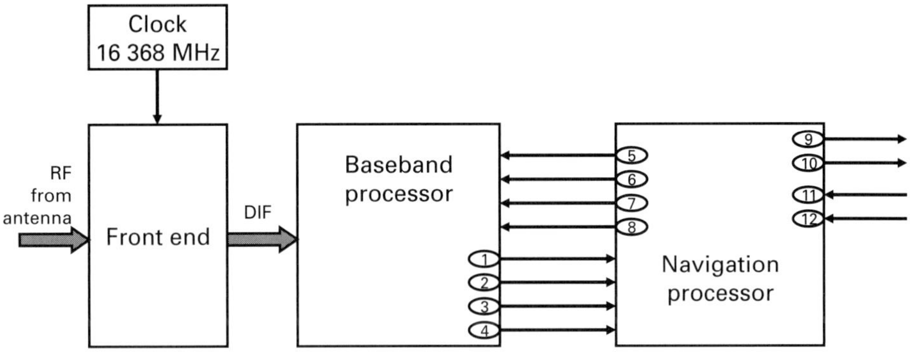

Figure 5.2 Functional diagram of a receiver.

chunks of the signal, it cannot align the internal signal replica with the satellite signal with an accuracy better than half the sample.

The GPS L1 C/A code main lobe has a bandwidth of 2 MHz. For good performance, a minimum sampling rate of 4 megasamples per second (Mps) would be required; 2.5 Mps may work for processing a GPS signal, but the signal quality would be rather poor.

If we consider a relatively high 16 Mps sampling rate, the interval between samples would be defined by 16 000 samples per millisecond. Each millisecond contains 1023 code chips of the GPS L1 C/A signal. One chip length is about 300 m, (the speed of light divided by the duration of the chip), and each chip contains slightly more than 15 samples, so the distance between the samples is about 19 m. The code phase measurement accuracy would not be much better than about 10 m, and on multiplying this code error by the DOP, we can estimate the positioning accuracy for a pure snapshot receiver.

For a receiver with code tracking, the accuracy depends on the receiver bandwidth, which is closely related to the receiver code tracking loop. The code phase tracking accuracy can be on the order of decimeters, much more than normally required for mobile applications.

To enhance accuracy further, we can use carrier phase, which has a 19 cm wavelength for GPS L1 C/A, or 9.5 cm half-length to compare with a 300 m code chip length. A receiver can measure carrier phase with an accuracy limited by the carrier phase errors and the receiver clock phase noise. These measurements allow coordinates to be determined with potentially millimeter accuracy in post-processing. Even for mobile devices, the capability to use carrier phase for positioning can improve accuracy from meters to centimeters in real time.

The digitized signal from the front end goes into a baseband processor (Figure 5.2). The function of the baseband processor is to find the GNSS signal and analyze it. The components in the baseband processor may vary depending on the function and price of the receiver.

The radio signal at the input of the baseband processor is in digitized form. We have described a GNSS signal from one satellite in Chapter 2 as follows:

$$A = A_0 \sin\left((\omega_0 + \omega_D)t + \varphi\right) \cdot D \cdot B, \tag{5.2}$$

where ω_0 is the signal central angular frequency, ω_D is the Doppler angular frequency, D is a spread code, and B is a navigation message. We now modify this equation in order to describe a signal from multiple satellites. Code and carrier phase measurements exist only as relative measurements, so they can be taken in relation to the receiver-generated signal replica. We need to remember though that the receiver time frame is not fixed, so this would not provide any extra information. The signal from multiple satellites can then be presented as follows:

$$A = \sum_{i}^{N} A_i = \sum_{i}^{N} \{A_{0_i} \sin((\omega_0 + \omega_{D_i})t + \varphi_i) \cdot D_i \cdot B_i\}, \tag{5.3}$$

where N is the number of satellites.

One purpose of the baseband processor is to obtain the following values:

(1) the Doppler shift,

$$f_{D_i} = \omega_{D_i}/2\pi,$$

of the signal from a nominal frequency $f_0 = \omega_0/2\pi$;

(2) the code phase between the satellite signal and the receiver code replica,

$$\Delta B_i = B_i - B_R;$$

(3) the carrier phase between the satellite signal and the receiver carrier replica,

$$\Delta\varphi_i = \varphi_i - \varphi_R;$$

(4) the encoded navigation message, D_i;

(5) the amplitude of the signal, A_{0_i}, for each satellite in relation to a noise floor.

The receiver-generated replica supplies the code and carrier references B_R and φ_R in the receiver time frame.

Additional information can be derived from the signal, for example the statistics of the carrier and code fluctuations with time.

The baseband processor functionality is encapsulated by *acquisition* and *tracking*. The purpose of acquisition is to initialize signal tracking by finding approximate values for the Doppler shift f_{D_i} and the code phase ΔB_i in (5.3). These approximate values are then transferred to the tracking. The values ΔB_i and f_{D_i} are refined in the code tracking loop. The carrier tracking loop, where carrier phase $\Delta\varphi_i$ is defined, may also be included in the tracking process.

The final operation is to record a navigation message D_i after an operation called *bit synchronization*.

For some applications, a baseband processor can be reduced to an acquisition part only. These applications apply snapshot positioning if only short snapshots of the GNSS signal are available. In this case, the ephemeris information normally derived from the contents of a navigation message should be supplied from outside the receiver. A positioning algorithm should also resolve the ambiguity in the code measurements, because spreading code repeats itself. For example, a GPS L1 C/A code repeats itself approximately every 300 km. Therefore, by measuring code phase only, the receiver cannot find out to which 300 km interval, within a distance to a satellite, the measured phase belongs. With longer codes, this is less of a problem. The methods of acquisition and tracking for GPS receivers are well developed and described in the literature [1]–[6].

5.1.2 Receiver specification

When a potential user is looking to buy a receiver, he or she first looks at the specification (or spec) to see if the receiver will fit the required application. The application defines not only the values given at specific positions within the specification, but also the contents of the specification itself.

5.1.2.1 Specification parameters

We first need to identify a potential user, whether this is a system integrator, an end user of a receiver, or an end user of an integrated system. An end user may be interested in coordinates and how quickly and reliably they can be estimated, whereas a system integrator, who, for example, uses an OEM component to make an end product, may be interested in the receiver raw data. An OEM component is here defined as a component that is sold to a system integrator by an original equipment manufacturer with the purpose of being integrated into a system integrator's product. The OEM component brand does not need to be shown on the final product. The system integrator may not be interested in the positioning accuracy, but rather in the accuracy of the raw data measurements, particularly if a navigation processor functionality lies outside an OEM component. An end user of an integrated system may just be interested in the derivative information. For example, a person who buys a car does not usually have access to the specification of the receiver implemented in the car's navigation system.

We will consider a specification in detail from both the end user's and the system integrator's perspectives.

5.1.2.1.1 Accuracy

Accuracy can be specified by various parameters. If the parameter is not specified, one has to assume that it is most likely a root mean square (RMS), i.e. the square root of a mean of the squares of the errors. One usually specifies the RMS of the horizontal errors as follows:

$$RMS = \sqrt{\frac{\sum_{i=1}^{n} (r_i - r_{true})^2}{n}}, \tag{5.4}$$

where $r_i = \sqrt{x_i^2 + y_i^2}$ and r_{true} denotes the true position coordinates.

Instead of the true position, one can consider the population mean. In that case, one obtains the value of the standard deviation:

$$\sigma_x = \sqrt{\frac{\sum_{i=1}^{n} (x_i - \hat{x})^2}{n}}. \tag{5.5}$$

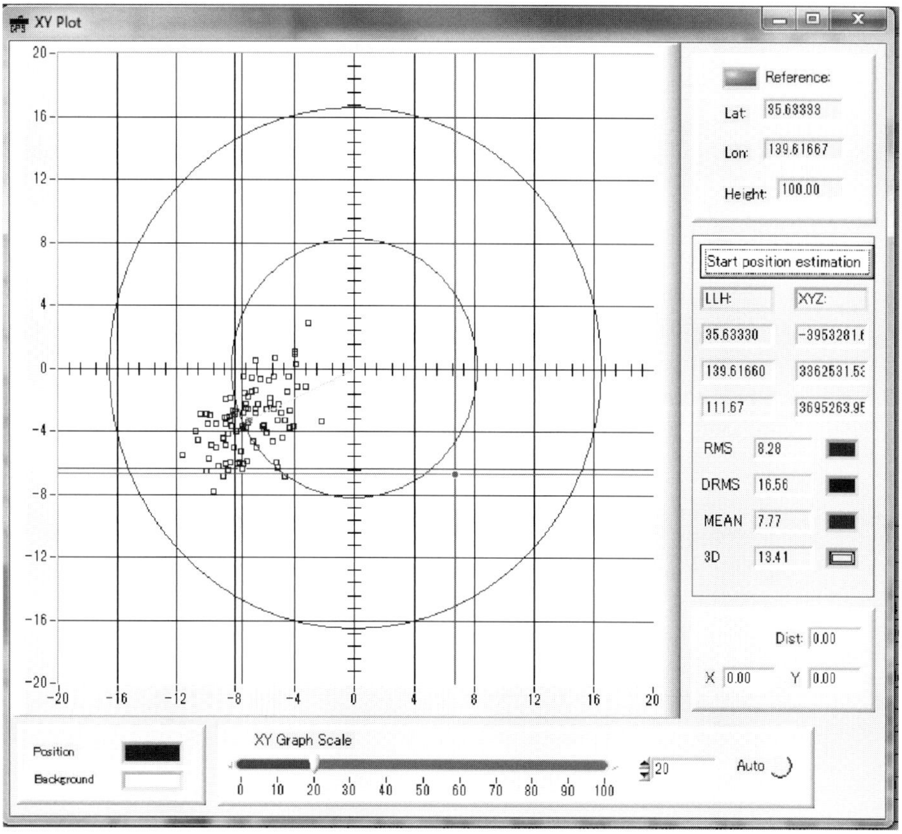

Figure 5.3 RMS and 2DRMS outputs of a software receiver.

For a normal distribution, the RMS ellipse contains about 68.27% of all errors. If the distribution is not normal, one can measure the percentage of the errors inside the RMS circle by measuring an area under the probability density function (PDF) curve.

The twice distance root mean square, or 2DRMS, is a value twice the RMS of the horizontal errors. For a normal distribution, the 2DRMS circle contains about 95.45% of all errors. Note that $\pm 3\sigma$ contains 99.73% of all errors.

Figure 5.3 shows an example of positioning output from an iPRx software receiver. The bull's-eye plot shows RMS and 2DRMS statistics of the position estimates.

The circular error probable (CEP), a term that came from the field of ballistics, is defined as the radius of a circle centered at the true position that encloses 50% of all estimates of horizontal position. The spherical error probable (SEP) is defined as the radius of a sphere (centered at the true position) that encloses 50% of all position estimates. These values can be directly converted to each other using linear error conversion factors.

The accuracy of a receiver can be specified depending on its capabilities for stand-alone, differential, and RTK modes. The receiver can also have the capability to receive

Table 5.1. Accuracy spec example

Function	Value	Low end (RMS)	High end (RMS)
Standalone	coordinate	7 m	1 m
	velocity	n/a	0.05 m/s
Differential	coordinate	n/a	$\pm$0.25 m + 1 ppm
RTK	coordinate	n/a	$\pm$10 mm + 1 ppm
WAAS	coordinate	1.5 m	0.7 m
Raw data	code phase	n/a	30 cm
	carrier phase	n/a	10 mm

signals from space-based augmentation systems (SBAS) such as WAAS, EGNOS, MSAS, or Luch (see Chapter 2). In this case, the name of the SBAS and the achievable accuracy are also specified.

Accuracy can be specified separately for coordinates, velocities, and other measurements, including raw data, such as code and carrier phases. Table 5.1 gives examples of specification values, but these values do not describe a specific receiver.

The velocity accuracy does not change much as a function of receiver mode, and it is basically the same in standalone, differential, RTK, or WAAS mode. This is because the velocity measurement came from a Doppler measurement in the tracking loop and was not obtained by coordinate differentiation. Therefore the velocities are not affected by the error budget of the code and phase measurements. Snapshot receivers have low accuracy and may have tracking loops absent, in which case attaining a meaningful estimate of velocity is problematic.

5.1.2.1.2 *Sensitivity*

The sensitivity describes the ability of a receiver to hear a signal. Normally, a GNSS signal when it reaches the surface of the Earth is already below the noise floor. It can be recognized by a receiver only because of a spread code in the signal, which resonates with a replica receiver. Additional signal attenuation, caused by various obstructions, such as buildings or vegetation, makes the acquisition of a signal by a normal receiver difficult. Sensitivity is a very important specification parameter for mobile devices. A high-sensitivity receiver uses various methods and algorithms to enhance the sensitivity, and ultra-high sensitivity requires massive parallel correlation, featuring from 500 000 to one million hardware correlators. Such receivers can even make a positioning with a GNSS signal indoors. Examples of sensitivity specification parameters are given in Table 5.2.

Signal power is measured in decibels (dB), which is the logarithmic ratio of power to a specific reference level. A GNSS signal is measured in decibels with a reference level of either 1 W (dBW) or 1 mW (dBm), and

$$P_{(dBm)} = P_{(dBW)} + 30. \tag{5.6}$$

Table 5.2. Sensitivity spec example

Value (dBm)	GPS signal power near the Earth surface	High sensitivity	Ultra-high sensitivity
Acquisition	−130	−145	−160
Tracking		−160	n/a

The logarithmic scale is used for two main reasons. First, it makes it possible to work with numbers over a very large dynamic range. Second, multiplication of ratios may be substituted by an addition operation. Consequently, the difference between normal and high sensitivity, which does not look very large in decibels, is really huge. If the standard signal strength on the surface of the Earth is specified at about −130 dBm, the ultra-high-sensitivity receiver can acquire signal indoors with signal power −160 dBm, which is about 30 dB lower. This 30 dB corresponds to a power ratio of 1000.

5.1.2.1.3 *Systems and frequencies*

A receiver can receive signals from one or many GNSS, and can work with a signal on one or more than one frequencies. For mobile applications, from a user's point of view the more systems the receiver supports the better. We will look at this from the manufacturer's point of view in Chapter 10. Additional systems provide more satellites, which is especially useful in urban canyons or obstructed environments. Support for GPS is a necessary feature of all receivers.

Related to the number of supported systems is the parameter that describes the number of channels. Normally, 12 parallel channels are required for a system, and there should be no more than 12 satellites on the sky at any time (see Figure 5.4). Therefore, if the receiver supports, for example, GPS, GLONASS, and Galileo, it has to have 36 channels. Additional channels may be useful to allocate SBAS; a receiver may, for example, support an SBAS ranging service, a correction service, and an integrity service.

A second frequency can be used to remove ionospheric error in the receiver. It may not be useful for all mobile applications (i) because the specified accuracy of the device may be above this error level, and (ii) because ionospheric error compensation can also be provided as part of the assistance information from an external source. Mobile applications will not benefit from more than two frequencies.

5.1.2.1.4 *Time to first fix*

The time to first fix (TTFF) describes the time interval from the moment the user starts the receiver operation until the first position output from the receiver is received. TTFF is usually defined for various start-up modes, for example "hot," "warm," and "cold" starts. The definitions of start-up modes, however, are not standardized and depend on the technology implemented in the specific receiver. They are usually defined by the balance of information content available in a receiver and that which is required to be received from a broadcast navigation message. We associate all definitions to a situation when a receiver operates in autonomous mode without a data link to a server.

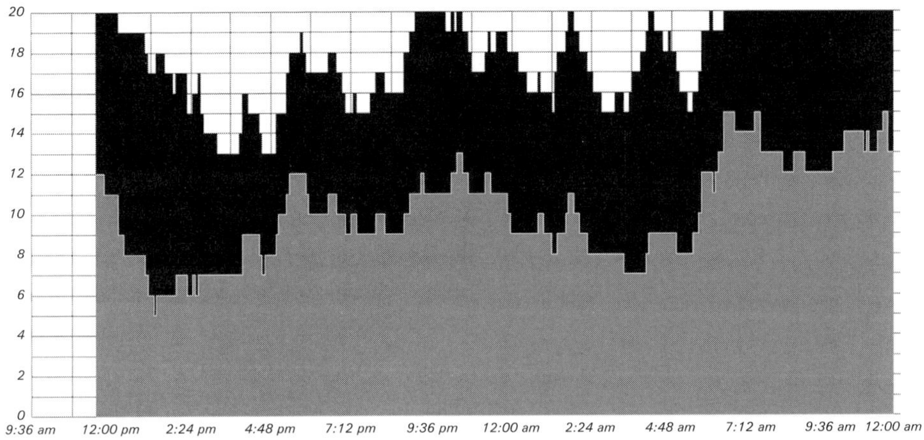

Figure 5.4 GPS and GLONASS satellite visibility over 24 hours.

(1) *Cold start* defines normal receiver start, when no a-priori information is available about satellite ephemerides and time.

(2) *Warm start* defines a start when satellite orbit information is available, possibly from a previously acquired navigation message. This situation can happen if satellites were blocked from view and then reacquired. This mode requires only the time mark from a navigation message to be received. For a GPS L1 C/A signal this would require up to 6 s tracking, during which time one frame with Z-Count should definitely be received. For a GLONASS SP signal, this time is 2 s. Warm start can also be associated with a situation when satellite ephemerides are supplied to the receiver from a data link.

(3) *Hot start* defines a start when orbit and time information are available. This situation can occur if at least one satellite was tracking continuously, while others were blocked from view and then reacquired. Hot start can also be associated with a situation when satellite ephemerides and time marks are supplied to the receiver from a data link.

(4) We define a *cool start* as a start-up in the so-called BGPS mode [7], which we discuss in detail in Chapter 9. This occurs when satellite orbit information is available and time is resolved without reading a broadcast navigation message.

Typical values for TTFF are given in Table 5.3.

5.1.2.1.5 *Interface*

Interfaces must be specified in terms of a physical interface, i.e. available connectors, output data, including data formats and available data output rates, and input/output data, such as aiding information, assistance information, and corrections.

Physical interfaces can include RS232, USB, LAN, and other connectors. We have already discussed data content which can be made available from a receiver. Additionally, a receiver may output the precise time as one pulse per second (1 PPS). The update rate for output data can range from 1–1000 Hz.

Table 5.3. TTFF spec example

Type of start	Typical TTFF (s)	Limits (s)
Hot	1	~10^{-3}
Warm	15	6
Cool	1	~10^{-3}
Cold	45	36

The data format may include NMEA 0183 for positioning output, RTCM 104 for output or input correction data, and RINEX for raw data output. Some files may not be provided directly as an output from a receiver, for example RINEX, but rather be created after the session from raw data proprietary formats. In this case special utilities for file conversion should be supplied by a manufacturer. Inputs may support differential corrections, for example in the RTCM 104 format, assist information, such as ephemerides, inertial navigation system (INS) aiding, and so on.

5.1.2.2 Spec specifics for main application fields

Now we have considered general specification parameters, we look at different application areas from a specification perspective and at mobile receiver spec in detail.

The main civil application areas that would define different specifications are mobile, geodetic, geophysical, and aviation applications. Let us consider how an application area shapes a specification.

5.1.2.2.1 Geodetic applications

Geodetic applications require high-accuracy raw data, which should include code phase, carrier phase, Doppler, and signal strength measurements. A receiver must be able to output raw data. It must also output satellite ephemerides. A manufacturer should provide software to convert the proprietary receiver output format into RINEX format.

Usually, a geodetic receiver is a multi-frequency receiver. Obtaining measurements over more than one frequency allows the formation of combinations of ionosphere-free observables, which consequently do not contain ionospheric errors. Although this is usual, it is not essential. The ionosphere-free combinations are more noisy than single-frequency observables. If the network is of high density, the single-frequency observables may work even better than ionosphere-free combinations. This consideration is important for an RTK network.

Another essential parameter is the code–carrier coherency. As we describe in this chapter, the code and carrier tracking loops should be well synchronized for some algorithms to work, especially in the case of coherent tracking.

5.1.2.2.2 Geophysical applications

Geophysical GNSS applications are related to the measurements of the quantities pertaining to the GNSS signal. In particular, one may need to measure signal scintillation, which is caused by the ionosphere. Such measurements provide a large amount of information about the conditions of the atmosphere, and may include signs of geophysical disturbances, such as earthquakes, in the days before they happen. For details on GNSS geophysical applications, see [8].

In order to obtain the required geophysical application measurements, one needs to include additional requirements in a specification to ensure code and carrier quality. For example, one should look at carrier phase noise and at code–carrier divergence. Code and carrier coherency is less important because positioning in such applications is not required.

5.1.2.2.3 *Aviation applications*

The parameters that are considered to be essential and very sensitive for cellular phone applications, such as price, size, and power consumption, are not that important for aviation applications. The sensitivity of the price parameter goes from the level of a few dollars to hundreds or thousands of dollars, that of size goes from centimeters to tens of centimeters, and that of mass from grams to kilograms. There are, however, new requirements, which have not been considered for other fields. These additional requirements for aviation applications are related to safety, and include integrity. The safety requirements not only bring extra parameters into the specification, but also change how the specification is related to a receiver. For example, aviation applications require receivers to undergo a certification process that ensures that the specification is adequate and is met by manufacturers. The other essential parameter for aviation is code phase accuracy.

An aviation receiver may require an input and an output to provide data exchange for integration with the INS. For details on aviation applications and integration with the INS, see [8]. For both geodetic and aviation applications, a wider bandwidth of the receiver front end is required in order to keep all signal information, restore the shape of the code chip more accurately, and as a result have a better shaped autocorrelation function.

5.1.2.2.4 *Mobile applications*

Up to now, the specifications for various mobile applications have varied as much as the applications themselves. They may include high sensitivity and short TTFF and almost completely disregard accuracy. If the receiver is intended for seamless indoor/outdoor positioning, then, along with high sensitivity and short TTFF, it must be able to use assist information. For example, for cellular phones, high sensitivity and the capability to use assist information are required in order to provide indoor positioning. Other applications may be less concerned with high sensitivity because they would require higher accuracy, and up to now these two parameters have not gone well together.

The accuracy standard for general mobile applications came from E911 requirements, which were determined by the limitations of snapshot positioning. We will look at snapshot positioning in detail in the following chapters. Here we just establish the fact that snapshot positioning is a positioning which is achieved using only a snapshot of code measurement obtained by an acquisition process without tracking. Therefore the receiver is not able to obtain high-accuracy code measurements. Additionally, snapshot measurements can be done indoors, therefore using a multipath instead of direct measurements, exactly those signals that we are trying to get rid of in the case of normal positioning. The accuracy of such measurements would be limited to about 50 m RMS.

5.1.2.3 Evaluation of parameters

Specification issues are closely related to issues of certification. For mobile applications, the certification process may be implemented by emergency positioning functions, similar to E911. The certification in this case would define the receiver parameters and what the user might expect from a receiver's performance. In general, certification may include a receiver technical certification, a receiver performance certification, and a receiver security certification [9].

OEM specifications, modules, and chips which are integrated into the final product always must be evaluated, even when detailed specifications are available. Evaluation kits are essential when an OEM or system integrator intends to add something to the purchased component. For example, in the case of cellular phone application, if a navigation processor is implemented in a device's general processor, the system integrator may be interested only in the quality of raw data. The navigation processor may not be a part of the OEM components at all in this case. However, for some applications, it may not be enough just to look at the specification raw data accuracy parameters in order to estimate the available accuracy. For example, even very accurate code measurements may not necessarily result in accurate positioning if the measurements are not properly synchronized on different channels, or if the noise reduction technique in tracking loops is not in agreement with the coherent tracking if implemented, and so on.

For these purposes, simulator equipment and test methodology became essential. We consider these subjects in Chapter 11.

5.1.3 GNSS receiver design

5.1.3.1 Hardware and generic receivers

This chapter considers conventional GNSS receivers. These receivers implement digital signal processing, and in most cases can be classified as software defined radio (SDR) receivers. In Chapter 7 we show that conventional receivers can be often classified as SDR, and that the difference between them and software GNSS receivers now only exists in baseband implementation specifics. By describing in this chapter a conventional GNSS receiver, we in fact describe a generic receiver. Chapter 6 deliberates on the specifics of receiver implementation on a general processor.

5.1.3.1.1 Receiver functional model

In [8], we consider a functional model of a GNSS receiver. It comprises the following blocks (see Figure 5.2):

(1) an RF front end, which is an essential component;
(2) a baseband processor, which is usually included in the receiver definition:
(3) a navigation processor, which is generally an optional component of a receiver.

This functional model is especially handy when we consider what is called a *software receiver*. In this case, the first block front end (with clock) is the only hardware component; other blocks reside on a general processor. For conventional receivers, blocks are implemented in the hardware. However, as we will see, the hardware includes programmable hardware and embedded microprocessors.

Today, especially for mobile applications, a receiver can be defined as a unit which combines a front end and a baseband processor [8]. Sometimes, a navigation processor is not located inside a mobile device. In many cases, even if a receiver has a navigation processor onboard, it is not used when the receiver outputs raw data for post-processing (geodetic tasks) or processing in a remote location (reverse positioning, including assisted GPS (AGPS), differential, and VRS applications).

Nowadays a front end alone is sometimes marketed as a GNSS receiver, but in [8], we define a receiver as a front end plus a baseband processor.

We can line up a set of arguments in favor of leaving a navigation processor out of the definition for a GNSS receiver as follows.

(1) All functions of a navigation processor can be implemented in the device's general processor.

(2) A mobile receiver, especially one used in cellular phone applications, often uses a host device processor for positioning functions. The receiver may not include a navigation processor as such, and positioning application design is often carried out separately from the receiver design.

(3) It is especially convenient to exclude a navigation processor from such devices as cellular phones, smart phones, and other devices with communication functions, because the assist information, its collection, as well as navigation and communication function synchronization and time sharing, are the responsibility of the device's processor.

(4) There is a different system concept for many applications, such as fleet management, for example. In these applications, the mobile receiver sends the information, i.e. code estimates and possibly other data, from the baseband processor to the control center, which calculates the rover position and sends it back to the user or other customer.

This reverse system can be extended to the case when data from the front end itself are sent to the control center or written into the memory for further processing.

5.1.3.1.2 *Receiver structural model*

When we consider a conventional receiver, a structural model becomes more useful than a functional one (Figure 5.5).

We consider here one of the possible implementations: one in which the navigation processor cannot be separated from the baseband functions that are implemented on the microprocessor. However, the reasoning from Section 5.1.3.1.1 about removing the navigation processor still applies. When considering the structural model, this means that one can implement a microprocessor which has a lower power consumption, is cheaper, and has a smaller footprint. For example, removing navigation functions from a microprocessor may reduce its price tenfold, for example from \$50 to \$5.

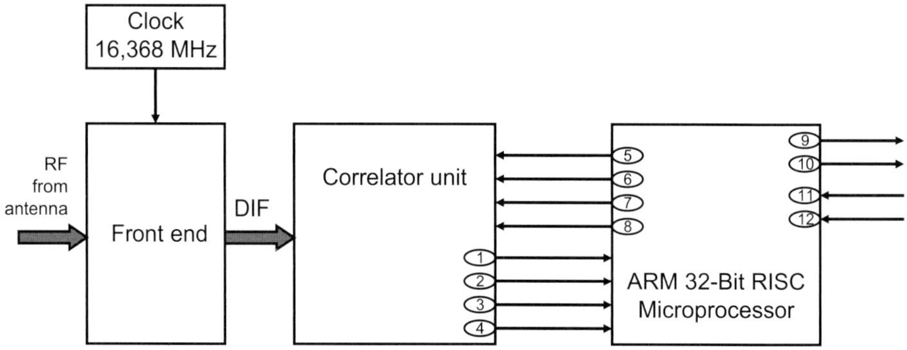

Figure 5.5 Receiver structural diagram.

This is especially important when we consider multi-system receivers, which have much more load on a processor, mostly in terms of baseband operation.

Correlators in hardware receivers can be realized either in hardware or in software. In fact, correlators were first realized as software, then, with the development of silicon technology, they became hardware based [10].

5.2 Receiver components

5.2.1 Correlators

5.2.1.1 Signal acquisition

The RF signal is down-converted from the L1 radio frequency to the baseband intermediate frequency (IF), hence the name *baseband processor*. The specific implementation of the down-conversion process can take the form of a single down-conversion operation to IF (a.k.a. heterodyne architecture), down-conversion to IF in two steps (a.k.a. super-heterodyne architecture), or direct conversion to IF $= 0$ (a.k.a. homodyne architecture). For example, in the case of a heterodyne architecture, for GPS, a 1542 MHz RF is down-converted to an IF of, for example, 4.92 MHz. After down-conversion, the IF signal is digitized and quantized. We consider down-conversion and digitization operations in detail in Chapter 7. The digitized intermediate frequency (DIF) signal goes to the baseband processor of our functional model or to the correlator unit on our structural model. Correlators are the main workhorses of the baseband processor.

The first task the correlators should achieve is signal acquisition, the purpose of which is to find an encoded spread code signal in the incoming DIF signal. Acquisition begins by looking for a particular satellite signal

$$A_i = A_{0_i} \sin\left((\omega_0 + \omega_{D_i})t + \varphi_i\right) \cdot B_i, \tag{5.7}$$

where A_i is the ith satellite signal and B_i is the spread code. Acquisition ignores the existence of a navigation message. However, the navigation message may interfere with coherent acquisition, which is essential for high sensitivity, and we will look at this later in this chapter.

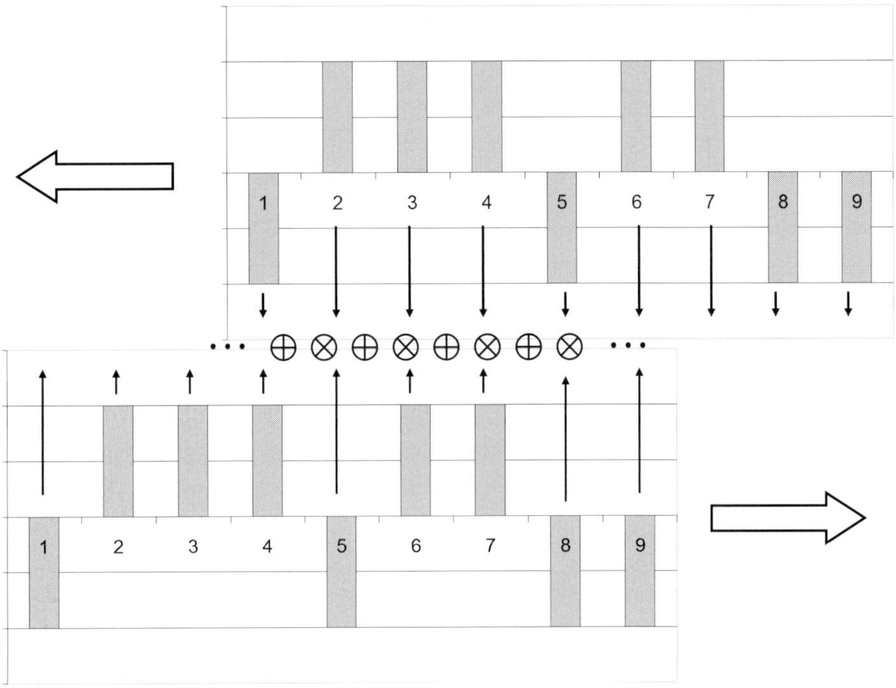

Figure 5.6 Schematic representation of convolution process.

As a first step, acquisition removes the carrier phase from the signal by mixing it with the carrier replica. We know the nominal value of the signal carrier f_0 as it is transmitted from all satellites. The Doppler shift f_{D_i} is mostly defined by satellite movement, which is approximately 4 km/s for a GPS satellite and could be up to 800 m/s along the line-of-sight (LOS) to the receiver. The Doppler shift is then in the range ± 6 kHz for a static or low-dynamics user and in the range ± 10 kHz for a high-dynamics user. This limits the search range of the carrier frequency of the incoming signal. Additional Doppler changes may be introduced during signal propagation through the Earth's atmosphere. A drift in the satellite and receiver clocks also manifests itself as a Doppler shift of the incoming signal. Therefore, we need to make a search within the possible Doppler range for the correct frequency of the incoming signal.

After the carrier frequency is removed, we check whether the remaining signal contains a satellite PRN spread code. For that purpose, we generate a replica of the satellite spread code in a receiver and check whether there is a correlation between the replica and the incoming signal. To do that, we look at the convolution between the incoming signal and the replica (Figure 5.6).

We do not know how far the satellite is from the receiver antenna, so we need to search through all possible shifts between the incoming signal and the replica code. Therefore, the search is conducted in two-dimensional space: code delay and Doppler shift to the nominal frequency (Figure 5.7).

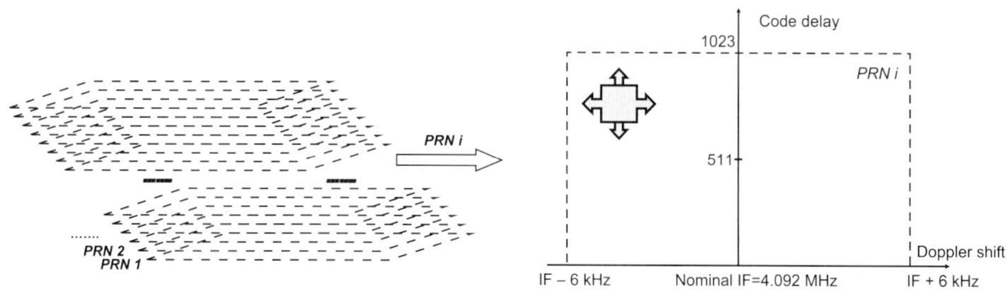

Figure 5.7 Two-dimensional search area for acquisition.

The search range for the Doppler shift is divided into a number of frequency bins. The idea of the acquisition search is to multiply the incoming signal with the locally generated carrier replica for all possible frequencies (defined by the number of bins) and delays. The correct bin and delay would give the maximum result of the multiplication. The search is conducted by sequentially multiplying the DIF signal, which is a digitized and down-converted RF signal from the satellites, by a carrier replica generated on a frequency from each bin. By means of this multiplication, the DIF signal is converted to a baseband signal (i.e. the carrier is removed from the signal), which is then compared with the receiver-generated replica of the spread code by shifting and multiplying. A maximum of the correlation peak indicates the candidate for the correct code phase (code delay). The search in code delay should be carried out along the entire code length, which is 1023 chips for GPS L1 C/A.

The properties of the spread code define the shape of autocorrelation function. The shift between the replica spread code and the incoming baseband signal can be conducted in steps of 0.5 chip. Figure 5.8 shows the typical correlation output for a search area for GPS C/A and GLONASS SP code for an iPRx receiver (only four channels are shown for each constellation).

Correlators can be realized in programming hardware such as a field-programmable gate array (FPGA) (Figure 5.9). Even when correlators are realized in application-specific integrated circuits (ASIC), it is likely that a prototype had been made first using an FPGA. Figure 5.9 shows an example of the realization of a correlator using an FPGA for GPS C/A code with 1 bit resolution. The GPS C/A code can be generated by shift registers. Some other codes, called memory codes (see Chapter 2), cannot be generated by shift registers and have to be stored in memory. A delay circuit facilitates a search function along the code delay axis in Figure 5.7. An example of delay circuit implementation using an FPGA is presented in Figure 5.10. The carrier replica is mixed with the DIF signal from the front end. If the frequency of the replica matches the frequency of the incoming satellite signal, the carrier is removed from the product. A correlator provides a bit-by-bit multiplication of the incoming signal, which is down-converted and stripped from the carrier, and the replica code. The result is accumulated for some period and then fed to the microprocessor, which works with it to find whether there is a signal present. The accumulated product has a different bit width from the codes.

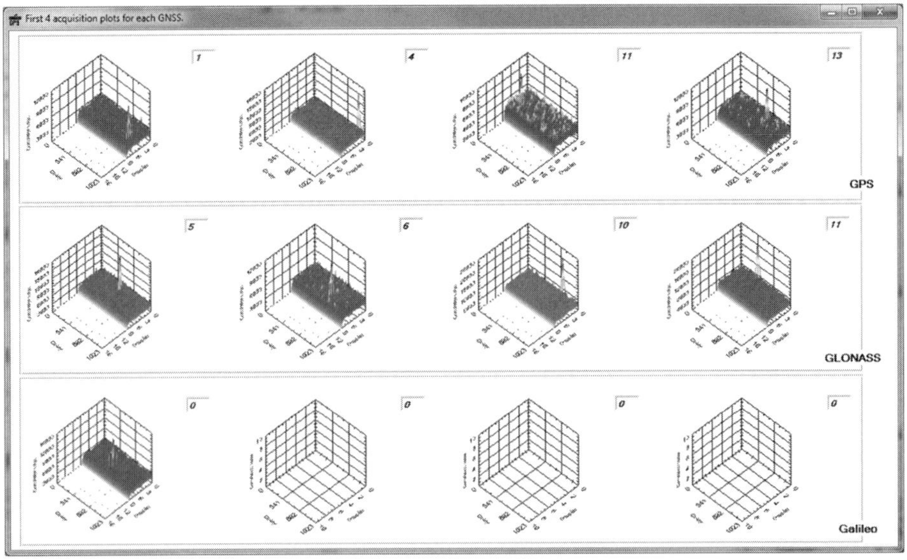

Figure 5.8 iPRx acquisition panel.

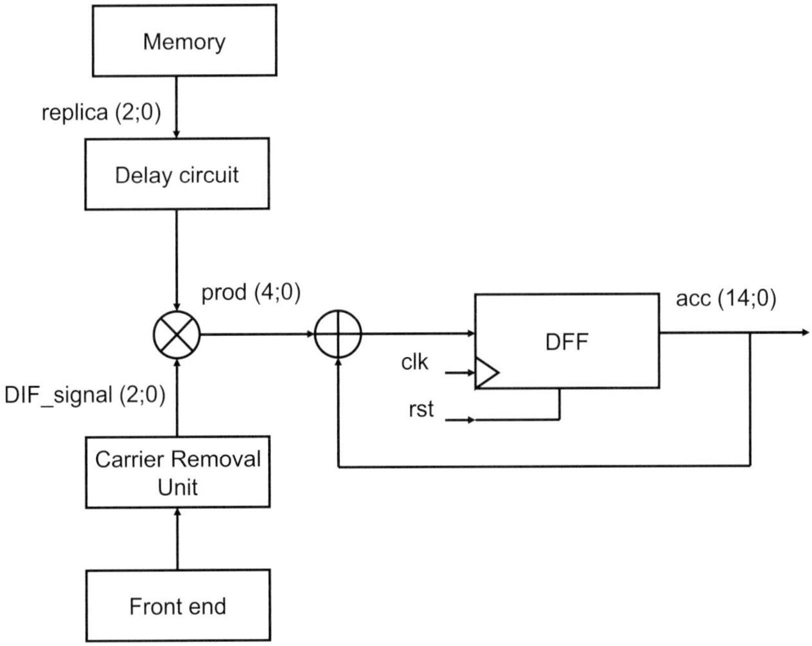

Figure 5.9 Correlator.

Once the output of the correlation has been accumulated, a microprocessor estimates whether the satellite signal is present. The decision is usually based on the set thresholds and probabilities for Type I and Type II errors, which are errors of wrong and missed detection. The threshold cannot be set once and for all. It depends first of all on signal

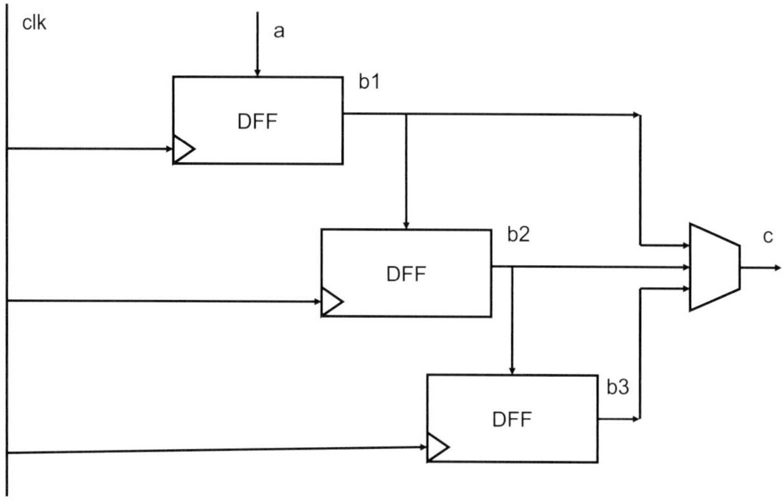

Figure 5.10 Delay circuit.

strength, and therefore may vary with different antennas and in different environments. In order to make this process automatic, we need to estimate a noise floor during each session. This can be achieved using the results of the acquisition of a satellite which is not present in a signal [4], or via the acquisition of a signal with non-assigned PRN.

5.2.1.2 Massive parallel correlation

A microprocessor reads the output of a D-type flip-flop (DFF) every millisecond for a GPS C/A signal. One correlator can process one code phase combination of signal and replica at a time. In generic GPS receivers there are three or five correlators for a channel. That means each of them needs to go through all combinations of code phase. Instead, a large number of correlators can be implemented in order to cover larger parts of the search area quickly, working on the same signal in parallel, so each of the correlators will work only with a small number of code phase combinations [10].

Increasing the number of correlators would increase the load on an embedded processor to the point that it would not be feasible to service all the correlators in real time. In modern high-sensitivity receivers, there may be hundreds of thousands of correlators. Therefore, the part of the baseband functionality handled by an embedded processor is implemented also in the digital hardware. In a receiver featuring massive parallel correlation, almost all baseband functionality is implemented in the hardware. This is very handy for cellular phone applications, because the technology facilitates high sensitivity and also means that the rest of the microprocessor functionality can be put into the host processor. This is possible because the remaining functionality does not lead to high processor load. Any sensitive intellectual property (IP) can be handled through providing an additional functionality for the host processor through the application programming interface (API). A microprocessor may be removed, with the corresponding functions being transferred to a device host processor. Massive parallel correlation was implemented first by Global Locate, Inc., which was then acquired by Broadcom Corporation for $143 million. The technology allowed a receiver to achieve unprecedented sensitivity, whereby a receiver was able to make a positioning indoors. The prototype used a high-accuracy clock to compensate for the absence of some of the assist information. Global Locate has also developed a worldwide GPS reference network with patented long-term orbit (LTO) solutions to provide assistance when a network connection is not available.

Based on Global Locate and similar technology [10], massive parallel correlation can be characterized by the following properties.

(1) It uses hundreds of thousands of correlators.
(2) Those baseband processor functions related to signal acquisition are transferred to a hardware, including frequency discrimination and data bit extraction.
(3) The results from accumulators (DFF in Figure 5.9) are values with various bit widths, and specially developed algorithms are required to optimize the processing of these results over a massive scale in the hardware.

The implementation of massive parallel correlation lets the load on the baseband processor decrease, up to the point when its functionality can be moved to a host processor, at the same time decreasing acquisition time and increasing sensitivity.

5.2.1.3 Coherent signal integration

Let us suppose that we have to find a GPS signal in n milliseconds of DIF signal. We can do it in two ways. We can process chunks of signal in code length sequences (1 ms for GPS L1 C/A code), then sum the results of n acquisitions. This is called non-coherent integration. Alternatively, we can use an n millisecond replica, in which case we do not just dump the data, but instead continue to accumulate them on the accumulator register acc(;), for all n milliseconds; see Figure 5.9. The bit width of the register may be increased accordingly, as it must handle larger numbers. The technology of coherent integration was developed for high-sensitivity receivers, and is given in detail in [10].

The strength of a signal can be characterized by the signal-to-noise ratio (SNR), defined as the ratio of the post-correlation signal power to the noise power, i.e.

$$SNR = 10\log\frac{P_S}{P_N} = 10\log\left(\frac{S}{\sigma_N}\right)^2 \quad [\text{dB}], \tag{5.8}$$

where S is the amplitude of the correlation peak and σ_N is the standard deviation of the noise. In the case of coherent integration, the SNR is much higher for the same signal length in milliseconds.

Figure 5.11 shows an output from an iPRx receiver for non-coherent and coherent integration of a 2 ms GPS signal.

There are, however, a few issues that limit the application of coherent integration.

(1) A single bit of navigation message in a GPS C/A signal has a length of 20 complete spread code sequences, and therefore occupies 20 ms of the signal

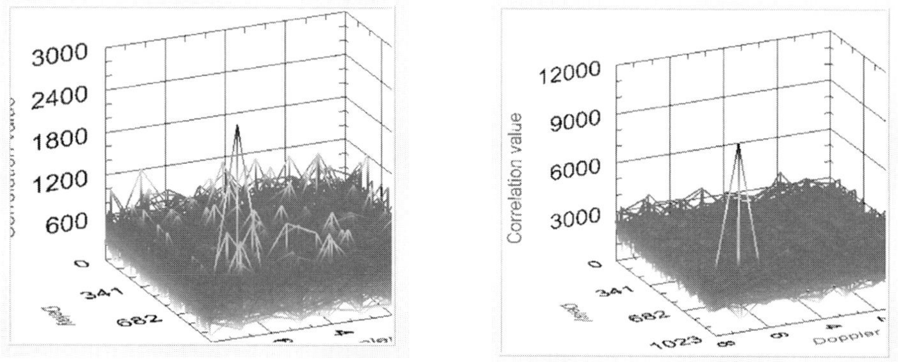

Figure 5.11 Acquisition panel of iPRx receiver showing (a) non-coherent and (b) coherent integration.

length. If the incoming signal has a navigation bit transition, the polarity of the codes in that bit will be changed, and the overall results from the coherent integration will be degraded. For the price of the processing time, we can integrate, for example, an incoming 20 ms signal with two 20 ms replicas, one with bit transition and another without. This is not a good approach for hardware, because it requires extra circuitry to facilitate a bit transition and effectively cuts the number of correlators in half, as two correlators now will be processing the same code phase–Doppler cell with and without bit transition.

(2) The instability of the receiver clock and the Doppler shift from the satellite and receiver antenna movement will affect the appearance of spread code in the incoming signal when compared with an ideal replica. It is possible to construct a replica that will compensate for these changes, but the price in terms of processing load may be prohibitive for most implementations. The various algorithms of such compensation are considered in [11] in detail.

5.2.1.4 Frequency resolution

If a receiver uses tracking loops, they place additional requirements on acquisition. Low-end receivers, and especially receivers for cellular phones, may implement an acquisition only in the baseband processor. In this case, they implement positioning using only snapshots of the GNSS signal. Even if a receiver implements signal tracking, it can still start working in a snapshot mode using acquisition results only to improve TTFF. The accuracy of code delays in acquisition is basically limited by the DIF sampling rate. We cannot overlap an incoming baseband signal with a receiver code replica with an accuracy better than that of the sample, which effectively limits range measurements. Additional propagation and receiver errors will give us an estimate of the user range error. The product of the range error and the DOP (Chapter 3) gives us the user positioning error. We can see now why the specification for the E911 service is limited to 50 m, when we normally get much better performance from a GNSS receiver. A receiver has to implement tracking loops to improve accuracy. If a baseband processor implements tracking loops, they must be initialized with code phases and Doppler frequency estimates from the acquisition.

In order for tracking loops to operate, the frequency from the acquisition should be defined with a certain accuracy. A possible frequency resolution for any signal depends on its length in samples [12]. We can introduce a digital angular frequency, defined in radians per sample, as follows:

$$\tilde{\omega} = \frac{2\pi f}{f_S}, \tag{5.9}$$

where f_S is the sampling frequency. The frequency resolution we can obtain from K samples can be defined as follows:

$$\Delta\tilde{\omega} = \frac{2\pi}{K}. \tag{5.10}$$

We can therefore define Δf using $\Delta\tilde{\omega}$ and then express it via the length of the processed signal chunk, i.e.

$$\Delta f = \frac{f_S}{K} = \frac{1}{KT_S} = \frac{1}{\Delta T}, \tag{5.11}$$

where T_S is the sampling interval and ΔT is the duration of the processed signal length in seconds. This puts certain restrictions on the minimum signal length. These restrictions, which are based on the signal-to-noise threshold required to find a code phase, are much lighter. To start signal tracking, the estimates of both frequency and code phase are required. While at least 20 ms of signal is required to find a frequency with sufficient resolution, only a few milliseconds are required for code phase estimation.

5.2.2 Receiver channel functions

Usually, most baseband functions, including loop discriminators, are allocated to a microprocessor.

5.2.2.1 Tracking loop theory

At the acquisition stage, a receiver finds the approximate values for the signal frequency and code phase. Then these values are transferred to tracking loops, which refine these values to a greater accuracy and keep them updated. We consider a baseband processor which implements two tracking loops, one for carrier tracking, and another for code tracking. Both tracking loops implement the concept of phase-locked loop (PLL). Details of code and carrier tracking loop theory for GPS applications are given in [1]–[6], and a more general discussion of tracking loop theory can be found in [13] and [14]. Here we describe one specific implementation.

The tracking loops should operate in parallel as they are using one another's information in order to wipe out the carrier from the baseband signal and the spread code from the carrier. This structure is shown in Figure 5.12. The general idea is that the code tracking loop estimates the code delay and applies it to remove the code from the

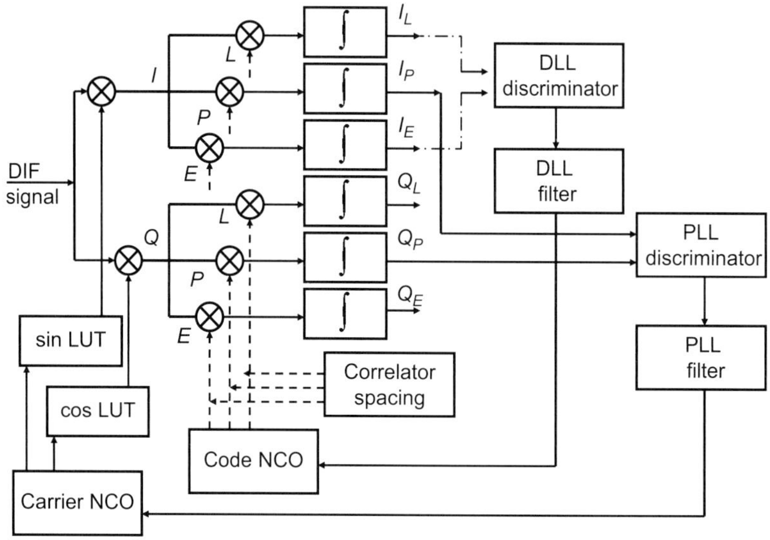

Figure 5.12 Tracking loops. After [1]–[3].

DIF signal. This codeless DIF signal is then tracked by the carrier tracking loop, which in turn estimates the carrier phase and applies it to the original DIF signal in order to remove the carrier from it. The converted (to baseband) signal is tracked by the code tracking loop.

The carrier tracking loop is designed as a PLL. In PLL phase, a comparator compares the phases of a locally generated carrier with the carrier of the incoming signal and adjusts the phase by applying the feedback in such a manner that the phase difference is kept to a minimum. The code tracking loop, also called a delay-locked loop (DLL), is also designed to keep the phase difference between the two codes small. This is achieved by keeping the cross-correlation between the local replica and the incoming baseband signal at a maximum. The implementation of these two loops differs in terms of the mechanism of comparison and the feedback implementation. These mechanisms are defined by code and phase discriminators.

Let's look at the tracking loops in Figure 5.12 in more detail. The code and carrier tracking loops are implemented as DLL and PLL, respectively. A digitized IF signal from a front end is mixed with a carrier phase replica.

In this implementation, we combine code and carrier tracking loops [3]. The carrier tracking loop is implemented as a Costas loop by multiplying the incoming signal by sine and cosine components of the local carrier replica. The specifics of the Costas loop is in that it is not affected by phase transitions. For GNSS, the phase transition is due to navigation bits. After the multiplication we have in-phase (I) and quadrature (Q) arms in the Costas loop. The I-arm outputs the signal modulation. It is represented by a navigation message bit sequence. If we remove the navigation message, then the I-arm output from the carrier tracking loop should be zero. The I and Q outputs are integrated usually over a period equal to the code length.

The carrier numerically controlled oscillator (NCO) drives the sine and cosine lookup table (LUT). The output of the LUTs comprises two carrier phase replicas, shifted by 180° relative to each other.

Resulting from these multiplications are two instances of incoming satellite signal, stripped of the carrier. One instance is in phase and the other is quadrature. Each instance is mixed with three code replicas, shifted, relative to each other, prompt, early, and late. The mixing can be interpreted as a multiplication, or XOR, operation, depending on the implementation (which is the same as the previously mentioned operation of mixing the DIF with a carrier replica).

This mixing provides information about the correlation. The value of this correlation is described by an autocorrelation curve. The precision of the curve, i.e. how close it is to an ideal autocorrelation function, depends on the front-end bandwidth [8]. For example, for GPS L1 C/A, a 20 MHz front-end bandwidth yields an autocorrelation shape very close to ideal. For 6 MHz, it is reasonably close, and for 2 MHz the similarity is much less prominent. The shape is changed because with narrow bandwidth we keep only the main lobe; the information on the side lobes is lost. Only by keeping all the information, including side lobes, can we restore the chip shape completely as an ideal rectangular shape without losing any information, and therefore the autocorrelation function will retain its ideal shape as well.

The correlation information is retained and accumulated by the accumulators. Then the tracking loop needs to estimate the size of the shift between the signal and the replicas based on this information about correlation; this is achieved using discriminators. The choice of discriminator is determined by a particular function that describes how the discriminator output depends on the input value, i.e. the value accumulated by the accumulators.

The output of the discriminator is fed to the feedback loop, which drives the code NCO, which is then adjusted in such a way that the generated code replica tends to move closer to the incoming signal code, minimizing the shift between them. The feedback filter can be driven by this shift itself or by its derivatives.

The filter order can be defined as the highest power in the denominator of the loop transfer function.

In a similar way, the carrier tracking loop operates in parallel with the code tracking loop. Carrier tracking feedback can be organized with carrier phase, in which case the loop is PLL.

As the PLL's order is increased, it tends to compensate for an instantaneous change in the next higher derivative of the input. The code loop order is usually smaller than that of the carrier loop, because the carrier loop provides internal aid to the code loop and compensates for most of the dynamics in it. Therefore, the code loop bandwidth is smaller than the carrier loop bandwidth [3].

Alternatively, the feedback loop can be driven by frequency rather than phase. In this case, the carrier tracking loop is FLL; FLL is more stable than PLL, and can be pulled in more easily.

We can see the behavior of I and Q outputs in the tracking panel of an iPRx software receiver. One of the advantages of the software receiver is that it can provide access to the baseband processor functionality. The I-arm outputs the demodulated data symbol, which is in fact a navigation message data sequence. In an IQ plot, we can see that the code tracking loop mostly affects the I-arm behavior, whereas the Q-arm shows a carrier tracking error. The following example (Figure 5.13) is taken from [8]. The figure shows the output of I and Q signals from prompt and early channels in the iPRx receiver panel. The iPRx IQ graph shows the dynamics of the tracking loops pull-in process. We can see that if DLL is locked and PLL is not locked, the IQ spots start to rotate. If, on the contrary, the PLL is locked, and DLL is not, the IQ spots converge without rotation.

As discussed, the error for each loop is calculated by a discriminator, which describes the error as a function of the shift between the received signal and the replica. A loop filter is in charge of how this error is handled. There are three types of filters [3]. Normally, receivers employ second-order feedback loops (Figure 5.14). In this case, NCO, which supplies the frequency for the replica's code and carrier, is controlled using information about error change. The first-order filter controls the NCO using information about error value, and the third-order filter controls it using the error rate of change.

In order to keep the phase difference to a minimum, we construct carrier discriminators and generate feedback according to their values. A discriminator is characterized by its value as a function of the phase difference. The phase difference should be within certain limits in order that the discriminator can be used to maintain a lock on the signal. Most common discriminators for carrier tracking loops can be defined as follows [3]:

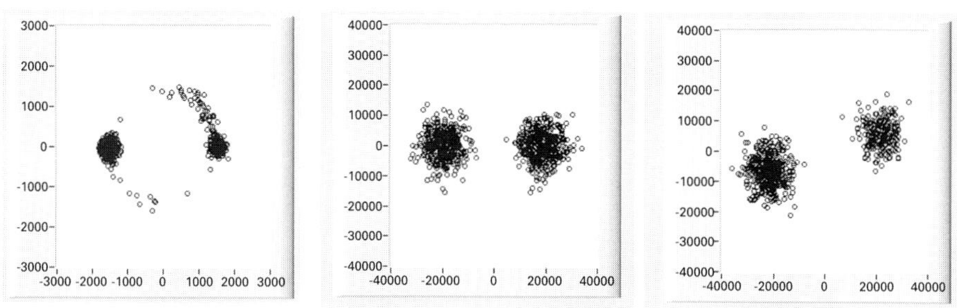

Figure 5.13 IQ plot behavior for (a) locked DLL and locking PLL, (b) locked DLL and PLL, (c) locked DLL and FLL.

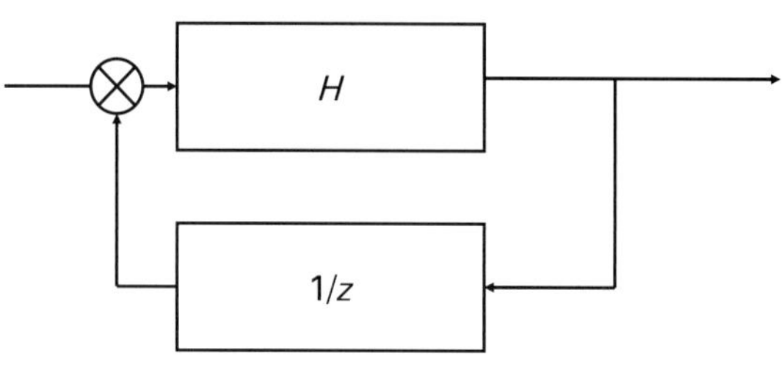

Figure 5.14 Second-order feedback loop.

$$D_{PLL1} = Q_P \times I_P, \tag{5.12}$$

$$D_{PLL2} = Q_P \times Sign(I_P), \tag{5.13}$$

$$D_{PLL3} = \arctan\left(\frac{Q_P}{I_P}\right). \tag{5.14}$$

These discriminators have the following performance [3]. DiscriminatorD_1, defined by (5.12), has a near-optimal performance at low SNR with output phase error defined as the sine of the doubled phase difference, $\sin(2\cdot\Delta\varphi)$. Discriminator D_2, defined by (5.13), has a near-optimal performance at high SNR with output phase error defined as the sine of the phase difference, $\sin\Delta\varphi$. Discriminator D_3, defined by (5.14), has an optimal performance; it requires more computations, and is usually realized using lookup tables. Its output phase error is defined as the phase difference, $\Delta\varphi$.

To provide code tracking functionality, the in-phase (I) and quadrature (Q) arms are multiplied by local replicas of the code. These replicas are generated as prompt, earlier, and delayed versions. The maximum number of such replicas implemented directly (without oversampling) is restricted by the number of samples within a chip length. The results of multiplications of these replicas by the incoming baseband signal depend on the shape of the spread code autocorrelation function. An autocorrelation function for GPS C/A code is shown in Figure 5.15, together with three code replicas.

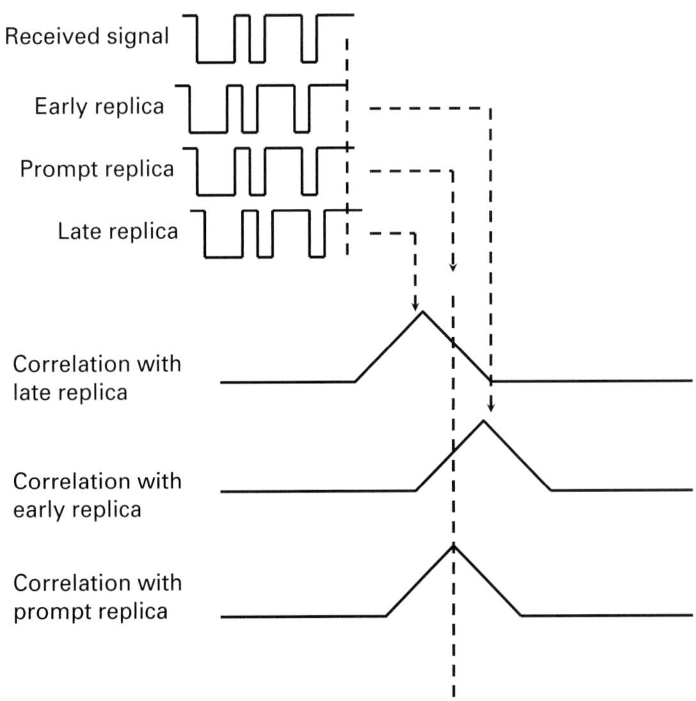

Figure 5.15 Autocorrelation function for various replicas for a GPS C/A signal.

A prompt replica is generated initially in accordance with the code phase estimation from the acquisition step. Early and late replicas are generated with some shift in code phase, called correlator spacing. This phase shift can be set in the range from half the chip length to much smaller values. Smaller correlator spacing between the replicas make the tracking loop more accurate, but can get out of lock more easily due to user dynamics. At some point, decreasing the correlator spacing further may not improve the performance. This is because a real baseband signal has a trapezoidal rather than a rectangular shape, and the shape of the autocorrelation function is defined by the front-end bandwidth [1].

A code discriminator provides the tracking loop with feedback on the code delay value. The purpose of this discriminator is to generate feedback in such a way that locally generated prompt replicas will experience minimum shift with an incoming signal. A code discriminator can be calculated in various ways. We will consider a few code discriminators here [3],[6]. The early-minus-late discriminator requires minimum computer resources, because Q_i values are not computed, i.e.

$$D_{DLL1} = I_E - I_L. \tag{5.15}$$

This is a coherent discriminator, which means that it requires coherent carrier tracking in order to operate. The discriminators discussed in the following are non-coherent.

The early-minus-late power discriminator is defined as follows:

$$D_{DLL2} = {}^1\!/_2\left(I_E^2 + Q_E^2 - I_L^2 - Q_L^2\right). \tag{5.16}$$

The dot product uses all outputs thus:

$$D_{DLL3} = I_P(I_E - I_L) + Q_P(Q_E - Q_L). \tag{5.17}$$

The correlator spacing can be adjusted during receiver operation as a function of signal quality.

To reduce the computational load on, various approximations were developed [3]. The *Robertson approximation* is expressed as follows:

$$\sqrt{I^2 + Q^2} \simeq \mathrm{MAX}\left(|I| + {}^1\!/_2|Q|, {}^1\!/_2|I| + |Q|\right). \tag{5.18}$$

Receivers normally use much more than three correlators. The price of correlators is so low that they are basically free for hardware receivers. For software receivers, extra correlators are extremely expensive in terms of computational load. However, the usefulness of more than five correlators is limited to specific applications, such as multipath and interference mitigation [3].

As previously mentioned, a baseband processor can also implement frequency tracking, which can be considered as a differential carrier phase tracking. Basically, all frequency-locked loop (FLL) discriminators suffer from a bit sign change. The *decision-directed cross-product discriminator* [1] is created out of two discriminators to detect and compensate for the sign change:

$$D_f = (I_{i-1}Q_i - I_iQ_{i-1})\mathrm{sign}(I_{i-1}I_i + Q_{i-1}Q_i). \tag{5.19}$$

Frequency tracking is used mostly either at the beginning of tracking or under special conditions, such as interference, when phase tracking is difficult.

For a comparison of BPSK(1) and BOC(1,1) autocorrelation functions, front-end bandwidth effects on these functions, and more details, see [8].

5.2.2.2 Tracking loop implementation

Here we discuss the tracking loops implementation, based mainly on one developed for GPS and described in Jet Propulsion Laboratory reports, in particular [15] and summed up in detail in [16]. This implementation is illustrated in Figure 5.16. The accumulators supply I and Q outputs every millisecond. The tracking loops operate at a rate of 50 Hz, which is a particularly convenient choice due to the 20 ms length of a navigation bit. The tracking loop output is provided at a rate of 1 Hz.

The carrier NCO provides assistance to the code NCO; the difference between the adjusted code NCO and the code-only NCO is accumulated, and then used to drive the code phase measurements.

Code measurements are adjusted through line fit, whereas carrier measurements are adjusted through curve fit. We now consider the specifics of such an implementation in greater detail.

5.2.2.2.1 PLL-aided DLL

Carrier tracking error is much smaller than code tracking error. We can use the output from the tracking loop to drive the code tracking loop. Code and carrier tracking loops follow changes in the Doppler shift in a similar manner. The carrier aiding to the code tracking is in a way similar to carrier smoothing, which we discussed in Chapter 3, but on the baseband processor level. Using carrier aiding, we can remove LOS dynamics from the code loop. This allows us in particular to narrow a code loop bandwidth and hence improve accuracy. Carrier aiding to DLL is achieved as follows. The carrier frequency is scaled to track the code and adjusted for IF. The scaling factor is required because the Doppler frequency shift caused by the LOS dynamics is different for code and carrier. The Doppler shift is proportional to the frequency (see Chapter 8 for Doppler calculations). For the same satellite-to-rover relative dynamics, the Doppler shift of the spread code is much smaller than the Doppler shift of the L1 carrier. The down-conversion process does not change the value of the Doppler shift, so the Doppler shift is the same for IF as for RF.

However, the code and carrier diverge due to the opposite signs of an error caused by the ionosphere. Therefore, the code tracking loop should be periodically adjusted. This adjustment can be done using an averaged code error, which compensates for the code tracking error if it is accumulated by using carrier tracking only.

An algorithm of carrier aiding is presented in Figure 5.17. Feedback from the DLL is used to monitor the tracking quality. Note that DLL aiding could also be arranged using orbit information from ephemerides, but carrier aiding is better because it includes clock error and thus avoids errors inherited from clock error estimation in positioning algorithms, which are worse due to cross-links with other channels and DOP.

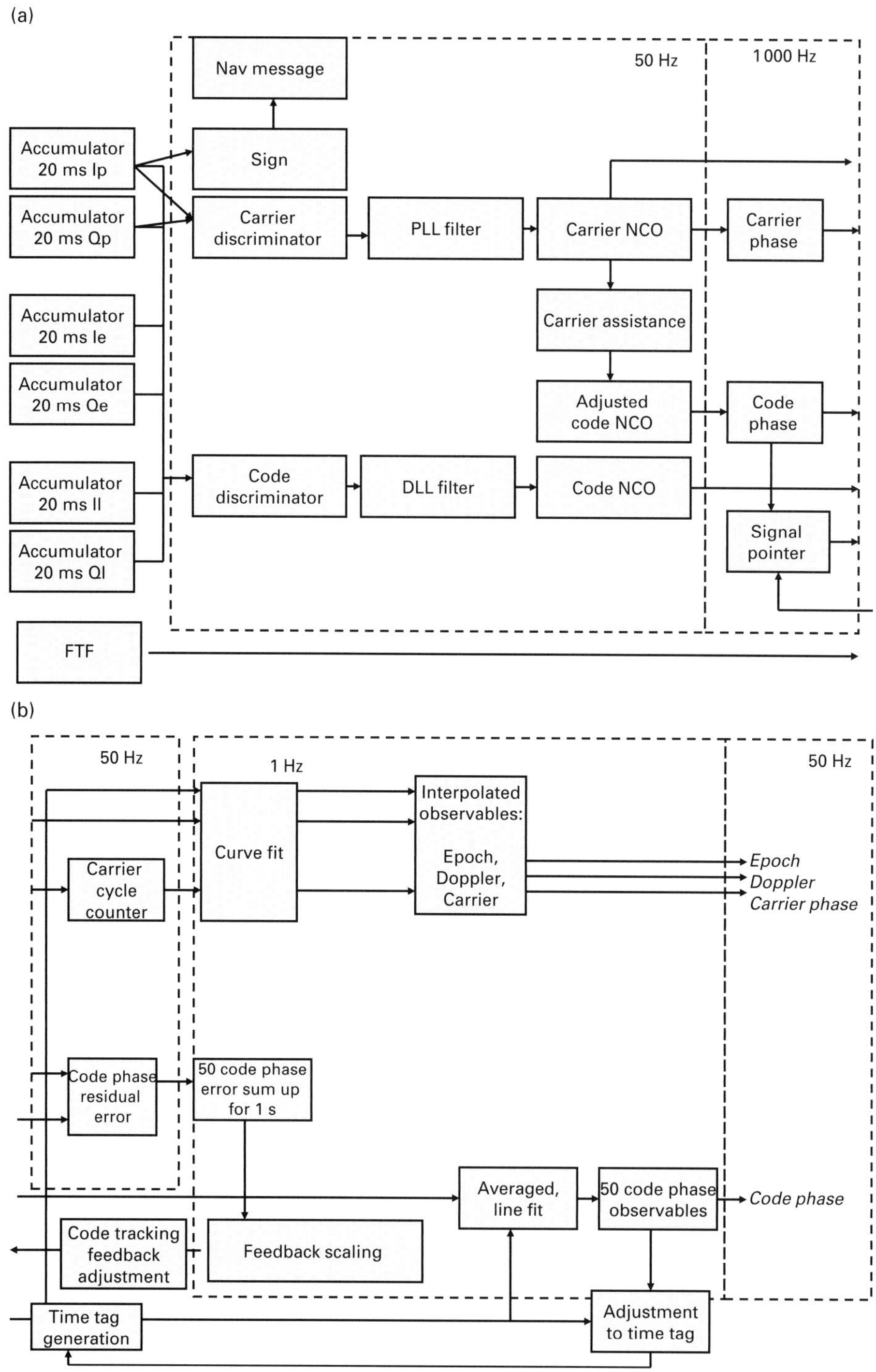

Figure 5.16 Tracking loop implementation. After [15],[16].

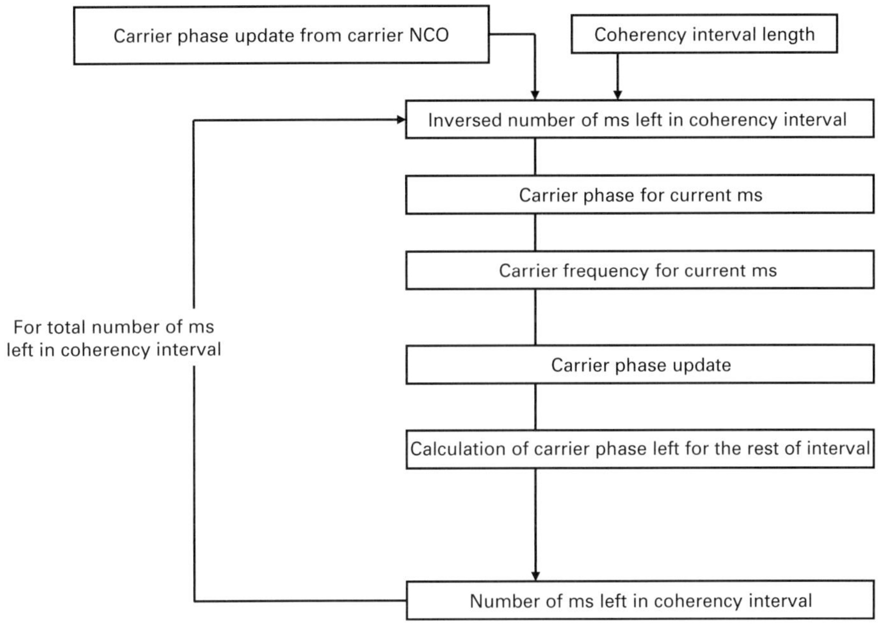

Figure 5.17 Carrier-aiding flowchart.

5.2.2.2.2 *Coherent tracking with 20 ms coherency interval*

Coherent tracking allows a significant improvement of the tracking sensitivity.

For GPS C/A code, one can implement coherent tracking with a 20 ms coherency interval. In this case, I and Q prompt, early, and late values are accumulated over a 20 ms interval. Tracking starts in a non-coherent mode and is then transferred to a coherent mode. This transfer can be performed at the moment when lock is achieved and the receiver starts to collect the navigation message. At this moment, tracking loop parameters for PLL and DLL can be changed to the different values. In particular, a narrow bandwidth for tracking loops can be implemented. Tracking loop parameters depend on clock type. For more precise clocks, such as an oven-controlled crystal oscillator (OCXO), a narrower bandwidth can be implemented by default within the tracking loop.

Coherent tracking must be used together with carrier aiding in DLL. It allows for a much greater sensitivity to the satellite signal and an increase in tracking accuracy.

Filters and other functions can run at rates of 50 Hz; NCO, accumulators, and lock detectors should still operate at 1000 Hz.

One could also expect that coherent tracking would allow for much more accurate tracking, and consequently for better positioning accuracy. However, positioning accuracy in a coherent mode may become slightly worse than in a non-coherent mode unless special measures are taken. The accuracy degradation can be caused by a smaller interpolation interval in the coherent mode and fewer stochastic characteristics of DLL tracking due to carrier aiding. It also depends on the strength of the DLL feedback to PLL-aided DLL. However, the signal strength is greatly improved (see Figure 5.18).

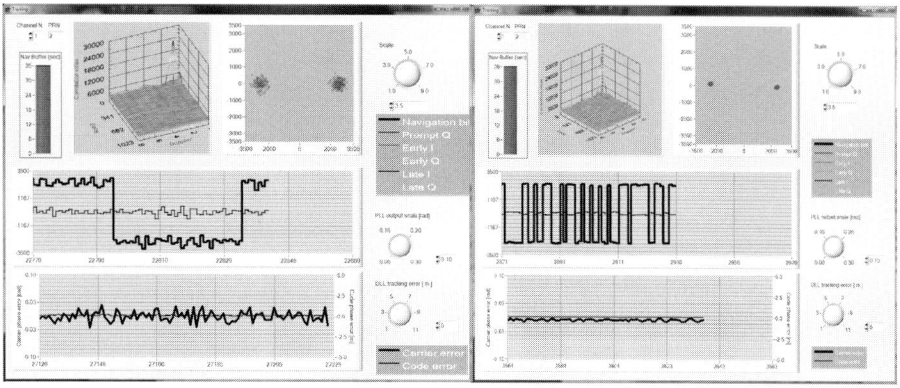

Figure 5.18 Comparison of (a) non-coherent and (b) 20 ms coherent tracking.

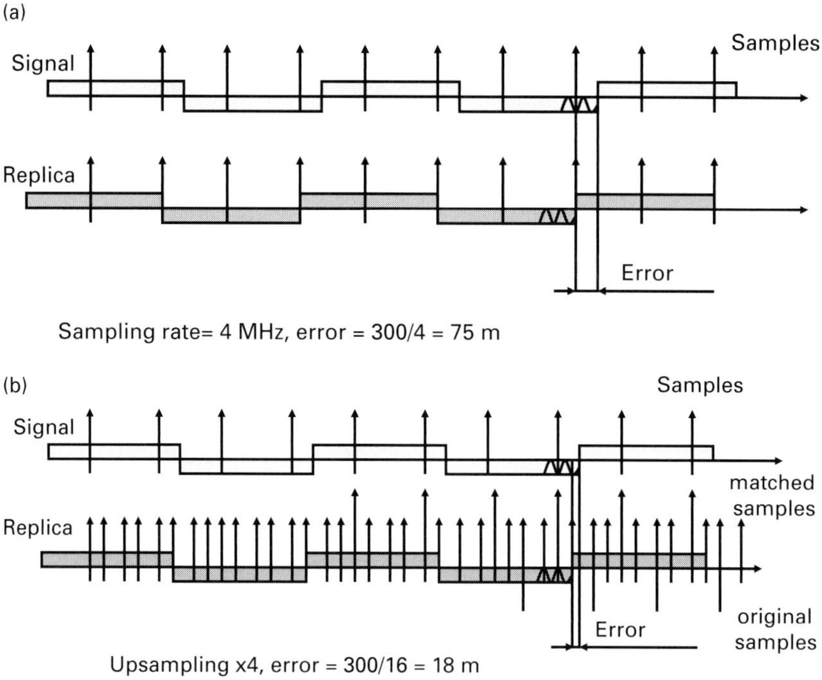

Figure 5.19 Tracking with (a) normal vs. (b) up-sampled code replica.

Figure 5.19 shows how the tracking accuracy can be improved by up-sampling replica code. By aligning up-sampled replica code, we can measure the code phase with an accuracy defined by the signal bandwidth rather than the sampling of a receiver front end. In fact, the sampling rate of the front end does not affect accuracy if the sampling rate satisfies the Nyquist criterion. The same approach can be used to improve the acquisition accuracy, if the signal length allows it. Tests show that up-sampled signals can be precisely tracked in the case of coherent tracking, which would allow for a higher observable accuracy.

Up to 20 ms of coherent tracking can be achieved without taking into account the satellite–receiver dynamics. Coherent tracking with a 20 ms interval can also be achieved with TCXO on standard mobile devices.

5.2.2.2.3 *Coherent tracking with 1 s coherency interval*

An additional increase in the coherency interval would improve sensitivity further. An increase in the coherence interval requires a high clock stability, so we use an off-the-shelf iPRx software receiver with OCXO as an example in this section. A high-accuracy clock, such as OCXO, may not be a practical solution for most mobile applications, but an alternative solution can be achieved with frequency aiding from a cellular phone network.

The main challenge in the design of coherent tracking is constructing the carrier tracking loop. It is possible to remove feedback in the code tracking loop if we use an OCXO clock and Doppler aiding. In Chapter 8, Section 8.2.1 discusses in detail how Doppler aiding is organized for acquisition. Similarly, a Doppler shift is calculated to aid tracking. The code tracking loop is less sensitive to inaccuracies in Doppler aiding due to low code frequency. This frequency should account for relative rover movement and for receiver clock instability. For the carrier tracking loop, we cannot completely remove the tracking loop feedback in order to substitute it with a calculated frequency. Even a slight error in our knowledge about the receiver OCXO frequency will result in tracking loops losing the signal lock.

Alternatively, we could use the carrier tracking loop with aiding. The problem is that usually tracking loops are not aware of the fact that replicas are driven by external information within the coherence interval. This results in incorrect behavior of the tracking loop. In order to use the tracking loop along with external information, it is necessary to make the tracking loop aware of that information, so it must be constructed using a model of carrier replica driven by external information. One of the possible implementations of such a loop is based on a Kalman filter.

If the receiver is static for the period of observation, and its approximate coordinates are known, for example from the previous positioning, then no external information is required for Doppler aiding. If the receiver is on the moving vehicle, information about the vehicle's motion may be necessary. This information can come, for example, from the INS.

Our particular implementation, which is realized in the iPRx receiver, uses one channel to track changes in clock frequency. This channel could be assigned to a particular satellite, for example one of the QZSS satellites, or to a satellite with the highest elevation or highest carrier-to-noise ratio. Frequency assistance from a cellular phone network can also substitute for this channel. The information update on clock frequency comes from this channel non-coherent carrier tracking loop. The adjusted clock is then used to keep the carrier loop on other channels locked within the coherency interval, when no feedback from the loop is available to align the signal with a replica. The alignment between the incoming signal and the code and carrier replicas within the coherency interval is achieved using Doppler aiding only and an estimated clock frequency drift. The clock drift estimation can come either from the

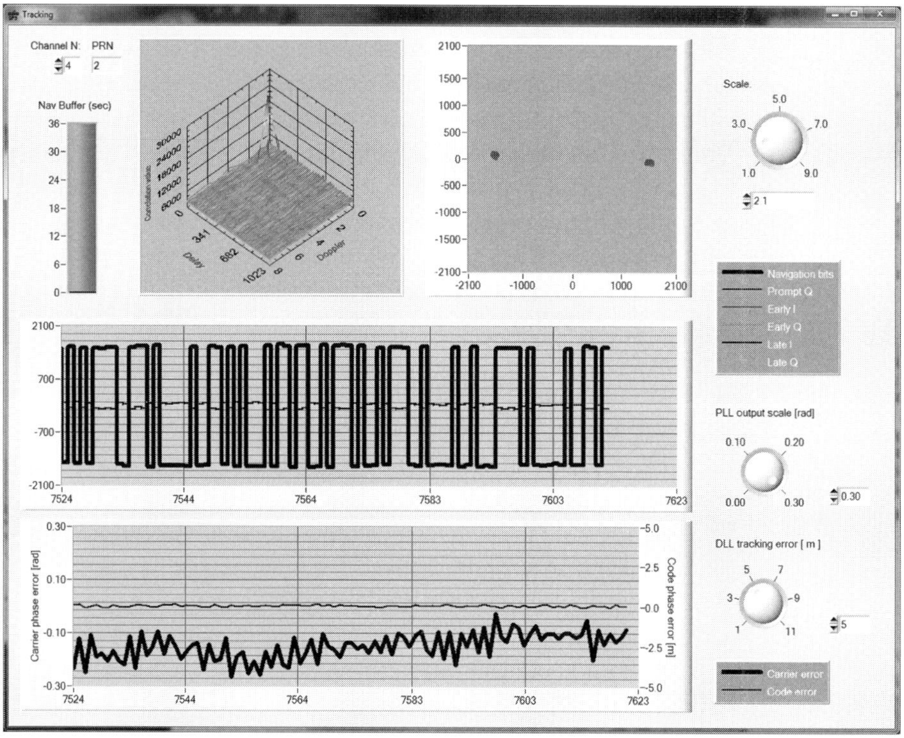

Figure 5.20 Coherent tracking with 1 s coherency interval.

non-coherent channel or from the coherent carrier tracking loop itself. In this case, the loop filter should be designed to include this additional variable.

Code tracking loop feedback can be substituted by external aiding information. In this case, signal alignment with the code replica within the coherency interval can be achieved using external Doppler information only and a stable enough clock. The coherency interval in this case is limited to tens of milliseconds.

To keep a lock on the carrier, we need to retain the carrier tracking loop. For coherent tracking with up to a 1 s interval it is practically impossible to exclude code tracking loop feedback. Even a slight shift in the clock from the nominal frequency will result in divergence if no feedback is available in the loop.

The implemented technique allows coherent tracking to be extended up to 1 s, and possibly more. Figure 5.20 shows tracking with a 1 s coherency interval.

5.2.2.3 Lock detectors

Lock detectors play a very significant role in receiver design. In order to be able to start to use results from the tracking loop and to start collecting measurements, we need to make sure that the tracking loop is actually locked on the satellite signal.

It is relatively easy to see whether the signal is locked by observing the I and Q components in the receiver panel. There are various ways to automate this process.

We follow [2] to devise such an algorithm. The lock detector is designed by an analysis based on a carrier-to-noise C/N_0 estimation. This is because code lock is a necessary condition to achieve a good C/N_0. In the implementation discussed here, code lock is impossible without carrier lock. A lock detector may be defined as follows:

$$\hat{\mu}_P = \frac{1}{K}\sum_{i=1}^{K} P_i, \tag{5.20}$$

where P_i is the normalized power for ith chip length interval and K is the total number of intervals. For a GPS C/A signal, the lock detector measurement $\hat{\mu}_P$ is taken over 50 intervals, which gives a 1 s length of averaging interval in total.

The normalized power P_i at each interval is calculated as the ratio of narrow-band power to a wide-band power,

$$P = \frac{P_N}{P_W}, \tag{5.21}$$

where wide-band power,

$$P_W = \sum_{j=1}^{M}\left(I_j^2 + Q_j^2\right), \tag{5.22}$$

and narrow-band power,

$$P_N = \left(\sum_{j=1}^{M} I_j\right)^2 + \left(\sum_{j=1}^{M} Q_j\right)^2, \tag{5.23}$$

are both computed over M samples.

The signal lock detector measurement $\hat{\mu}_P$ is compared against a threshold, and a decision about the lock is made.

The lock detectors continue to work after the racking loops are locked in order to indicate the time at which the lock is lost.

The same $\hat{\mu}_P$ measurements can be used to estimate C/N_0 as follows:

$$\frac{\widehat{C}}{N_0} = 10\log_{10}\left(\frac{1}{T}\frac{\hat{\mu}_P - 1}{M - \hat{\mu}_P}\right). \tag{5.24}$$

5.2.2.4 Bit synchronization

After the tracking loop is locked, it starts to output data for measurements. The purpose of bit synchronization is to align with the navigation message data bits in order to be able to read the binary navigation message from the signal. We describe here a method given in [2].

We create a cell counter with the number of cells equal to the number of chips in 1 bit of navigation message. When the tracking loops are locked, the cell counter counts each time the in-phase component changes its sign. We can actually see navigation message bits as outputs of the in-phase component (see Figure 5.18). The cell counter switches

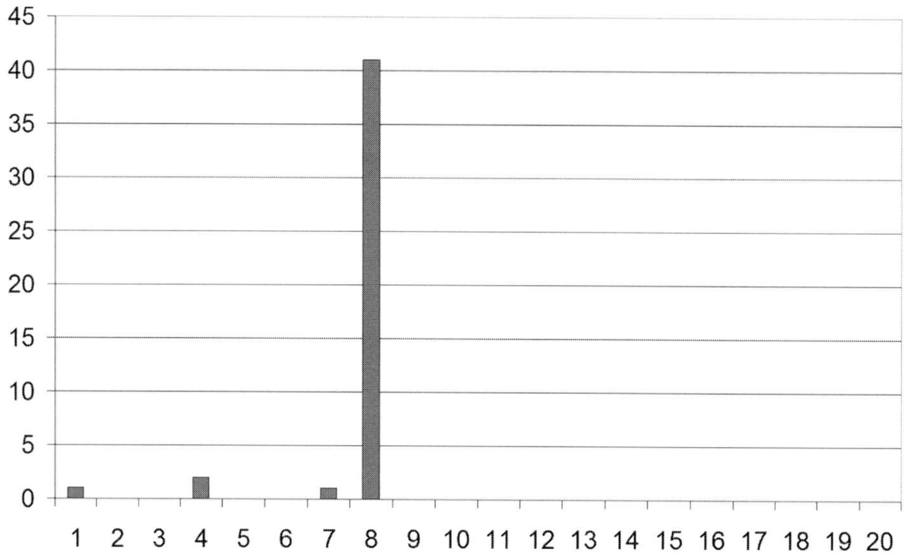

Figure 5.21 Bit synchronization histogram.

cells in intervals of equal code length, so the sign of the in-phase component can change between cells only. Every time the change of sign occurs, the counter increases the number in the corresponding cell. The resulting bit synchronization histogram will show the number of such changes in each cell. If the tracking loops are properly locked, all changes should be counted in the same cell (see Figure 5.21). There are two thresholds set. The bit synchronization is successful if one cell exceeds the upper threshold. The bit synchronization has failed if either two cells exceed the bottom threshold, or the tracking loop lock detectors indicate loss of lock. If bit synchronization is successful, the bit synchronization counter should not be used further. Also, it should be checked that all changes of in-phase output sign occur at the time indicated by the bit synchronization counter.

5.2.2.5 Measurements

The baseband processor should provide the following measurements:

- code phases or pseudoranges,
- Doppler shift frequency,
- signal-to-noise measurements,
- carrier phase measurements,
- decoded navigation message.

Carrier frequency estimates are provided by the carrier tracking loop. The Doppler shift frequency is calculated as the difference between those estimates for each satellite and central IF signal frequency. These Doppler estimates include front-end clock drift. If we want to exclude this drift, further adjustments should be made after a positioning algorithm estimates the antenna coordinates along with the clock error.

Carrier phase measurements also come directly from the carrier tracking loops. Code phase measurements, however, can be calculated only after at least one frame of navigation message is decoded, because the code phase measurements from different satellites must be aligned before they are made available in the output.

What are the benefits of implementing the various enhancements in code and carrier tracking that we have discussed in this chapter? Figure 5.22 shows various code and carrier quality measurements for "usual" (on the left) and enhanced tracking (on the right) from the off-the-shelf ARAMIS^{TM} receiver. In particular, we use carrier noise and code-carrier diversion as criteria. We can see that the usual, straightforward implementation can hardly be used for high-accuracy positioning or for any other application which requires a fine analysis of a GNSS signal. However, for cellular phone applications and a wide range of other mobile applications, where tracking mostly serves to deliver a navigation message and resolve the ambiguity of pseudoranges, these enhancements are not essential, in particular because they come at the price of additional power consumption, larger device size, and increased cost.

5.3 GPS/GLONASS receiver

In this section, we highlight the specifics of the implementation of the discussed methods and algorithms to GLONASS signal processing.

The bit synchronization process should be modified in order to account for meander. When the first meander border is located, the meander is removed and then bit synchronization starts.

Coherent tracking in GLONASS mode is implemented by removing the meander sequence from results accumulated from the output of correlators.

In order to search for a GPS L1 C/A signal, we need to go through a number of shifts defined by the code length and the length of the shift in chips, i.e.

$$n = \frac{N}{d}, \tag{5.25}$$

where d is the correlator spacing. For acquisition of BPSK, we can conduct a search choosing d equal to a maximum value of 0.5. The number of chips for GPS L1 C/A is equal to 1024, and for GLONASS it is equal to 511. Therefore, the initial search area is as follows:

$$\begin{aligned} n_{L1C/A} &= 2,048 \\ n_{SP} &= 1,022 \end{aligned} \tag{5.26}$$

where $n_{L1C/A}$ is the searching space for GPS, n_{SP} is the searching space for GLONASS.

Figure 5.23 shows a multi-system-receiver diagram. In this chapter we consider only GPS and GLONASS receiver because of two reasons:

(1) at the time of writing (2013), GPS and GLONASS were the only two operational GNSS;

(2) GPS and GLONASS L1 signals have different central frequencies, whereas, for example, Galileo is on the same frequency as GPS.

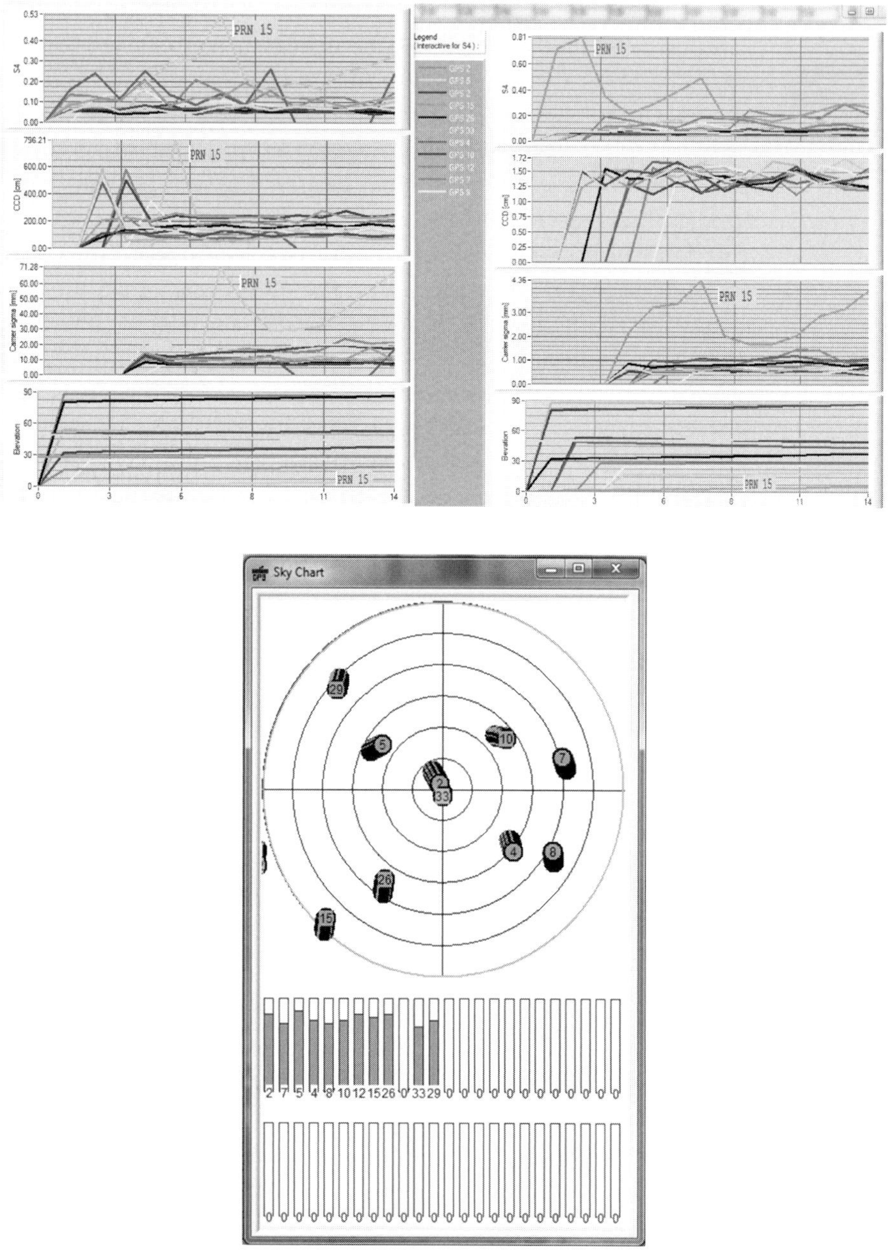

Figure 5.22 Quality measurements for "usual" and enhanced tracking (ARAMIS^{TM} panel).

The second issue is important, because it means that the receiver has to use two separate front ends for these two systems. The correlator unit may also be different because we have to search over a wider frequency region, but with only one code. That makes the search procedure and the design of the correlators significantly different for these two systems. On the contrary, although the secondary codes mean that the correlation algorithms are also different, the logic of the search is similar for GPS,

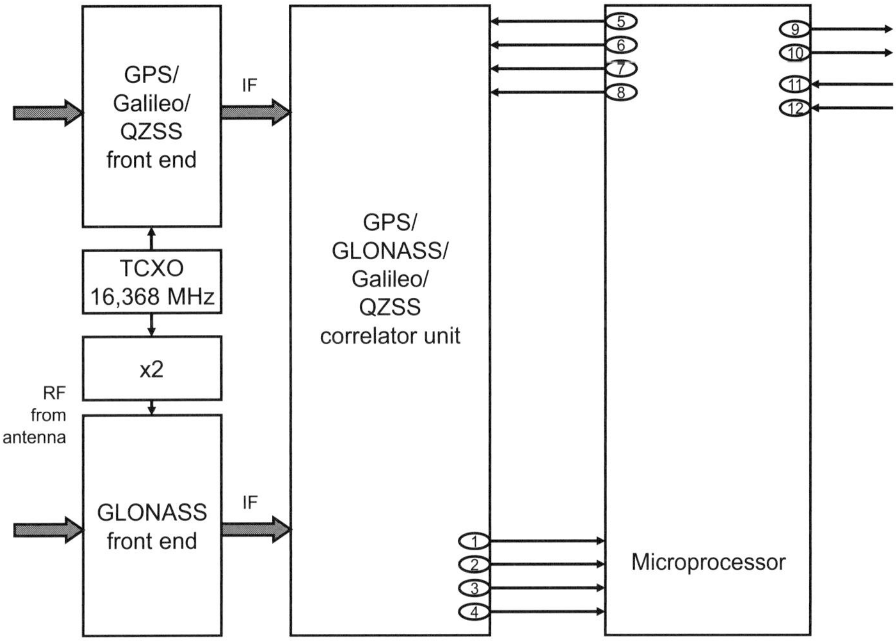

Figure 5.23 Multi-system-receiver implementation.

Galileo, and BeiDou L1 signals. The new GLONASS L1 CDMA and L3 signals are also similar in that respect to GPS.

As previously mentioned, the FDMA component of the GLONASS code design implies a wider bandwidth, which defines, according to the Nyquist criterion, a higher sample rate. As we can see in Figure 5.23, a 16 368 MHz clock (for example) should be doubled, for example using both rising and falling edges, to 32 736 MHz. This higher sampling rate defines the minimum requirements for the digital hardware and embedded processor power. Again, for other L1 signals this is not the case, and they can all be processed with the same sampling. Compromising on the sampling rate would lead to a reduction in accuracy due to the side lobes of the signal being cut off. This can be done only if the original signal is at first filtered to the narrower bandwidth that corresponds to new sampling rate in order to avoid aliasing errors.

For new signals, such as GPS L1C or Galileo E1, a receiver can track the complete signal or the BOC (1,1) component only. It is not possible to track the complete signal with a 1-bit resolution receiver. For mobile receivers, which feature a narrow front-end bandwidth, BOC tracking is basically the same as tracking legacy signals (GPS L1 C/A, GLONASS SP).

References

[1] J. J. Spilker, Fundamentals of signal tracking theory, in *Global Positioning System: Theory and Applications, Vol. I*, B. W. Parkinson and J. J. Spilker, Eds. Washington, D.C.: American Institute of Aeronautics and Astronautics Inc., 1996.

[2] A. J. Van Dierendonck, GPS receivers, in *Global Positioning System: Theory and Applications, Vol. I*, B. W. Parkinson and J. J. Spilker, Eds. Washington, D.C.: American Institute of Aeronautics and Astronautics Inc., 1996.

[3] P. W. Ward and J. W. Betz, Satellite signal acquisition, tracking, and data demodulation, in *Understanding GPS, Principles and Applications*, 2nd edn., E. Kaplan and C. Hegarty, Eds. Boston, MA: Artech House, 2006.

[4] J. Tsui, *Fundamentals of Global Positioning System Receivers: A Software Approach*. New York: Wiley, 2000.

[5] T. Pany, *Navigation Signal Processing for GNSS Software Receivers*. Boston, MA: Artech House, 2010.

[6] A. Schmid, *Advanced Galileo and GPS Receiver Techniques, Enhanced Sensitivity and Improved Accuracy*. New York: Nova Science Publishers, Inc., 2009.

[7] I. Petrovski, T. Tsujii, and H. Hojo, First AGPS – now BGPS. Instantaneous precise positioning anywhere, *GPS World*, vol. 19, pp. 42–48, Nov. 2008.

[8] I. Petrovski and T. Tsujii, *Digital Satellite Navigation and Geophysics: A Practical Guide with GNSS Signal Simulator and Receiver Laboratory*. Cambridge: Cambridge University Press, 2012.

[9] J. McNeff, GPS receiver specifications, *Inside GNSS*, pp. 50–56, May/Jun. 2012.

[10] F. van Diggelen, *A-GPS: Assisted GPS, GNSS, and SBAS*. Boston, MA: Artech House, 2009.

[11] N. I. Ziedan, *GNSS Receivers for Weak Signals*. Boston, MA: Artech House, 2006.

[12] S. J. Orfanidis, *Optimum Signal Processing*, 2nd edn. New York: McGraw-Hill Publishing Company, 1988.

[13] R.E. Best, *Phase-Locked Loops: Design, Simulation, and Applications*, 5th edn. New York: McGraw-Hill, 2003.

[14] F. M. Gardner, *Phaselock Techniques*, 2nd edn. New York: John Wiley and Sons, Inc., 1979.

[15] J. B. Thomas, *An analysis of digital phase-locked loops*, JPL, Pasadena, CA, Publ. 89–02, 1989.

[16] D. Doberstein, *Fundamentals of GPS Receivers: A Hardware Approach*. Berlin: Springer Science+Business Media, 2012.

6 Receiver implementation on a general processor

6.1 Development of the "software approach"

In this chapter we discuss a receiver with a baseband processor implemented on a general processor, which is usually referred to as a *software receiver*. GPS software receivers originated from the concept of software-defined radio (SDR), which appeared some time before 1995 in the communication field [1].

In fact, almost from the beginning all GPS receivers were SDR receivers in a general sense. An SDR receiver, as it is defined by Joe Mitola who created the SDR concept, is one that can be reprogrammed and reconfigured through software.

In the GPS field, the concept of the software receiver was assumed to be somewhat different from SDR. It originated in works by Akos [2] and Tsui [3]. The GPS software receiver concept is narrower than that of SDR in the sense that all signal processing is implemented completely in the software. This approach provided a unique perspective on our understanding of GNSS receivers and sparked a great deal of research in the receiver field that was previously available only to receiver manufacturers and large research labs. We discuss GNSS receivers from the perspective of a general SDR concept in Chapter 10.

In this chapter, we discuss software receivers from the point of view, common in the GPS field, in which everything in the receiver after the front-end block (see the functional model in Figure 5.3) is implemented in the software. This idea has developed in two main directions: software receiver implementation in (i) personal computers and (ii) embedded processors for mobile devices, such as cellular phones. The first software receivers were implemented in powerful general-purpose computers.

The very first GPS software receiver was developed at the Jet Propulsion Laboratory (JPL) in Pasadena, CA, and was introduced in 1997 by Wu *et al.* [4]. This receiver had an architecture in which "all GPS-specific signal processing typically implemented in hardware has been moved to the software."

The first commercial off-the-shelf real-time GPS software receiver was developed by the Swedish company NordNav. The NordNav GPS receiver operated under the Windows operating system (OS) and had a user-friendly graphical user interface (GUI), which allowed access to major parameters, and some visualization, such as sky charts, positioning on the map, views of correlators, and signal spectra. The receiver has been very popular for R&D applications. It also provided researchers with access to a digitized intermediate frequency (DIF) signal. Many of the researchers were using DIF

Figure 6.1 The author presenting his software receiver for pseudolite applications during the ENC 2006 conference in Manchester, UK.

recordings from a NordNav USB front end in various in-house solutions for software receivers. In particular, the author was using the NordNav front end together with a post-processing receiver written in MathWorks MATLAB language. The receiver was able to acquire and track in order to receive pseudolite signals (see Figure 6.1) [5]. The NordNav USB front end was built around a SiGe front-end module.

In 2007, CSR (Cambridge Silicon Radio), a multinational fabless semiconductor company, acquired NordNav for \$40 million, plus an additional up to \$35 million, on condition that some expectations were met [6]. By that time, NordNav also had a software version for embedded processors. This event showed that software GPS technology developed for general processors could be applied to mobile devices, with algorithms moved to embedded processors. This particular acquisition maybe did not meet the expectations at that time, but the technology has significantly matured since then.

The development of PC-based and embedded software receivers, which had hitherto been linked, split up at approximately that point. The NordNav software receiver had been developed as a test bed for embedded software receivers. Today, PC-based software receivers are usually intended for R&D and special purposes. A multi-frequency, multi-constellation software receiver, which was developed by the German company IFEN, is suitable for high-end users in R&D, for example those who are working on creating the Galileo system [7]. Another example of a high-end software receiver is the ionospheric scintillation monitor ARAMIS™ from iP-Solutions, which was developed for R&D and geophysical applications [8]–[11]. The first free real-time software receiver was made available, also by iP-Solutions, to accompany a textbook on GNSS applications for geophysics and navigation [8].

A different, PC-based, approach was to create receivers in MATLAB. It should be noted that MATLAB-based receivers are rather slow; they are not really suitable for the purposes of navigation, positioning, or even research. However, they provide an excellent tool for education. The first commercial MATLAB receiver was developed by DataFusion [12], and the first free MATLAB software receiver was developed and released by the Prof. Borre school [13]. This publication sparked huge interest within the GNSS community by providing the necessary tools to the research community. Further, the MATLAB GPS receiver was modified to work on the open-source Scilab platform and extended to support GLONASS [14].

The development of the embedded software receiver took a different route. The functionality of the PC-based software receiver was exchanged for smaller memory and reduced CPU requirements. Even mainstream companies successful in the mass production of GNSS receivers for mobile devices are now looking at the software implementation. What does the future hold for software receivers? Will they occupy the high-end receiver market, the mass market, or will they be forever left in academia?

6.2 Software receiver design

6.2.1 Baseband processor implementation

In this section we look at software receiver design and how it is different from a conventional or "hardware" receiver. It is necessary to underline that in this book we are not talking about a "hardware" receiver in the sense of analog implementation, an approach abandoned almost immediately for GPS.

When we talk about software receivers, as compared with conventional receivers, we shall consider only a baseband processor. The baseband processor implementation is the only difference between a software and a conventional receiver (Figure 6.2).

In the case of a conventional receiver, all the baseband processor functions can be implemented on different platforms, which may include FPGA, ASIC, SoC, and a dedicated embedded processor. The only essential difference between "hardware" and

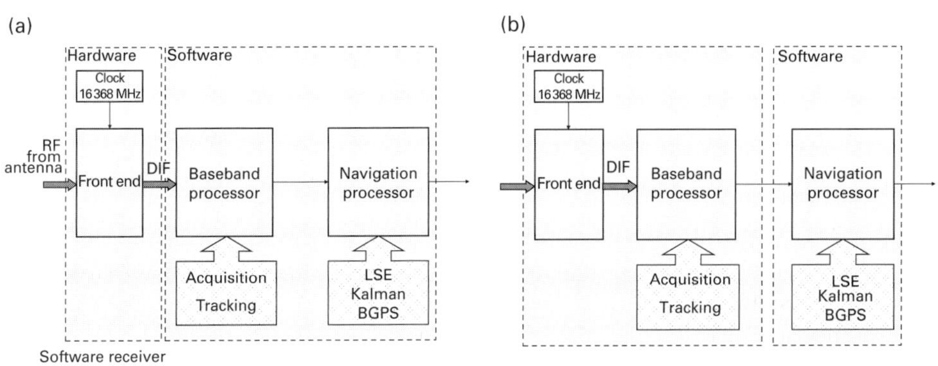

Figure 6.2 (a) Software receiver vs. (b) conventional receiver functional diagrams.

software implementation is the possible number of parallel threads. In the dedicated hardware, the number of correlators working in parallel can reach hundreds of thousands, whereas the number of physical parallel threads on a general computer processor is limited to just a few, for example eight on the i7 Intel processor. Apart from this limitation, there is no algorithm in the "hardware" receiver that cannot be successfully implemented in the software, and vice versa.

As a result of baseband implementation in the software, all acquisition and tracking loops are directly accessible to a user. Figure 6.3 shows a graphical output from the baseband processor of an iPRx receiver which was developed by the author.

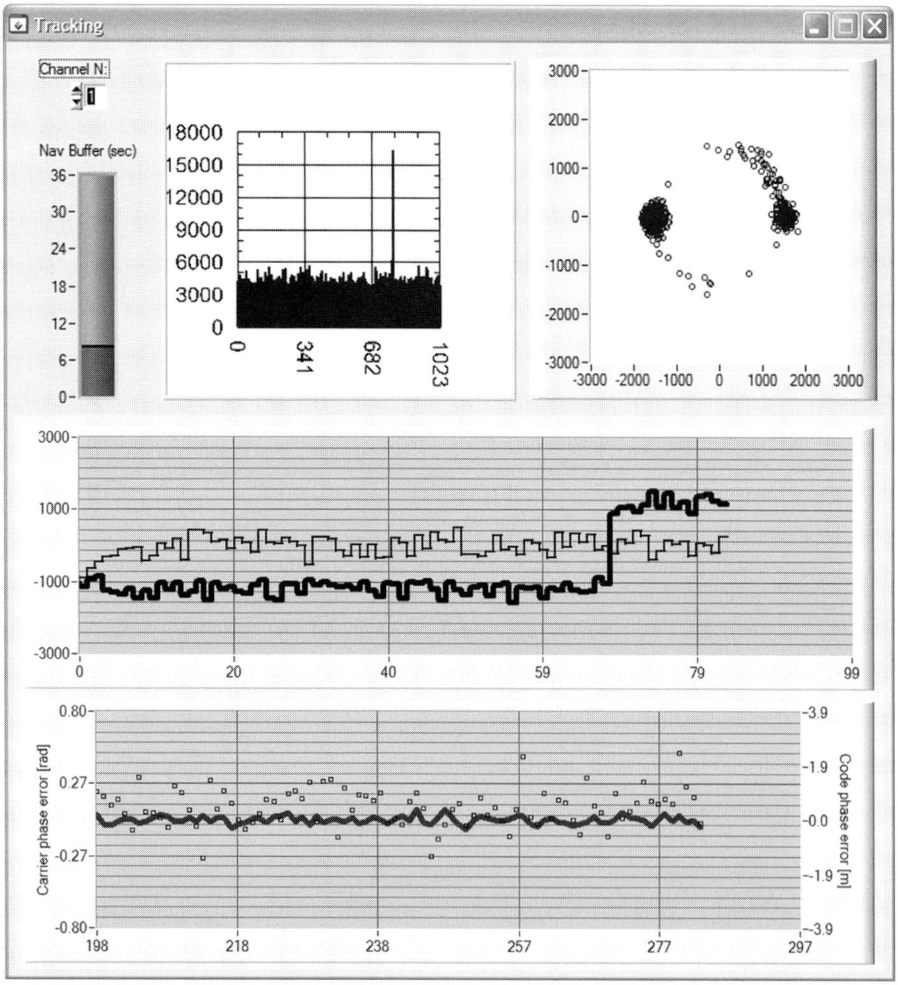

Figure 6.3 Interface of real-time iPRx software receiver, which visualizes the baseband processing inside the receiver.

6.2.2 Acquisition implementation

A software receiver employs the same tracking and navigation algorithms and methods as conventional receivers. However, acquisition methods for a software receiver are different from those applicable to conventional receivers. Conventional receivers apply multiple correlators which work on signals in parallel, whereas a software receiver has to work through the acquisition sequentially. The search can be done using a brute-force method, but this would take too long for a software receiver, so a number of methods can be used to optimize the speed of this search. One of the most widely accepted methods for acquisition in software receivers is based on the fast Fourier transform (FFT). The flow chart of the parallel code phase search algorithm with circular correlation is presented in Figure 6.4. There are a number of acquisition algorithms that could be used, but the circular correlation algorithm [3],[15], which we discuss in this section, is currently the optimal choice for a software receiver.

The acquisition algorithm discussed in Chapter 5 needs to multiply a signal that has been converted to IF (intermediate frequency) by the code replica and then by the carrier replica. We can significantly speed up the algorithm using FFT. The result of the multiplication of the incoming signal by the spread code with each possible delay can be transferred to a frequency domain in one step and analyzed there. If the replica spread code has the same delay as that in the incoming signal, the multiplication will strip the incoming signal of the spread code. The spectrum of the signal resulting from the multiplication will show a peak at the frequency corresponding to the IF plus signal Doppler shift. FFT allows for a computationally efficient transfer from the time domain to the frequency domain.

In order to use FFT, we need the number of signal samples be equal to 2^n. Otherwise, FFT performance will be the same as that of the standard digital Fourier transform (DFT), which is significantly slower. Therefore, in order to enjoy the advantages of

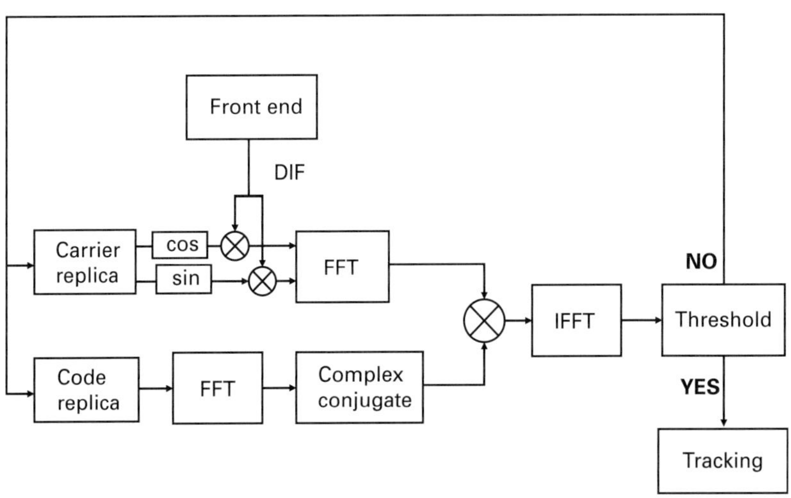

Figure 6.4 Acquisition algorithm with FFT.

FFT, we need to pad out the samples with zeroes. This will, however, cause a slight decrease in correlation performance.

We can devise an even quicker algorithm if we go back to the initial method and strip the incoming IF signal of the carrier first and then search for the code delay in the frequency domain (Figure 6.4). When we strip the carrier, we have to assume a certain value for the Doppler shift. This assumption cannot be made precisely, so we have to repeat the carrier stripping process a few times to get a number of replica versions. To this end, the frequency range within which we shall search for the Doppler shift is divided into a number of bins. How large the search area should be is defined by the rover dynamics and the accuracy of the front-end clock. We repeat an FFT process sequentially through all the frequency bins, the number of which can be chosen to be from 18 to 72 depending on the width of the frequency bin and the number of bins. In the case of the previous algorithm, when we strip the signal of code first, we had to repeat FFT for 1023 code possible code delays. Returning to the second algorithm, we employ FFT to transfer both the replica and the incoming baseband signal to the frequency domain, and compare them in there. After finding the correlation in the frequency domain, inverse FFT gives the correlation versus delay in the time domain. Overall, the algorithm can be presented as follows:

$$IFFT\big(FFT(S_i) \times FFT(B_j)\big) = S_i \otimes B_j, \tag{6.1}$$

where FFT ($IFFT$) is the direct (inverse) fast Fourier transform operation, S_i is the incoming baseband signal, and B_j is the spread code replica generated in the receiver, where j is the PRN number and i is the number of frequency bins.

The left-hand side of equation (6.1) gives our exact circular correlation algorithm as it is depicted in Figure 6.4. The right-hand side defines a convolution between two signals, which is implemented as sample by sample shifting and multiplication of two signals in the time domain. The resultant vector should have a distinct peak at a shift equal to the code phase. This type of algorithm is referred to as circular because it will look through all the possible delays between the incoming baseband signal and the receiver code replica by shifting them as in a circular buffer. Although the FFT acquisition algorithm is most often discussed in relation to a software receiver, it can be implemented on digital hardware as well.

6.3 Advantages of software receivers

6.3.1 Software receiver advantages for mobile applications

We list the most important advantages that a software receiver may bring to mobile applications.

6.3.1.1 Potential reduction of required hardware

A mobile device with a software receiver no longer needs to incorporate a hardware baseband processor. The only hardware component in the receiver is the front end. This allows a reduction in:

- cost,
- size,
- weight,
- power consumption.

In turn, reduction in power consumption would allow longer operation of the device, which works on batteries.

This requires implementation of the baseband processor functionality on the host processor; the baseband processor functions of a software receiver may then be transferred to a general processor. For a snapshot receiver, the load on the processor may be minimized because the receiver works during short periods of time.

The software receiver solution could become almost universal and unified across a wide range of products. For products that have a processor into which the receiver software can be ported, a manufacturer would need to add only a front-end module and an antenna.

Furthermore, combining all the functionality in the same processor allows for better synchronization with other device functions, for example better time sharing with phone operation in a cellular phone device.

However, in many applications a dedicated embedded processor may still be required. In particular, if a receiver implements tracking, then, depending on the implementation of tracking loops, an interruption rate from 1000 Hz to 50 Hz may be required. In that case, it would be difficult to combine this functionality with cellular phone functions.

6.3.1.2 Upgradeability

This refers to the procedure of implementing new algorithms in a mobile device after the device has been sold to the customer. This can be done in a similar way to upgrading application software on a PC. The aforementioned application of new signals can be achieved even when the device is already with a customer.

6.3.1.3 Bug fixing

Serious bug fixes can be made without recalling the product. The baseband processor algorithms can be fixed by software patches. A mobile device with a hardware receiver, on the other hand, would have to be recalled if the baseband processor were implemented on FPGA or ASIC, or would need to undergo a firmware upgrade in the case of an embedded processor and possibly FPGA. A firmware upgrade is more difficult to automate than a software upgrade.

6.3.1.4 Reduction of new product development cycle

It becomes much less expensive to introduce a new product version for a software receiver, and consequently the product development cycle is significantly shorter.

6.3.1.5 Adaptability to new signals

When a front end permits acquisition of a new signal, the baseband and navigation processor functions can be adapted for the new signal on the fly.

Figure 6.5 The iP-Solutions USB front end for the GNSS real-time software receiver.

For example, a few years ago the author and his colleagues developed the GPS/GLONASS software receiver, for which all the baseband processor functionality was written in the software. Figure 6.5 shows the USB front end, as built by the author, for the receiver. The USB front end was built around a MAX2769 chip from Maxim Integrated. When BeiDou and the first Galileo satellites became available, only modifications in the software were required to make the receiver work with new GNSS. Of course, new signals may increase the load on the processor due to additional modulations, secondary codes, or wider bandwidth, for example. This, however, can be compensated for using various computational tricks, which are considered in the following sections.

6.3.1.6 Change of receiver type

It is possible to make a receiver even more flexible and adaptive by changing its type on the fly. For example, a high-sensitivity receiver can become a much more accurate GIS receiver just by the user clicking on a key. This can be achieved by downloading either new software from the Internet or memory onto this or an external device.

6.3.1.7 Third-party product involvement

The preceding function also opens doors for third-party application developers. Certain types of functionality for receivers can be purchased from third-party companies; some is even provided on a free basis. No special open architecture would be required, because all algorithms operate in the device host processor and require no special interaction with the front end.

6.3.2 Software receiver advantages for high-end applications

Software receivers engaged in R&D, geophysics, and education also have significant advantages over conventional receivers.

6.3.2.1 Flexibility

A receiver can be easily modified in order to accommodate any changes in the signal or navigation message. This makes a software receiver a very attractive option during the design of new satellite navigation systems. Developers can modify the signal and evaluate receiver performance during the real field test.

A researcher may take advantage of changing and testing algorithms inside a receiver during their development. For example, it would be a time consuming and possibly expensive process to test a new version of a signal processing algorithm implemented inside a conventional receiver. For a software receiver, such a process is reduced to routine software debugging. This is not limited to receiver developers, but can extend to researchers in various fields. For example, the ARAMIS™ ionospheric scintillation monitoring software receiver was modified at the level of the baseband processor by researchers to improve the quality of code and carrier phase observables, implement additional outputs from tracking loops, and accommodate additional requirements specific to the ionospheric scintillation monitor [9–11].

6.3.2.2 Access to baseband processor

A software receiver provides unprecedented access to the inside of the baseband processor. A user can see the behavior of acquisition and tracking loops in real time, which gives a unique insight to the physical processes. This function makes the software receiver an exceptionally attractive tool for academia [8].

6.3.2.3 RF signal post-processing

A software receiver permits one to record and keep RF records, which may be processed again, possibly with different algorithms, including acquisition and tracking algorithms [16]. This may be essential for many applications when repeating tests would be too expensive, such as those involving airborne and space-borne receivers. Wu *et al.* [4] presents such an example, in which GPS RF signals for LEO satellite were recorded and post-processed with a JPL MicroGPS receiver.

Another important application is related to the recording of a GNSS RF signal for ionospheric scintillation research [8]. These events may be unique because solar activity has a period of 11 years. The researchers can process recorded signals in post-processing mode with various algorithms and tracking loops, improving the accuracy of the measurements and the stability of the tracking.

6.4 Real-time implementation

Applying the term “real time” to a receiver looks like a tautology. However, due to the huge impact which the slow MATLAB, and now Scilab, receivers had on the educational field, they always come to mind first when someone talks about software receivers [3],[13],[17]. These MATLAB/Scilab receivers are extremely useful for academic purposes, though their slow speed renders them almost useless for practical applications. If processing takes too long, the receiver becomes difficult to use, so near-real-time receivers also become of interest if they can operate if not with live satellites, then at least quickly [18].

Apart from these extremely important, but relatively forgiving, educational applications, any practical application requires real-time operation. These applications can greatly benefit from a software receiver approach. They include geophysical applications, R&D, and mobile applications.

As mentioned at the beginning of this chapter, the first software receiver available for the community was the commercial real-time receiver from NordNav developed in C++. At the time of writing, commercial real-time GNSS receivers are available from IFEN and iP-Solutions. The iP-Solutions iPRx receiver is also available as a free academic version [8].

The term “real time” is often quite incorrectly understood as referring only to speed of operation. In the case of a software receiver, however, “real time” comprises:

(1) concurrency;
(2) speed of operation.

6.4.1 Concurrency

Concurrency means that front-end operation, acquisition, tracking, positioning, and possibly other tasks should be working in parallel. In a commonly used MATLAB receiver, including one developed by the author [5], the signal logging, acquisition, and

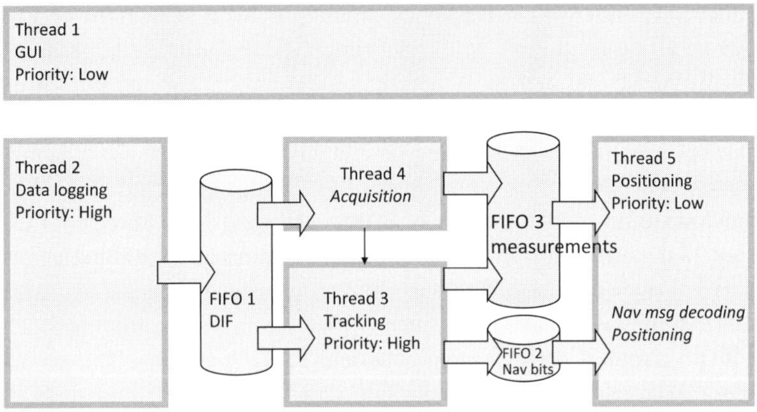

Figure 6.6 Main parallel threads in a software receiver.

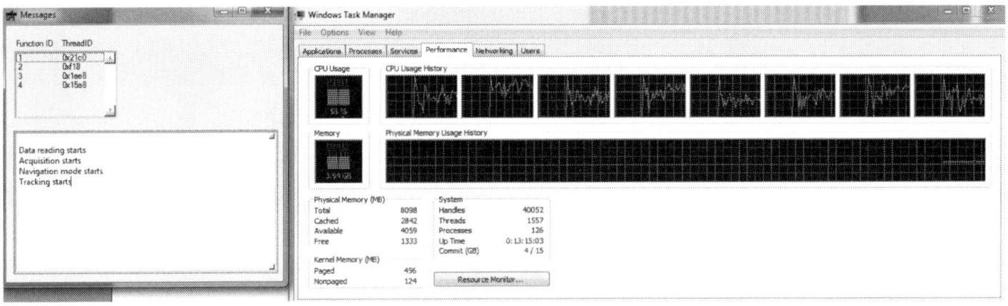

Figure 6.7 Software and physical CPU threads.

tracking are consecutive tasks. This makes perfect sense from an educational point of view, because it follows the signal processing logic. However, concurrency is required in order for the receiver to work continuously, without interruption of the positioning task and reacquisition, and without accuracy degradation.

Figure 6.6 shows the necessary threads in a receiver. There are a minimum of five main threads. Figure 6.7 shows the threads with ID on the iPRx software receiver panel along with the physical parallel threads in the Task Manager window.

A graphical user interface (GUI) (Thread 1 in Figure 6.6) is in charge of communication with a user, taking commands, and providing visualization. This low-priority thread may be thinner for some applications, and for some it may not be necessary at all. It may, however, put a heavy load on a computer, for example for the iPRx receiver, when data tracking loops are collected each cycle and plotted with an update rate of up to 1000 Hz, as shown in Figure 6.3.

In a PC-based receiver, the thread priority can be set up in the software from low to high with many levels.

Data logging is a thread (see Thread 2 in Figure 6.6) responsible for accepting DIF data from a front end. As the sampling rate increases, so the load becomes heavier on this thread. The GPS C/A sampling rate is defined by the front-end clock. For most applications, the front-end clock is implemented as a temperature-compensated crystal oscillator (TCXO) with a frequency around 16 MHz. The front end is a real-time component, and it cannot stop its data supply. Therefore in order to handle these data in non-real-time OS, such as Windows, various buffers must be implemented.

FIFO 1 in Figure 6.6 is a First In First Out buffer, which keeps the DIF signal after it arrives from the front end. A higher sampling rate requires the use of larger buffers. For GLONASS applications a sampling rate of ~16 Mps is not sufficient due to the wide bandwidth occupied by a number of FDMA channels. In addition, a GLONASS signal comes from a different front-end channel and requires additional buffers, which should be larger than those allocated for the GPS signal. The logging thread has a high priority because it interfaces the non-real-time OS with the real-time front end.

FIFO 1 supplies data to acquisition and tracking threads. The acquisition thread (Thread 3 in Figure 6.6) takes the DIF data from FIFO 1 and finds a signal. The tracking thread (Thread 4 in Figure 6.6) has a high priority because it must operate at the same speed as the incoming signal. Depending on the receiver design, the tracking thread may be omitted. In that case, the measurements come from the acquisition thread. A navigation message is accessible only if a tracking thread is present.

A navigation thread (Thread 5 in Figure 6.6) may also be absent if the receiver is used only to collect raw data, such as code and carrier phases, Doppler, and signal-to-noise ratios. Further, these measurements can be used for finding positions using third-party positioning software, for example geodetic software such as Bernese GPS.

For embedded processing, the structure of the threads and their priorities may be different, especially because the embedded processor operates in a real-time mode.

As for any rule, there are exceptions for concurrency. One possible exception is a snapshot receiver, which may operate sequentially, i.e. it may take a snapshot of the data, record it into a memory, make an acquisition, then make AGPS (Chapter 8) or BGPS (Chapter 9) positioning, and proceed to the next snapshot.

6.4.2 Bottlenecks in GNSS signal processing

The most time-consuming operation in a software receiver is related to correlators operating in tracking loops. We have considered tracking loops in detail in Chapter 5. Each correlator must operate by processing all samples; how many are processed is defined by a sampling rate. A GPS L1 C/A signal is usually sampled by a front end with a sampling rate of ~16 Mps. The minimum sampling rate for a GPS L1 C/A signal is about 5 Mps. In the case of tracking, a receiver must operate with at least three or sometimes five (e.g. double delta discriminator for aviation receiver specification) correlators for each channel. Also there are two channels, in-phase and quadrature (see Figure 5.13). This means that a twelve-channel receiver should make

$$\sim 16\,000\,000 \times 12 \times 8 = 1\,536\,000\,000 \text{ multiplications per second.}$$

Multiplication is a very expensive operation on its own in terms of processor recourses. The correlators also use some additional operations required to sum the results of the multiplications (see Figure 5.6). The approximate number of such operations in our example is 1 152 000 000.

There are many tricks of trade available to optimize GNSS signal processing. Some of these tricks are applicable for general processors, and some are suitable for both general processors and embedded processors.

Some of these methods are trade-offs between requirements of resources, such as memory and processor load, and processing speed. Continuing processor development allows for further ease in the requirements of the software, at the same time meeting real-time constraints.

6.4.3 Algorithmic methods used to speed up processing

6.4.3.1 Early-minus-late discriminator

In Chapter 5 we considered various discriminators, and in Figure 5.15 we presented correlation curves between a received signal and receiver prompt, early, and late replicas. Comparison of Figure 5.15 and Figure 6.8 demonstrates that we can substitute the early and late replicas with one early-minus-late replica.

Early-minus-late discriminators are particularly important for software receivers because they do not require an output from the prompt branch. Each of the correlator branches, early, late, and prompt, requires a number of operations proportional to the number of samples. The calculation of I and Q values presents a computational bottleneck for the software receiver. Thus, removing one of the branches allows a significant decrease (of one-third) in computational load. Furthermore, some of the quadrature components may not be required, depending on the choice of discriminators.

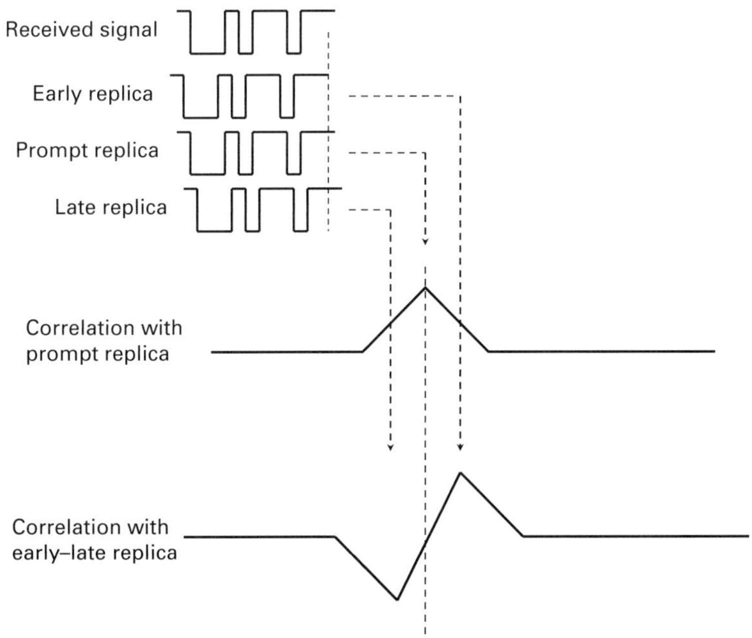

Figure 6.8 Using an early-minus-late replica.

In our example, removing each branch allows a gain of about 192 000 000 multiplications and the same number of summations as in our example.

6.4.3.2 Signal decimation

We have stated that the load on the processor is defined by the sampling rate, which in turn is defined by the receiver front-end clock. For GPS, it is usual to use a 16 MHz TCXO. However, a GPS L1 C/A signal bandwidth is 2 MHz. So, according to the Nyquist criterion, the sampling rate should exceed twice the bandwidth. For practical applications, 2.5 times the bandwidth actually works better. An excess in the sampling rate would not provide additional information about the signal. It improves the signal-to-noise ratio, because of the additional samples, but greater accuracy will not be achieved, with the exception of a snapshot receiver. In that case, the accuracy may be defined by the sampling rate unless special algorithms which compensate for the absence of tracking loops are implemented.

Therefore, 16 Mps of the GPS L1 C/A signal can be down-sampled to 5 Mps, or even 4 Mps, equivalent to a four-fold reduction in load. The catch is that the signal should be filtered prior to decimation. Decimation without filtering would lead to signal aliasing, and even signal acquisition would be impossible. Some front ends allow filter parameters to be changed in the front end, in which case the filtering procedure in the software can be omitted.

Having said that, we must now consider how the sampling rate affects the signal processing in more detail. Figure 2.4 shows the spectrum of the pulse train. As we have discussed in Chapter 2, the GNSS code sequence has a similar spectrum (moved to the L1 band by the carrier). As we have learned in Chapter 2, the chip length defines the spectrum. The main lobe in the case of the GPS C/A code, which has a chip length of 300 m, is about 2 MHz, which is in general good enough to allow us to get a position with an accuracy required for mobile applications.

Once again, the Fourier transform would give us an ultimate tool with which to establish a correspondence between the signal representation in the frequency and time domains. The signal representation in Figure 6.9 shows that the chip (pulse train) in the time domain is represented by a spectrum that contains a few lobes. If we cut off all but the main lobe, the shape of the chip, which can be restored from this limited bandwidth, is not correct. As we include more side lobes, a more accurate chip shape can be restored (see Figure 6.9). In turn, this more accurate shape allows us to narrow the autocorrelation function, which is depicted in the third column. Therefore, a 20 MHz bandwidth is actually the maximum bandwidth requirement, for an L1 C/A signal, which would allow a complete restoration of the code chip shape and therefore the best possible autocorrelation function. The 20 MHz bandwidth is also a requirement for high-end aviation receivers.

6.4.4 Hardware-dependent methods

Hardware-related methods include using special CPU instructions. These allow the use of a single instruction on multiple data (SIMD). This instruction is used in x86 processors. Heckler and Garrison [19] provide a detailed explanation of the application of these instructions with source code examples in C++.

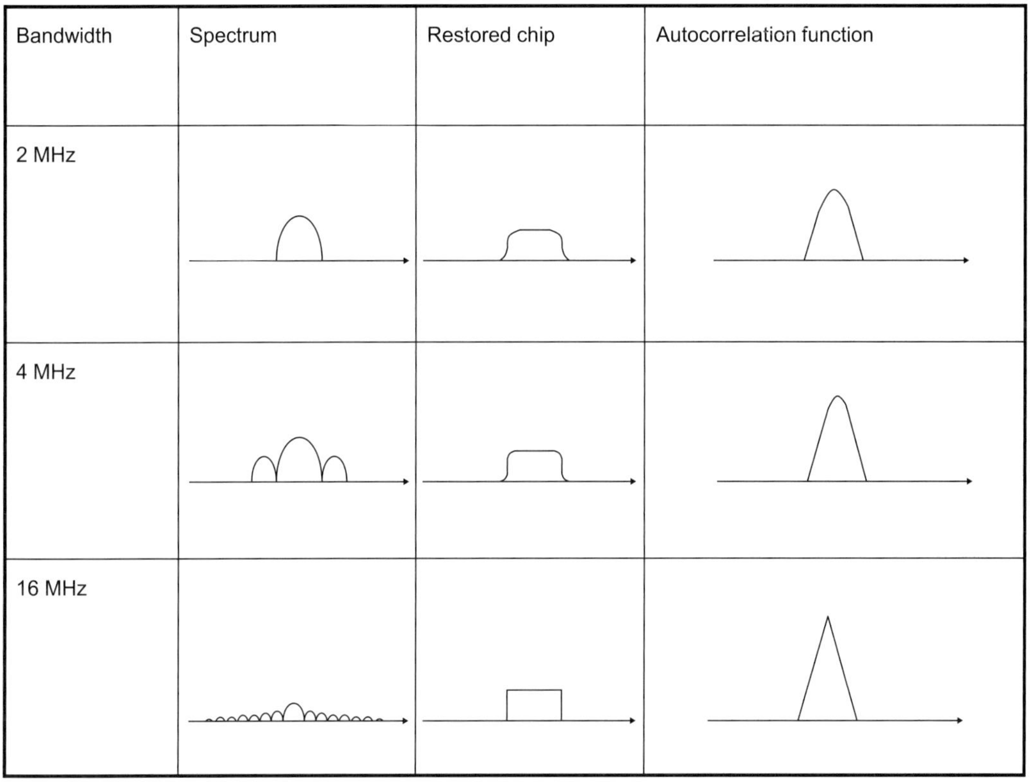

Figure 6.9 Quality of signal recovery dependence on receiver bandwidth.

An alternative hardware-related method is to use a graphic processing unit (GPU) to assist the CPU. The graphic card on any PC features a dedicated processor unit. We can compare a top CPU with a top GPU: in 2013, a high-end PC with an Intel Core i7-2600K CPU can provide eight physical threads and has a memory bandwidth of about 20 GB/s and a 3.4 GHz clock, whereas GTX 590 has 1024 CUDA (computer unified device architecture) cores, a 327 GB/s memory bandwidth, and a 1.2 GHz processor clock.

Originally a GPU had a narrow specialization, with hardware optimized for specific tasks. Gradually, software engineers working on applications gained more and more access to the GPU through the API. At the same time, GPUs provide more and more flexibility for programming.

In 2003, a research team from Stanford University created the first widely accepted programming model to extend C with data-parallel constructs and adopt the concept of stream computing with GPU [20].

The graphic card manufacturer NVIDIA developed a solution with GPU-accelerated libraries that allows software engineers to develop parallel applications using C++. The approach was facilitated by developing CUDA parallel computing architecture to access GPU power on NVIDIA video cards. This technology has been implemented in radio astronomical signal processing, which performs tasks similar to GNSS [21]. It was later proposed to use GPU for GPS signal processing [22].

New Microsoft AMP (accelerated massive parallelism) technology allows wider applications of GPU, which are coming also to mobile devices. This approach works for any discrete graphics card. The C++ AMP programming model includes multidimensional arrays, indexing, memory transfer, tiling, and a mathematical function library. Software engineers can use the C++ AMP library to control how data are moved from the CPU to the GPU and back to optimize performance.

A disadvantage of this approach is that not all general tasks can be easily transferred to GPU without loss. A processor spends time on processing and on data delivery to the processor. If data are in the cache memory, the gain in processing could be very high. GPS processing takes advantage of excessive cache use for manipulating data kept in caches. A GPU, however, has a much smaller cache than a CPU. Therefore a slightly different approach to the algorithms is required depending on whether they are processed in the CPU or the GPU. The use of a GPU makes the implementation of a software receiver on a PC similar to that on mobile devices with specialized digital hardware.

6.4.5 Software methods

6.4.5.1 "Bitwise processing – a paradigm for deriving parallel algorithms"[1]

An application of bytewise processing was introduced by L. Lamport in 1975, thus establishing a general approach whereby one regards the bits, bytes, or other subfields within a computer word as elements of an array of independent microprocessors. These microprocessors work on their subfields, synchronized through shift instructions and carry bits [23]. The idea was developed further by many programmers. Bitwise processing was specifically proposed for parallel processing in 1997 in a paper called "Bitwise processing – a paradigm for deriving parallel algorithms" [24].

We look at the idea behind this approach and consider a front-end supply baseband processor with a stream of 2-bit (for example) values. In a software general processor we are forced to represent these values by integers, floats, or even doubles, which have widths of 16, 32, or 64 bits. First, it is a waste of computer memory, because, for example, a 32 bit field is used to store 2 bits of data, therefore for this example we can reduce the size of buffers by a factor of 16. Second, if we can fit our 2-bit values into larger containers, such as doubles, and then process these containers instead of the separate values, we can reduce the number of operations by a factor of 16 as well.

Knuth described some algorithms in which we can work with bits in detail [23]. With respect to working with these bits, bytes, or other subfields within a computer word, Knuth refers to multibyte techniques, introduced by Leslie Lamport in 1975, and generalizes by combining with bitwise parallel processing.

[1] The section title quotes W. Koch's 1997 paper title [24].

The bitwise parallel processing technique was first applied to processing GPS signals by Ledvina [25]. The algorithms were further developed in [7],[26], and [27].

A disadvantage of this approach is that the signal processing algorithms become more opaque and less suitable for educational purposes.

The other disadvantage of these algorithms is that they may depend on processor architecture, i.e. have word length hardcoded into the algorithms. For example, the algorithms developed for a 32-bit processor will work on a 64-bit processor, but not necessarily vice versa. In order to port the general processor software to an embedded 16-bit or 8-bit processor, the algorithms may have to be rewritten. It is also much more difficult to modify them if needed.

6.4.5.2 Pre-calculation of replicas

If the bottleneck in the correlator subroutine is resolved, we may encounter the next time-consuming part of the program, which is the creation of code and carrier replicas for a specific code phase and Doppler shift frequency. During its operation, a receiver creates replicas for the code and carrier for the acquisition and tracking units. When the search moves to the next cell with a different frequency bin, the replica for the carrier must be created again. It would lessen the payload on a processor if replicas were calculated in advance and kept in memory. These replicas could also be used during tracking.

A code replica is mapped on the samples, kept in memory, and can be easily modified by shifting it in a circular buffer. However, the carrier replica is created for a specific Doppler shift. In tracking, frequency shifts are small and the carrier is recalculated each millisecond for each new frequency. Though it is possible to create a database and keep the replicas in a memory, this approach would create a trade-off, not only between speed and memory, but also between the speed of creating the replica and the speed of getting it from the data base. There also exist algorithms for more efficient generation of replicas by re-sampling [7].

A disadvantage of these methods is that they lead to excessive memory use if more than one frequency is used. It is even more difficult to implement for FDMA-based GLONASS signals, again due to the necessity to calculate these replicas for each bin in 14 frequency bands. The methods, however, may be realizable for a snapshot receiver, where only acquisition is required. This is specifically because the number of frequency bins in the case of acquisition is much less than required for tracking.

6.5 Applications of high-end real-time software receivers

In this section we look at the high-end real-time software receiver developed by iP-Solutions as an example of the high-end software receiver on a general processor and examine the specific application fields where it is implemented. The specifics of the receiver are discussed in [8], which also features a bundled academic version of the receiver.

6.5.1 Instant positioning

Software is an ideal tool for the development of algorithms and methods for a receiver. It allows a significant simplification of the stages of the receiver life cycle, which we discuss in detail in Chapter 11. Also, it reduces time to manufacturing by minimizing iterations at the research and development stage, and assists in making the design and validation stages risk free.

Indeed, a developer can have instant access to the test results of his or her ideas when they are implemented through software modifications, usually on high-level programming languages, such as those of the C-family. These tasks are much more expanded in time and much more complicated for a conventional receiver. Even in comparison with hardware description programming languages, which can be used in many cases for conventional receiver modifications, such as VHDL or Verilog, software programming languages have huge advantages in terms of simplicity of implementation due to sequential algorithm logic and ease of debugging.

iPRx was initially developed to test instant positioning algorithms, which are described in Chapter 9. At first it was a "lazy developer" solution for a receiver because the algorithms allow one to use predicted ephemerides and to resolve code phase ambiguities without a navigation message. That permitted one to make a positioning without tracking, using code phase data from acquisition only.

The iPRx implementations for other tasks have a completely different structure. The software implementation allows for much easier structural modifications, with re-use of large parts of the program.

6.5.2 Ionosphere monitoring

The iPRx receiver has been changed drastically on a structural level a few times, first to include tracking and concurrent processing, then to meet top-level specification requirements for geophysical applications. The result was a new Adaptive Receiver Applied for Monitoring of Ionospheric Scintillation – ARAMIS™.

A software receiver is an indispensable tool for geophysical applications. It is discussed in detail in [8]; here we list just a few of its important features.

(1) The important signals can be recorded. Even a receiver specialized for ionospheric scintillation monitoring may lose tracking; then the data are lost. The most important ionospheric events occur with an 11-year cycle. For a software receiver, the recorded data can be stored and processed over and over again.

(2) A software receiver allows one to access the signal processing and tracking loops core to modify existing, or implement custom, processing algorithms and methods.

(3) The recorded signal can be studied and processed many times with various algorithms or with various parameters, which can be modified interactively based on the results of the study.

(4) The software receiver has all the features of the conventional ionospheric scintillation monitor receiver.

(5) The software receiver allows implementation of adaptive tracking algorithms.
(6) The software receiver features a high level of visualization for parameters of interest.

6.5.3 Ultra-tightly coupled integration with INS [11]

The ARAMIS™ receiver has been used for ultra-tightly coupled integration with the INS (see Tsuji *et al.* [11]). The data from the INS are used inside the receiver baseband processor to assist tracking loops, in particular to remove the effects from vehicle dynamics. The satellite signal is transmitted on a certain frequency. This frequency is shifted from the expected amount by the Doppler shift, which is caused by satellite movement, vehicle movement, and satellite and receiver clock instability. The Doppler shift from satellite movement can be accounted for by using known ephemerides, and the receiver clock error can be minimized by means of a high-quality clock, such as OCXO. The only uncompensated part of the Doppler shift is due to the rover dynamics, which can be compensated for with information from the INS. The complete removal of all Doppler shifts allows us to narrow down the tracking loop bandwidth, which in turn allows for more accurate tracking. It also facilitates coherent tracking with a long coherency interval, which allows for higher sensitivity and reliability of tracking.

Access to its baseband processor allowed INS specialists to implement the iPRx receiver for ultra-tight INS integration with effort spent only on the integration part. The result was that ARAMIS™ was the only receiver that could function as an airborne ionospheric scintillation monitor. The ultra-tight integration with INS also allowed the realization of coherent tracking with a long coherency interval for dynamic vehicles.

6.5.4 Application in education

A software receiver is an indispensable tool for education for a few reasons. First, it brings receiver technology to anyone's desk, and allows anyone to start working with a receiver immediately. Second, a software receiver may serve as a tool for people who study subjects other than positioning or navigation, such as geophysics or geodesy. Finally, it is a great demonstration and visualization aid, which makes the study more interesting, more alive.

A software receiver is available for free in source code in MATLAB (Borre [13], Pany [7], and Johnson *et al.* [17]), on the open-source Scilab platform (Gavrilov [14]), and in a real-time version (Petrovski and Tsujii [8]).

References

[1] J. Mitola III, The software radio architecture, IEEE Commun. Mag., pp. 26–38, May 1995.
[2] D. Akos, A software radio approach to global navigation satellite system receiver design. Unpublished Ph.D. dissertation, Ohio University, Athens, OH, 1997.

[3] J. Tsui, *Fundamentals of Global Positioning System Receivers: A Software Approach*. Hoboken, NJ: Wiley, 2005.
[4] S. C. Wu, W. I. Bertiger, D. Kuang *et al.*, MicroGPS for low-cost orbit determination, JPL, Pasadena, CA, TDA Prog. Rep. 42–131, Nov. 15, 1997.
[5] I. Petrovski, K. Okano, and H. Torimoto, Application of pseudolites for high accuracy positioning in ITS, robotics and satellite navigation system test beds, in *Proc. European Navigation Conf., ENC-2006*, Royal Institute of Navigation, Manchester, UK, 2006.
[6] *GPS World*, vol.18, no. 2, p. 24, Feb. 2007.
[7] T. Pany, *Navigation Signal Processing for GNSS Software Receivers*. Boston, MA: Artech House, 2010.
[8] I. Petrovski and T. Tsujii, *Digital Satellite Navigation and Geophysics: A Practical Guide with GNSS Signal Simulator and Receiver Laboratory*. Cambridge: Cambridge University Press, 2012.
[9] T. Tsujii, T. Fujiwara, and T. Kubota, Flight test evaluation of INS-aided GPS tracking performance under equatorial ionospheric plasma bubbles, in *Proc. ION Pacific PNT*, Honolulu, HI, Apr. 22–25, 2013.
[10] P. Hoai, P. Supnithi, and T. Tsujii, Ionospheric scintillation monitoring using software GPS receiver at Chumphon Station, Thailand, in *Int. Tech. Conf. Circuits/Systems, Computers and Communications 2012 (ITC-CSCC 2012)*, Hokkaido, Japan, July 2012.
[11] T. Tsujii, T. Fujiwara, T. Kubota *et al.* , Measurement and simulation of equatorial ionospheric plasma bubbles to assess their impact on GNSS performance, *J. Korean Soc. Survey., Geodesy, Photogram. Cartog.*, vol. 30, no. 6–2, 2012.
[12] K. Krumvieda, P. Madhani, C. Cloman *et al.*, A complete IF software GPS receiver: a tutorial about the details, in *Proc. 14th Int. Tech. Meeting of the Satellite Division of The Institute of Navigation (ION GPS 2001)*, Salt Lake City, UT, Sept. 2001, pp. 789–829.
[13] K. Borre, D. Akos, N. Bertelsen, P. Rinder, and S. Jensen, *A Software-Defined GPS and Galileo Receiver: A Single Frequency Approach*. Boston, MA: Birkhäuser, 2007.
[14] A. Gavrilov, GLONASS software receiver, *J. Engineer*, vol. 9, Sept. 2012 (in Russian). Available at http://engbul.bmstu.ru/file/505590.html?__s=1 (last accessed Oct. 6, 2013).
[15] D. Akopian, A fast satellite acquisition method, in *Proc. 14th Int. Tech. Meeting of the Satellite Division of the Institute of Navigation (ION-GPS)*, Salt Lake City, UT, Sept. 11–14, 2001, pp. 2871–2881.
[16] I. Petrovski, B. Townsend, and T. Ebinuma, Testing multi-GNSS equipment, systems, simulators and the production pyramid, *Inside GNSS Mag.*, July–Aug., 2010.
[17] C. R. Johnson, Jr., W. A. Sethares, and A. G. Klein, *Software Receiver Design: Build Your Own Digital Communication System in Five Easy Steps*. Cambridge: Cambridge University Press, 2011.
[18] S. Gleason, M. Quigley, and P. Abbeel, A GPS software receiver, in *GNSS Applications and Methods*, S. Gleason and D. Gebre-Egziabher, Eds. Boston, MA: Artech House, 2009.
[19] G. Heckler and J. Garrison, SIMD correlator library for GNSS software receivers, *GPS Solutions*, vol. 10, no. 4, pp. 269–276, Nov. 2006.
[20] I. Buck, Stream computing on graphics hardware, Thesis, Stanford University, CA, Sept. 2006. Available at http://graphics.stanford.edu/~ianbuck/thesis.pdf (last accessed Oct. 6, 2013).
[21] C. Harris, K. Haines, and L. Staveley-Smith, GPU accelerated radio astronomy signal convolution, *Exp. Astron.*, vol. 22, no. 1–2, pp. 129–141, 2008.

[22] T. Hobiger, T. Gotoh, J. Amagai, Y. Koyama, and T. Kondo, A GPU based real-time GPS software receiver, *GPS Sol.*, vol. 14, no. 2, pp. 207–216, 2010.
[23] W. Koch, Bitwise processing – a paradigm for deriving parallel algorithms, in *Parallel Computing Technologies*, V. Malyshkin, Ed. Berlin: Springer, 1997.
[24] D. Knuth, *The Art of Computer Programming, Vol. 4A. Combinatorial Algorithms, Part 1*. Reading, MA: Addison-Wesley, 2011.
[25] B. M. Ledvina, M. L. Psiaki, S. P. Powell, and P. M. Kintner, A 12-channel real-time GPS L1 software receiver, in *Proc. Institute of Navigation National Tech. Meeting*, Anaheim, CA, Jan. 22–24, 2003, pp. 767–782.
[26] B. M. Ledvina, M. L. Psiaki, S. P. Powell, and P. M. Kintner, Bit-wise parallel algorithms for efficient software correlation applied to a GPS software receiver, *IEEE Trans. Wireless Commun.*, vol. 3, no. 5, Sept. 2004.
[27] J. Deng, R. Chen, and J. Wang, An enhanced bit-wise parallel algorithm for real-time GPS software receiver, *GPS Solutions*, vol. 14, pp. 133–139, 2010.

7 Common approach and common components

7.1 Common approach for receiver design

In Chapter 6 we discussed a "software approach" to GNSS receivers in comparison with a traditional "conventional" approach. In this chapter we consider a common approach taken in the implementation of a receiver, or more precisely in the implementation of a baseband processor. The "common components" refers then to the realization of those components, that are independent of the "conventional" or "software" approach. For example, a realization of an RF front end does not depend on whether a receiver is implemented with either a "conventional" or "software" approach in mind. The same is true for a navigation processor, as its implementation is always via software.

The difference between the "conventional" and the "software" approaches is that in the latter case the baseband processor is implemented in software on a general processor. This led to a definition of the "GPS software receiver," which is a more narrow definition than software-defined radio (SDR). We discuss the details in Chapter 10, Section 10.3. The software approach [1] effectively allowed researchers in the GPS field to look at the receiver from a different perspective. It was essential in order to change the approach to receiver development. It has also introduced a set of new (for this area) instruments, in particular fast Fourier transform.

This new approach has led to the development of a whole new field, which has now progressed as far as development for mass market. It also had an enormous impact on GNSS education. Every student studying in this field today is familiar with the software receiver concept. In order to distinguish between conventional and software receivers, a new term, the "hardware approach" has become useful. In particular, this has been used to describe algorithms and methods that have been developed for conventional receivers [2].

It is, however, important to understand that the difference between software and hardware receivers is in implementation only. The algorithms and methods which were developed for conventional receivers can all be successfully implemented in software receivers. The explanation for this is that all algorithms and methods for conventional receivers within a conventional approach are programmed into an embedded processor or an FPGA, which further could possibly be converted into ASIC. Therefore these algorithms can initially be described in programming language, such as VHDL or Verilog, for example.

Consequently, all these algorithms can also be programmed using a more flexible and powerful general processor environment, and languages such as C, C++, and C#.

The reason why all GNSS algorithms and methods work for the software implementation is because GPS receivers were originally designed with digital processing in mind, and basically all GNSS receivers today use digital signal processing. The last important advantage of the hardware approach is in its capability of massive parallel processing. Today this difference is diminishing with the development of massive parallel processing technology using GPU (see Chapter 6 for details).

Conversely, any algorithm that one can develop for a software receiver can be translated into hardware. Let's take for instance an acquisition algorithm based on FFT, which is the conventional algorithm for a software receiver to increase speed of acquisition. Today, FFT is routinely and very successfully implemented in FPGA.

In terms of programming, any software program written on C can be automatically translated into VHDL or Verilog languages for programming FPGA by special compilers. Furthermore, any FPGA can then be translated into ASIC. So we can see that any algorithms and methods in software can be transformed to ASIC, and any algorithms and methods developed for hardware can be implemented in the software, in most cases without losing their advantages. This gives one ground to consider a common approach to receiver design. The same algorithms and methods can be implemented in both software receivers and conventional receivers. The only difference will be in the implementation of the baseband processor; other components (the RF front end and the navigation processor) for both types of receivers are basically the same. Moreover, as we discussed in Chapter 6, the navigation processor is not necessarily part of a receiver. The receiver per se can be described as a front end plus a baseband processor.

In existing applications, one uses in fact the same front end for both types of receivers. For the software receiver, the front-end module must be added using an interface, which connects it to the host PC. Petrovski and Tsuji [3] describe the USB front end for a software receiver in detail.

This equivalence of "software" and "conventional" approaches is related to design, but not functionality, especially for educational and high-end geophysical applications. For example, for education, the software approach provides the enormous advantage of being able to observe directly GNSS signal processing in a baseband processor using a PC. This is not possible for conventional receivers, for which access to the inside of the receiver is limited by the navigation processor. Students could obtain raw data from a conventional receiver and then play around with positioning algorithms, studying and modifying them; for example, Strang and Borre [4] derive navigation algorithms and give examples in source code in MATLAB. This is possible because, as previously discussed, the navigation processor is, strictly speaking, outside the receiver.

Another example of where software receiver functionality makes a difference is in geophysical applications. In Chapter 6, we looked at iP-Solutions' ARAMIS^{TM} (adaptive receiver applied for monitoring ionospheric scintillation). One of the advantages of ARAMIS^{TM} is that the important signals can be recorded and then studied and processed many times with various algorithms. This advantage is also related to the fact that the baseband processor is open for programming.

7.2 Mobile antennas

The RF front end is connected to an antenna, which is responsible for converting electromagnetic waves from a GNSS satellite into an electrical signal.

Just a few words about semantics here. For high-accuracy applications, we usually speak of defining the coordinates of the antenna rather than those of a rover receiver or user. Even more precisely, we define the position of the antenna phase center. Proper antenna calibration is required to achieve millimeter-level accuracy with expensive geodetic antennas. For some mobile applications it may not be necessary to distinguish between the antenna, the rover receiver, or the user if the position accuracy is on the order of meters or less.

As an antenna connects hardware to the physical world, the impedance of the hardware elements should match that of the physical world. Impedance mismatch results in loss of signal power, which may result in greater power consumption, accuracy degradation, and loss of sensitivity, and, in some cases, to total operational failure. The physical world can be modeled by a radio signal propagating in free space. The impedance of the medium to the electromagnetic wave is calculated as the ratio of the electric field to the magnetic field, i.e.

$$Z \equiv \frac{E}{H} = \sqrt{\frac{\mu}{\varepsilon}}. \tag{7.1}$$

In free space, the radio-wave impedance Z_0 is therefore about 377 Ω (ohm). The impedance in the medium is a function of the refractive index:

$$Z = \frac{Z_0}{n}. \tag{7.2}$$

The requirement for most RF hardware circuits in the GNSS area is to have an impedance value equal to 50 Ω. Coaxial cables usually also have an impedance of 50 Ω. The 50 Ω standard was chosen in the 1930s as a compromise solution between 30 Ω (best power handling) and 77 Ω (lowest loss) for coaxial cables. The antenna must convert the radio signal impedance to a value of about 50 Ω, which can be easily matched to a front-end input or cable leading to a front end.

An omni-directional antenna transmits a signal that is equally distributed over all directions. Signal power is equally distributed over a sphere. Power per surface unit can be expressed as

$$P_0 = \frac{W_T}{4\pi r^2}, \tag{7.3}$$

where W_T is the transmitting antenna power, and r is the distance from the transmitting antenna phase center to a point at which the power is measured. Because of this distribution over the spherical surface, the dependence on distance for electromagnetic wave power follows the inverse square law.

An electric field induces a current in a receiving antenna. The current goes through the cable to the front end. The antenna aperture (or effective area) is defined as the ratio of antenna-produced power P to the power of received signal W_R, i.e.

$$S_A = \frac{P}{W_R}. \tag{7.4}$$

The antenna gain is defined by antenna aperture:

$$G_A = \frac{4\pi S_A}{\lambda^2}, \tag{7.5}$$

where G_A is the antenna gain, expressed as a ratio, S_A is the antenna effective area, and λ is the radio-wave wavelength. The antenna gain is therefore defined by how many squared wavelengths fit into the antenna effective area. If the signal frequency is changed, the aperture should also be changed in order to maintain the same gain.

The antenna aperture (effective area) can, as a rule of thumb, be estimated as one-half the physical antenna area:

$$S_A \approx 0.5 S_G, \tag{7.6}$$

where S_G is the physical (or geometrical) antenna area.

A directional antenna provides a gain in accordance with its antenna pattern. The antenna gain pattern defines the signal power transmitted by an antenna as a function of direction. The GNSS satellite transmitting antennas are directional with a narrow antenna pattern. Because the angle in which the satellite antenna is radiating is reduced, the directive gain goes up. For a hemispheric pattern, all energy will be concentrated in half of the original area. Therefore

$$P_0 = \frac{2 \cdot W_T}{4\pi r^2}. \tag{7.7}$$

This gives an extra 3 dB gain.

Receiver antennas have a hemispheric antenna pattern. Antenna gain is usually specified by its value in the direction of maximum gain in its pattern profile. The power generated in the receiving antenna can be defined as

$$P = \frac{W_T S_A}{4\pi r^2}. \tag{7.8}$$

The power produced by the antenna can be expressed via the current induced in the antenna (I) and the radiation resistance (R) as follows:

$$P = \frac{1}{2} R_{RAY} I^2. \tag{7.9}$$

For a simple dipole antenna, the radiation resistance can be expressed as follows [5]:

$$R_{RAY} = \frac{2\pi}{3} \sqrt{\frac{\mu_0}{\varepsilon_0}} \left(\frac{l}{\lambda}\right)^2, \tag{7.10}$$

where l is the length of the dipole. From the above equations we can see that small antennas operate less effectively, and that antennas of size on the order of the wavelength have better characteristics. If an antenna has a length of $\lambda/2$, a stationary current can be established in the antenna. When the length of antenna is $\lambda/4$, the antenna acts as a half-wave antenna, as it generates a symmetrical image in a conductor plane.

A GPS signal is right-hand circularly polarized (RHCP). An electric-field component of the electromagnetic wave is oriented perpendicular to the wave's path. When the signal is linearly polarized it means that electric-field component oscillates on the same plane in the same direction consistently. When the signal is circularly polarized, the electric-field component is rotating in the plane oriented perpendicular to the wave's path. A GPS transmitting antenna is polarized. To provide optimal reception, a receiving antenna should have the same type of polarization as the transmitting antenna.

For mobile applications, this feature of GNSS signals provides us with the means to mitigate multipath effects that lead to errors which are often dominant in a mobile device's error budget. A reflected GNSS signal changes polarization, and therefore becomes attenuated by an antenna with a different polarization. Therefore, such polarization allows a GNSS signal to be less affected by multipath effects. The second reflection, however, restores the original signal polarization, and a twice-reflected signal is not attenuated by the antenna.

Antenna designs for GNSS vary from thin microstrip patch antennas to large multipath-mitigating helical coils. Different applications place different requirements on antenna design. When choosing an antenna for a particular application, one should look at its gain pattern and multipath-mitigation characteristics. If carrier phase measurements are required, such as for reference stations or mobile devices capable of RTK positioning, the stability of the antenna phase center becomes important.

A simple patch antenna is presented in Figure 7.1. It consists of a very thin metal patch on a dielectric substrate, which is set on top of a preferably large metal ground plane. The height of the dielectric layer is usually chosen to be

$$h \leq 0.02\,\lambda_d, \tag{7.11}$$

where λ_d is the signal wavelength in the dielectric [6].

If we remove the ground plane, the patch antenna will operate as a simple dipole antenna. The ground plane doubles the antenna gain in the same way as described above for any directional antenna. The length of the patch (L) should optimally be equal to half the wavelength. The width (W) is less important; it can be chosen from the signal wavelength and dielectric parameters as follows [6]:

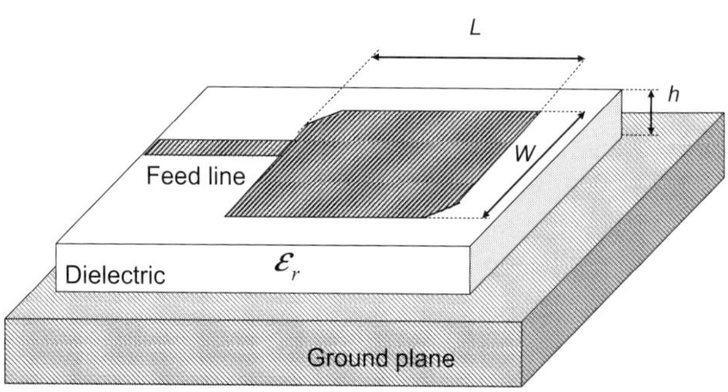

Figure 7.1 Patch antenna. After [3].

$$W = \frac{\lambda_0}{2}\left(\frac{\varepsilon_r + 1}{2}\right)^{-1/2}, \tag{7.12}$$

where λ_0 is the signal wavelength and ε_r is the dielectric constant of the dielectric substrate.

The patch length (L) for the specific dielectric can be calculated as follows [6]:

$$L = \frac{\lambda_d}{2} - 2\Delta l = \frac{\lambda_0}{2\sqrt{\varepsilon_{eff}}} - 2\Delta l, \tag{7.13}$$

where ε_{eff} is the effective dielectric constant and Δl is an edge extension correction term; i.e.

$$\varepsilon_{eff} = \frac{\varepsilon_r + 1}{2} + \frac{\varepsilon_r - 1}{2}\left(1 + \frac{12h}{w}\right)^{-1/2} \tag{7.14}$$

and

$$\Delta l = 0.412h\left(\frac{\varepsilon_{eff} + 0.3}{\varepsilon_{eff} - 0.258}\right)\frac{\frac{W}{h} + 0.264}{\frac{W}{h} + 0.8}. \tag{7.15}$$

The frequency of the induced wave will be higher than that in free space because of the dielectric. If a ceramic with a high dielectric constant is used for loading, the size of the patch antenna can be significantly reduced, to fit for example into cellular phones [7]. Further minimization can be achieved by using quarter-wave antennas. In comparison with a dipole antenna, a quarter-wave antenna's ground plane replaces the half-wave dipole null potential. In order to produce and receive circularly polarized signals, the antenna either has two feeds or has a rectangular (instead of square) shape, with one or two of its corners clipped.

An antenna is designed to work at a specific frequency. Therefore it can also be modeled as a pass-band filter. A GPS antenna usually has a bandwidth of about 2% of the signal center frequency. Therefore bandwidths for L1, L2, and L5 antennas are about 31.5 MHz, 24.6 MHz, and 23.5 MHz, respectively.

7.3 TCXO and bandwidth

A front-end clock facilitates the conversion of an RF to a DIF signal. Via the quality of this DIF signal, the clock's parameters therefore affect the baseband processor. They do not, however, affect the navigation processor directly. The signal replicas generated in the receiver, both for carrier and spread code, are not affected by this clock drift; but the drift *does* affect the carrier and spread code of the *incoming* signal. This difference affects acquisition and tracking, and may result in decreased signal acquisition capabilities and reduced accuracy of the tracking loops.

The receiver clock error comes from the receiver's internal time keeping. For a software receiver, this time is set initially to the time in the navigation processor. For receivers working with PCs, this time mark came from the host PC. If the receiver is operating in real-time mode, the receiver clock is set to the PC clock. If the receiver is operating in post-processing mode, the receiver clock is set to the time of the beginning of the data recording. This initial setup is, however, not necessary. It is only used in order to assist in acquisition and positioning.

After the receiver has acquired the signal (in either real-time or post-processing mode), the initial time is set to the time mark provided in the navigation message. After that, the time is kept by dead reckoning applied to the acquired signal code sequence. Therefore, the front-end clock may affect the signal quality but not the time keeping, because the time keeping essentially comes from the GNSS satellite.

The quality of the front-end clock affects the performance of a baseband processor. A front-end clock can usually be represented by a phase-locked loop (PLL) and an oscillator. A PLL is used in the front-end clock circuitry for two main reasons: (i) to generate frequencies other than that generated by an oscillator; (ii) to clean up noise from the noise frequency by removing short-term phase variations. Off-the-shelf GNSS modules provide control over the PLL, allowing it to be tuned to a specific user's requirements, for example oscillator frequency and hence the sampling rate.

The most simple type of oscillator is the voltage-controlled crystal oscillator (VCXO), which is stable over a range of ± 20 ppm. A simple model explaining the features of a VCXO can be constructed from a simple amplifier schematic [7]. A zero phase response of an open loop amplifier will be in close proximity to its oscillation frequency when it operates as an oscillator in a closed loop. Use of a variable capacitance (varicap) diode allows the creation of a simple VCXO from an amplifier (see Figure 7.2) by connecting its output to the input, as shown by the dotted line in the figure. The most commonly used clock in GNSS receivers is the temperature-compensated crystal oscillator (TCXO). A good-quality TCXO provides a user with up to 0.5 ppm stability and also provides low power consumption. A Rakon TCXO clock is shown as part of an iPRx receiver front end in Figure 7.3, as a stand-alone clock in Figure 7.3(a) and as part of front-end module in Figure 7.3(b).

For demanding applications, an oven-controlled crystal oscillator (OCXO) may be beneficial. In an OCXO an oscillator is contained inside a temperature-controlled enclosure, which maintains the crystal at a constant temperature, thus providing superior stability. An OCXO is essential for a number of applications in geophysics and aviation. For details on OCXO applications, a comparison with TCXO, and its effect on receiver performance, see Petrovski and Tsuji [3].

The specific requirements of a clock are defined by the GNSS signal. The minimum sampling frequency is defined by the Nyquist theorem, which states that the sampling frequency is defined by the signal bandwidth. For example, a GPS bandwidth is about 2 MHz; the corresponding minimum sampling frequency for this signal is 4 MHz. A lower sampling frequency may also work, though some signal information may be lost. The code bandwidth is defined by its chip rate. A narrow-band front end retains the part of the signal with spectrum in the main lobe only. We can narrow down the bandwidth further at the cost of signal degradation and, possibly, loss in positioning accuracy due to a degradation in autocorrelation function shape.

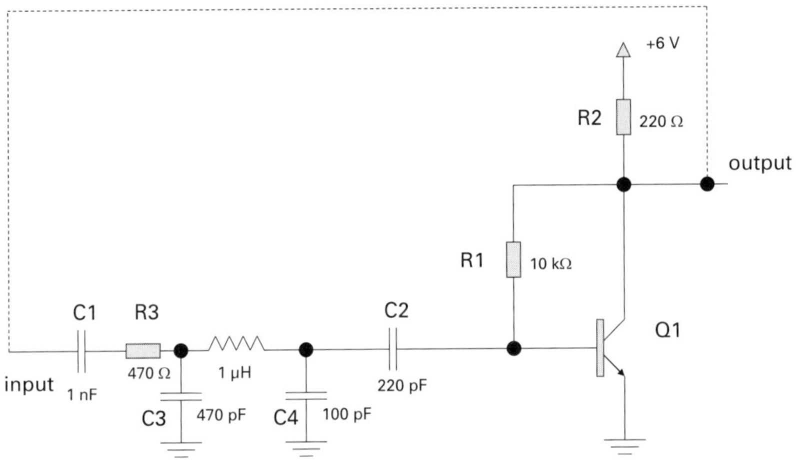

Figure 7.2 Creation of a VCXO from an amplifier. After [7].

(a) (b)

Figure 7.3 Rakon TCXO on iP-Solutions front end: (a) as standalone clock; (b) as part of front-end module.

Other higher-frequency components are removed from the signal as it passes through the front end. This process also removes out-of-band interference. However, due to an effective decrease in the sampling rate, it also decreases the resolution of the signal processing algorithms in the baseband processor. Wide-band front ends include several side lobes as well. This additional information can be useful for some applications, including multipath mitigation.

GPS L1C, L2C, and L5 front ends have their minimum bandwidths (4.092 MHz, 2.046 MHz, and 20.46 MHz, respectively) defined by the corresponding signal code design and chip rate.

The usual sampling frequency for GPS receivers is about 16 Mps (mega-samples per second). We need to note that samples per second and cycles per second (Hz) are different (see Figure 7.4 for an explanation).

The GLONASS L1 front-end bandwidth is defined not only by the GLONASS signal chip rate, but also by the frequency range, which contains L1 signals from all

Table 7.1. Achievable positioning accuracy for a receiver with 16 Mps sampling rate

	Direct measurements from the nearest sample	Code tracking		Carrier tracking
		GPS	GLONASS	
Achievable standalone positioning accuracy (RMS)	~ 25 m	~50 cm	~1 m	~10 cm

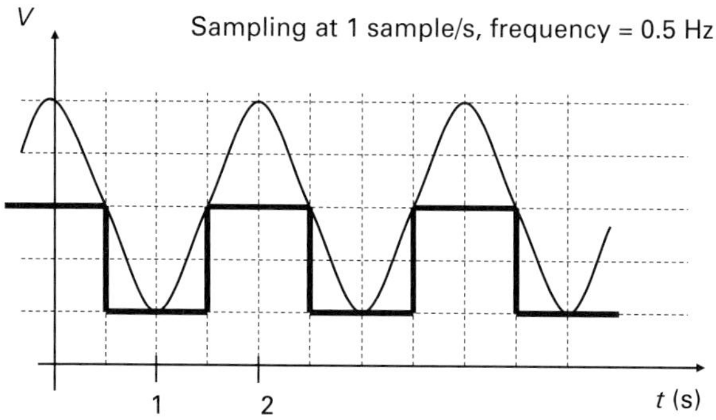

Figure 7.4 The difference between cycles per second and samples per second.

GLONASS satellites. Therefore a GLONASS L1 front end should have a minimum bandwidth of about 8 MHz. GPS and GLONASS receivers may work with the same TCXO, whereas GLONASS signals are sampled on both edges of clock, resulting in a 32 Mps effective sampling rate.

There is a trade-off between accuracy and processor load. Low-end GPS L1 receivers can work with a front end that provides narrower bandwidth, for example 1.8 MHz. Narrowing a bandwidth for such applications allows a baseband processor to operate at a lower sampling rate.

The sampling frequency ultimately defines the computational load on a processor in the case of a software receiver, and specifies requirements to the digital hardware in the case of a conventional receiver. Sometimes this leads to a trade-off between increasing power consumption, weight, and price (caused by an increased sampling rate), and decreasing accuracy (caused by a decreased sampling rate). This issue is more important for GLONASS rather than GPS due to GLONASS's wider overall bandwidth.

The misconception about the effect that the front-end bandwidth and the sampling rate have on positioning accuracy is very common, so let's consider an example. A typical GPS receiver analog-to-digital converter (ADC) has a sampling rate of 16 Mps. This yields a distance of about 19 m between samples. However, the accuracy with which we estimate the code phase is limited mostly by the code phase noise on the centimeter scale. As soon as a receiver has a carrier tracking loop, the achievable positioning accuracy is defined by carrier phase measurements with a possible accuracy in millimeters (see Table 7.1). The length of the carrier wave for L1 is about 19 cm, so most of the carrier measurements fall between the samples, but the receiver can restore them completely if the Nyquist criterion is satisfied.

7.4 Front end

There are a number of front-end chips available on the market for the L1 frequency, among them SiGE SE4162T 4110L, MAXIM MAX2769, Atmel ATR0601, ST STA5620, Nemerix NJ1006, and Texas Instruments TRF5101.

It is possible to use an L1 front end for GNSS signals on other frequencies by down-converting and filtering the incoming signal, providing that the front-end bandwidth fits the signal bandwidth.

An example of a USB front end for a software receiver based on the Rakon front-end module is considered in detail in [3].

Mobile devices usually have a low-noise amplifier (LNA) integrated in an RF front-end solution. For example, U-blox and MAXIM chips have an integrated LNA. However, for high-end receivers, the LNA is usually separated from the RF chip. Surface acoustic wave (SAW) filters are usually located outside the chip. Figure 7.5 shows such a filter located outside the MAXIM front-end chipset after LNA and before the mixer, which both are inside the chip.

7.4.1 Down-converter

The down-conversion process shifts the spectrum of the signal along the frequency axis. A signal from a particular satellite does not have its center at the signal center frequency; the LOS projection of the satellite velocity can be up to 800 m/s. The received signal frequency will be increased by the Doppler effect caused by this motion

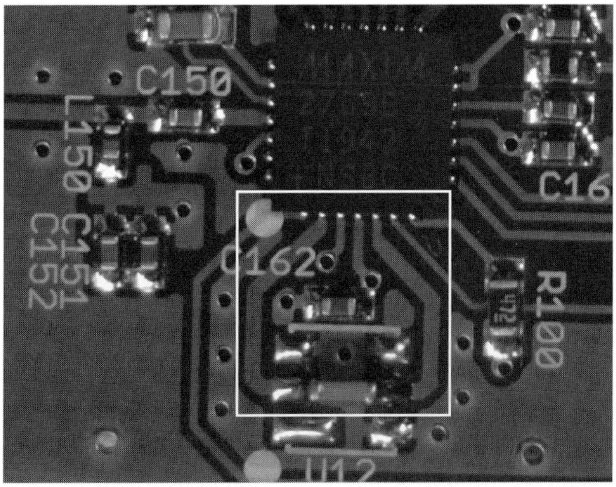

Figure 7.5 RF L1 filter located outside of MAXIM front-end chip.

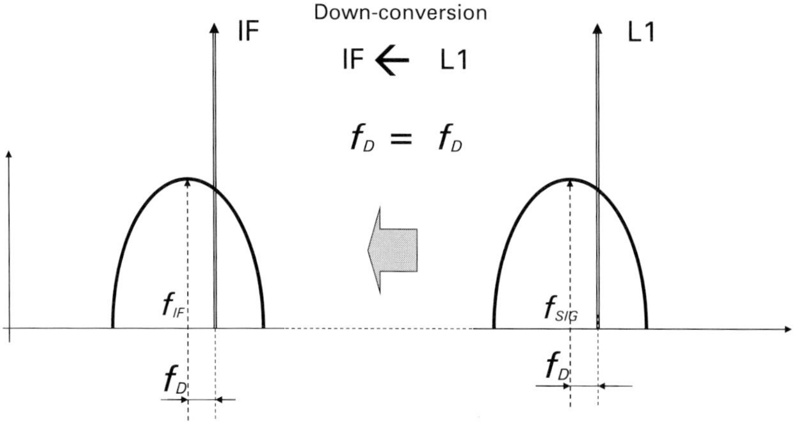

Figure 7.6 GPS L1 signal with Doppler shift.

if the receiver and satellite are converging, and decreased if they are moving apart. So we have

$$f_R = f_T - f_T \frac{v_{LOS}}{c}, \tag{7.16}$$

where f_R is the received signal frequency, f_T is the transmitted signal frequency, c is the speed of light, and v_{LOS} is the relative velocity between the satellite and the receiver along the line-of-sight. The Doppler shift is within 6 kHz for a low dynamic vehicle. The value of the Doppler shift is not changed by the down-conversion process (see Figure 7.6).

A mixer has a received RF signal and a low-frequency signal from a local oscillator (LO) on its input. A signal on the mixer output is the sum of two harmonics, one with frequency equal to the difference and another with frequency equal to the sum of the input signal frequencies. After filtering the upper signal out, we will have only one harmonic with the following frequency:

$$f_{IF} = f_R - f_{LO}. \tag{7.17}$$

Therefore, the IF of the received signal is the sum of the IF of the transmitted frequency and the Doppler frequency:

$$f_{IF} = f_{IFT} + f_D. \tag{7.18}$$

Let us look at this in detail, following [8]. In the circuit implementation, the product on the mixer output is represented by some complicated waveform, with main frequency described by (7.17):

$$x_{IF} = x_{RF} \cdot x_{LO} = \sin(\omega_{RF} t) \cdot x_{LO}. \tag{7.19}$$

A mixer can be constructed from diodes. The LO signal is large enough to control the diodes, which switch on and off depending on the sign of the LO wave. When the diode is off, the RF is not passed. As the result, the signal on the mixer output can be seen as a product of the incoming harmonic signal and a square wave with chip rate equal to double the LO frequency. The square wave can be expressed as a Fourier series as follows:

$$x_{LO} = \frac{4}{\pi}\left(\sin(\omega_{LO}t) - \frac{1}{3}\sin(3\omega_{LO}t) + \frac{1}{5}\sin(5\omega_{LO}t) - \cdots\right). \tag{7.20}$$

The other frequencies are filtered out by filters, and the output signal has an envelope with IF frequency defined by (7.17):

$$x_{IF} = \frac{2}{\pi}(\sin(\omega_{RF} + \omega_{LO})t + \sin(\omega_{RF} - \omega_{LO})t). \tag{7.21}$$

7.4.2 Analog-to-digital converter

At the final step, the IF signal should be digitized. Digitization comprises two processes: signal sampling and quantization. The sampling process of the band-limited analog signal in the receiver can be viewed as a multiplication of the incoming IF signal (after it is filtered by a band-pass or low-pass filter) by a periodic train of unit impulses. Figure 7.7 shows a spectrum representation of this process with sequential filtering.

The spectrum representation in Figure 7.7 can be expressed as a Fourier transform of the signal multiplication. The spectrum of the digitized DIF signal can be found via a Fourier transform of the DIF signal as follows:

$$X_d(f) = \sum_{n=-\infty}^{n=\infty} x_d(n)e^{-j2\pi fn}. \tag{7.22}$$

The resulting signal is a convolution of the IF signal spectrum and the impulse train spectrum and is expressed as follows

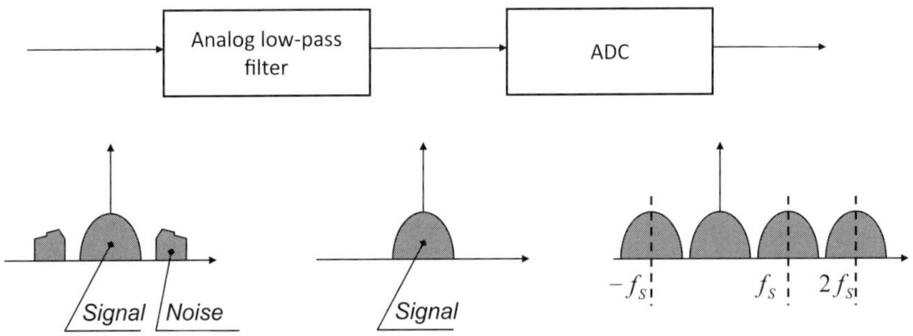

Figure 7.7 Spectrum representation of analog-to-digital conversion process with sequential filtering. After [9].

$$X_{DIF}(f) = \mathrm{F}\left[x(t)\sum_{n=-\infty}^{n=\infty}\delta(t-nT)\right] = X_{IF}(f)\otimes\left[\sum_{m=-\infty}^{m=\infty}\delta(f-mf_s)\right], \tag{7.23}$$

where X_{DIF} is the DIF signal spectrum on the ADC output, $x(t)$ is the analog IF signal on the ADC input, X_{IF} is the analog IF signal spectrum, T is the sampling period, f_s is the sampling frequency, and δ is the delta function. The resultant DIF signal has a spectrum consisting of repeated images of the spectrum of the analog IF signal (Figure 7.7). If the sampling frequency is smaller than the IF signal bandwidth, the spectrum lobes of the DIF images overlap, causing signal aliasing. The IF signal is digitized without loss of information only if this overlap does not occur. Then the signal can be restored from its spectrum via an inverse Fourier transform as follows:

$$x(n) = \int_{-1/2}^{1/2} X_d(f)e^{j2\pi fn}\,df. \tag{7.24}$$

The Nyquist requirements set to prevent signal alias define the minimum sampling frequency as follows:

$$f_N = 2\cdot B, \tag{7.25}$$

where B is the analog signal bandwidth. As we can see from (7.25), the Nyquist frequency can be defined by the signal bandwidth rather than the IF signal highest frequency, which is a sum of the central IF frequency and half the signal bandwidth. This is clear from the fact that the signal can be freely transformed along the frequency axis without distortion. Therefore, for the purpose of finding the Nyquist frequency without loss of generality we can consider an IF signal with zero central frequency. This Nyquist frequency sets the conditions at which we can restore the signal without losing information.

Regarding quantization, each sampled value can be presented by an N-bit word, which can be in one of 2^N states. Therefore, the analog IF signal can be represented by 2^N levels of the DIF signal. Most commercial receivers have 1- or 2-bit quantization. In particular, 1-bit quantization means that the analog signal is represented by two levels. In terms of hardware implementation, this means that one pin with two voltage states (high and low) is enough for the front-end output.

The frequency of the clock defines the sampling frequency, which is used to digitize an incoming baseband signal. The value of the signal is quantized using a certain number of levels. For conventional mobile receivers, 1- or 2-bit digitizing is enough; 1 bit means that there are two levels of signal, and 2-bit means four levels. A receiver can use 1-bit quantization without sacrificing the achievable accuracy. More than 2 bits may be required for receivers that have to deal with issues of interference or more generally with many signal sources in the same frequency range. In those cases, increasing the number of levels of quantization allows a distinction between either the signal and the interference signal or many signal sources.

7.5 Navigation processor

It was discussed in [3] that the navigation processor in modern applications can be placed outside the receiver and is no longer a necessary receiver component.

It is a requirement for mobile applications, in most cases, to output positioning information. In other cases, raw data from the baseband processor may be sent to a control center, which calculates mobile device coordinates. It may often be necessary to deliver some assistance information to the baseband processor and navigation processor if it is located on the mobile device.

The navigation processor can be embedded in a mobile device as a specialized processor or as software in the device general processor. The processor load and memory requirements for the navigation processor are relatively small. We consider the navigation processor regardless of its implementation as either a specialized or general processor in terms of its functions. The interface of the navigation processor on a mobile device may in general feature the following three information streams.

(1) Assist information from a data link, routed into the baseband processor. This is information used to assist satellite acquisition and tracking, if applicable.
(2) Another part of the assist information is required only for the navigation processor and can be separated from the baseband processor part. In a software receiver, which we considered in detail in Chapter 6, the baseband processor is realized in software and can also be placed in the device processor. In this case, the navigation processor can again be completely separated from the front end and the baseband processor, or it may even be placed outside the mobile receiver in the server or post-processing computer.
(3) A stream of output positioning data goes out of the navigation processor.

What kind of quality may we expect from the navigation processor? The navigation processor can be implemented in a handset or on a server. The power of the processing algorithm may vary as we go from simple to high-accuracy geodetic grade algorithms.

The main factor affecting the results is the quality of measurements delivered by the baseband processor. The baseband processor can be implemented on a hardware platform or on a processor. With advances in computer technology and signal processing, the difference between the software and hardware approaches is becoming superficial. If we look at the algorithms and methods developed for conventional receivers, we can hardly find one that cannot be successfully implemented in a software receiver.

There are, however, differences in implementation due to the capabilities of the hardware, although these differences continue to diminish as the OS of mobile devices and PCs converge rapidly.

The graphics capabilities of mobile devices are constantly increasing. As technology allows the use of GPU for GNSS signal processing, the capabilities of a software receiver move toward conventional implementation. Correspondingly, any method or algorithm developed for a software receiver can be programmed not only on a processor, but also on FPGA and consequently transferred to ASIC.

As a result, the achievable performance of a mobile device is ultimately defined by the antenna and the front end.

References

[1] J. Tsui, *Fundamentals of Global Positioning System Receivers: A Software Approach*. New York: John Wiley & Sons, 2000.

[2] D. Doberstein, *Fundamentals of GPS Receivers: A Hardware Approach*. Berlin: Springer Science+Business Media, 2012.

[3] I. Petrovski and T. Tsujii, *Digital Satellite Navigation and Geophysics: A Practical Guide with GNSS Signal Simulator and Receiver Laboratory*. Cambridge: Cambridge University Press, 2012.

[4] G. Strang and K. Borre, *Linear Algebra, Geodesy, and GPS*. Wellesley, MA: Wellesley-Cambridge Press, 1997.

[5] A. Moliton, *Basic Electromagnetism and Materials*. Berlin: Springer Science+Business Media, LLC, 2007.

[6] K. Chang, *RF and Microwave Wireless Systems*. New York: John Wiley & Sons, Inc., 2000.

[7] R. Lacoste, *Robert Lacoste's the Darker Side. Practical Applications for Electronic Design Concepts*. Burlington, MA: Elsevier Inc., 2010.

[8] A. Scott and R. Frobenius, *RF Measurements for Cellular Phones and Wireless Data Systems*. New York: John Wiley & Sons, Inc., 2008.

[9] R. G. Lyons, *Understanding Digital Signal Processing*, 3rd edn. Englewood Cliffs, NJ: Prentice Hall, 2011.

Part III

Mobile positioning at present and in the future

8 Positioning with data link: from AGPS[1] to RTK

8.1 Merging mobile and geodetic technologies

Positioning with data link is not the same as referenced positioning (a method that allows measurements from more than one receiver to be combined and processed together in order to enhance accuracy), which we have discussed in Chapter 4. In this chapter, we consider all possible external information that can be used to enhance receiver specification. This external information includes measurements from other receivers, but it also includes other information which can be used to improve not only accuracy, but also other parameters in the specification, such as TTFF and sensitivity.

It is very important for many applications to be able to provide instant positioning, i.e. to avoid the necessity of tracking a satellite signal and reading a navigation message. It takes up to 36 s to read a complete navigation message for a GPS L1 signal to ensure the decoding necessary for positioning data. If navigation message data are available through some other data link, it is still necessary to decode a time mark from the navigation message, which may require up to 6 s. BGPS (and AGPS before that) are very important for many applications because they allow instant positioning using just a snapshot of data.

It is often impossible to track a satellite signal indoors and therefore it becomes impossible to decode the navigation message, even partially. This is because moving indoors will change the multipath, and therefore the tracking will most likely be interrupted, even if it was possible in the first place.

When using the GPS function in cellular phone applications, the cellular phone cannot be used at the same time as GPS because of interference. This means that a user has to wait until all the data from the navigation message have been acquired.

In one sentence, the title of this chapter combines assisted GNSS (AGNSS) with carrier differential GNSS. The terms "carrier differential GNSS" and "AGNSS" came from different sides of the GNSS industry: the former has been developed as a precise positioning method in the satellite geodesy field, whereas the latter was developed much more recently for cellular phone applications. Both technologies use externally supplied information: in the first case, differential corrections to enhance accuracy, and,

[1] The correct term would be AGNSS, but AGPS is more traditional.

in the second, assist data, which are used to enhance the TTFF, sensitivity, and other parameters in the receiver specification.

Whether it is proper to combine these two technologies in one device depends on whether it would be feasible (i) to have one device that could benefit from both technologies and (ii) to provide all the information on the same data link.

This tendency to combine precise applications with mobile applications started some time ago [1]. In this chapter, we look at various information that is externally supplied in real time and discuss how this information is used in a receiver.

Usually, this subject is considered in relation to a specific field and therefore to a specific receiver type. The theory behind AGPS was first summarized and given to the scientific community by Frank van Diggelen, who authored various AGPS-related technologies, in Indoor GPS tutorials at ION GPS-2001. All this information has been extended and now comprises a textbook [2]. Reference [3] provides additional information on AGPS implementation.

In this chapter, we consider all this information as it is applied to the same receiver. This receiver has a front end, which can cost from $1 to $6 depending on type and quantity. The baseband and navigation processors can be implemented in general on a microprocessor, FPGA, ASIC, or a PC. As an example, we use an iPRx receiver, developed by the author, with an iP-Solutions USB front end. The software receiver and USB front end are described in detail in [4]. Both instant positioning and RTK tests were conducted on the same receiver. The receiver makes a position fix within 1 s using BGPS, which is discussed in detail in Chapter 9, after a *cool start*[2] with 25 m accuracy. After 36 s, during which time the receiver acquires the complete navigation message, the RTK positioning test yielded a few centimeters accuracy. The instant positioning and RTK tests were conducted with a software receiver using two different mobile solutions, a MAX2769 front-end chip and a Rakon GRM8650 front-end module. Both tests were successful, therefore we can conclude that *it is quite possible to combine AGNSS and DGNSS, even RTK functionality, in a mobile device*.

There are two different approaches taken for the application of external information. One concept is to use this information on the rover side. The other concept, which can be called *network-based* or sometimes *reversed positioning*, is to provide some data from a rover to the server and allow the server to calculate the rover's position. Reversed positioning allows us to implement geodetic techniques without load on a handset processor.

Reversed positioning can be used for cellular phones and also for fleet management, animal tracking systems, etc. It can be realized in two ways.

(1) A rover sends chunks of DIF records to a server, or keeps them in memory. We look at the latter approach in detail in Chapter 9. In this case, the rover receiver is only equipped with a front end, because all baseband functions are on the server side. This method is sometimes referred to as tracking with RF logs.

[2] The definition of cool start is given in Section 5.1.2.1.4.

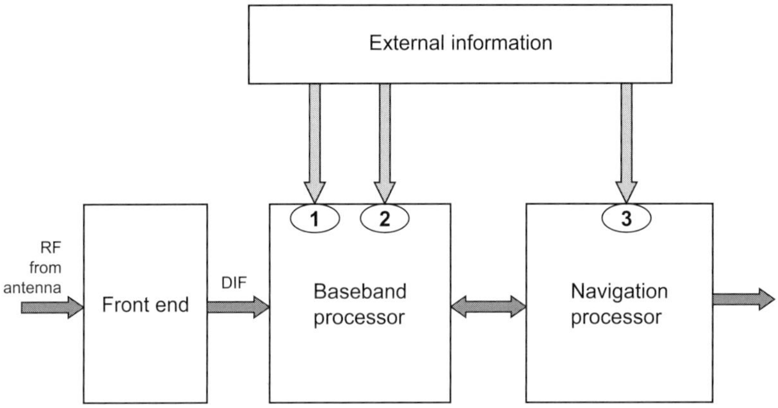

Figure 8.1 Receiver functional diagram.

(2) A rover sends to a server code phase measurements, which are processed on the server using all the available information at the server, such as navigation message and so on. This method is sometimes referred to as tracking with positioning logs. Note that in this case code phase measurement may not be converted to pseudoranges.

The rover estimate of the time at which it receives the signal is also required in both cases.

In rover-based positioning, the rover may also send some information to a server. For example, for open sky application in the case of VRS corrections, which we will consider later in this chapter, the rover may send its code-phase-based positioning. Using this position, the server calculates the carrier-phase corrections for this particular rover and sends them back. This allows an improvement in rover positioning accuracy from meters to centimeters.

Because we are using all the data within the same receiver, it is convenient to classify each area of external information by function. In this respect we look at the following external information for:

(1) the baseband processor, to assist in acquisition;
(2) the baseband processor, to assist in tracking;
(3) the navigation processor to assist in positioning.

See the functional receiver diagram, Figure 8.1.

8.2 Application of external information in the baseband processor

In Chapter 5 we discussed that acquisition needs to correlate the incoming signal with a replica created for a specific satellite, the Doppler frequency shift from the signal central frequency, and the code delay. This procedure is in fact a search procedure, which sweeps all possible Doppler frequencies and code delays in chips. The time required for acquisition depends on how large this search area is. It would be beneficial to put in

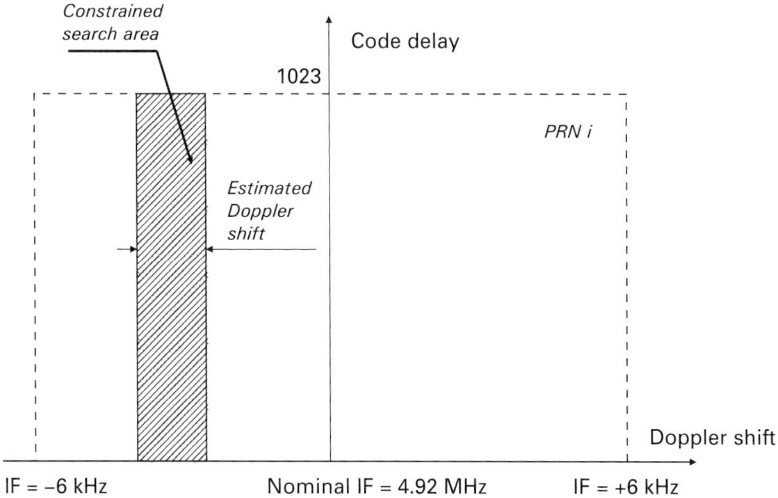

Figure 8.2 Acquisition area with Doppler constraints.

some constraints which would allow us to reduce this area and consequently the time required to conduct the search. In the case of parallel correlators realized on hardware, this would allow us to add more correlators to different parts of the same signal, such that their results can be summed up in order to provide higher sensitivity. For modern signals, the free correlators can be used to assist in the direct acquisition of tiered codes. Figure 8.2 portrays the idea of putting constraints onto a search area.

The receiver requirements define the design of the receiver. A need for ultra-high sensitivity in a mobile receiver would generally mean that massive parallel correlation is required in the baseband processor. If the baseband processor resides in the general processor rather than being implemented as a digital module, it can use, for example, FFT for acquisition. In that case, there would be a need for increased processor power. The sensitivity requirements define the length of the signal in samples to be processed and therefore the memory.

8.2.1 Doppler assistance in acquisition

A Doppler frequency shift appears in the satellite signal due to the relative movement between the satellite and the user.

To explain the Doppler effect, we can use the analogy of a swimmer in the sea. If the swimmer swims against the waves, the waves are encountered more frequently than if he swims along them. This means that, from the swimmer's point of view, the frequency of the wave becomes higher if he swims against the waves and lower if he swims along the line of the waves. The wave frequency in relation to the sea floor is midway between these two values. The difference in the frequency observed by the swimmer is the Doppler shift (Figure 8.3). The Doppler frequency shift can be calculated from satellite almanac or ephemeris data.

Figure 8.3 Doppler shift explained using the swimmer analogy. Based on artwork created by N. Petrovskaia.

To calculate the Doppler frequency shift, we need to know only the approximate rover position (the Doppler frequency will not change significantly as a function of the rover position error). The shift is calculated using a projection of the satellite and user velocities on a line-of-sight (LOS). The rule of thumb is that rover movement over 100 km distance would result in a change in satellite elevation in 1°, which would not affect the value of the projection significantly. This error in rover position may cause an error of 100 Hz in the Doppler frequency shift.

We also need to know the approximate time in order to calculate the satellite position. A GPS satellite velocity is approximately 4 km/s. An error in time of 1 s yields an error in angle proportional to ~4/20 000, and is completely negligible.

The Doppler frequency shift can be calculated by following steps (1)–(4).

(1) Estimate the satellite velocity from ephemerides. The algorithms are different for GLONASS than for the other GNSS because GLONASS uses tabular ephemerides. For GPS, Galileo, and BeiDou, the velocities can be found from the formulas in Keplerian parameters; for GLONASS, the velocity would be found by interpolation (see Chapter 1).

(2) Estimate the relative unit vector from the rover to a satellite.

(3) Estimate the projection of the velocity on the LOS velocity:

$$v_{LOS} = \left(v_x^{rov} - v_x^{sat}\right) \cdot u_x + \left(v_y^{rov} - v_y^{sat}\right) \cdot u_y + \left(v_z^{rov} - v_z^{sat}\right) \cdot u_z, \tag{8.1}$$

where $\vec{u}$ is the relative unit vector to the satellite, $\vec{v}^{rov}$ is the rover velocity, and $\vec{v}^{sat}$ is the satellite velocity.

(4) Finally the estimation of the Doppler shift is given by

$$D = f_{L1} \cdot \frac{v_{LOS}}{c}, \tag{8.2}$$

where f_{L1} is the L1 frequency and c is the speed of light.

When compared to the satellite dynamics, the user dynamics may be negligibly small in most cases. In cases when it cannot be neglected, for example for an airborne receiver, the receiver can be assisted by the inertial navigation system (INS).

This subject is beyond this book, and interested readers may consult Petrovski and Tsuji [4] for details about receiver INS integration and its methods and benefits, especially for aviation applications.

As a result of Doppler assistance, the acquisition search may be conducted for fewer Doppler bins, the number of which is defined by the error in Doppler compensation and unaccounted-for effects such as user dynamics, receiver clock error, and effects caused by signal propagation in the atmosphere.

Normally an acquisition search is conducted through all possible Doppler bins, for example 36 for GPS C/A. Using assist information such as predicted ephemeris, one can narrow down the search area from 36 bins to the three that are closest: the central bin, which contains the predicted Doppler frequency, and two adjusted bins.

The Doppler error coming from the satellite and receiver clocks can be expressed as follows:

$$D_{clock} = f_{L1} \cdot (e_{sat} + e_r). \tag{8.3}$$

Although the satellite clock error is known from the previous navigation message, it is found to be negligible due to the high stability of satellite atomic clocks.

The receiver front-end clock always experiences drift, and this cannot be distinguished from the Doppler frequency shift. The larger the clock drift, the greater number of adjusted bins should be added to the search area. An error of 0.5 ppm in a receiver clock leads to an error of about 1.5 kHz in the Doppler estimate.

Once one satellite is found, the Doppler drift from the receiver clock can be estimated and removed from the other channels, thus narrowing the search area to one bin.

As a first step, only one satellite should be acquired, and therefore the correlator resources can be reallocated to search for various receiver clock drift values. On the second run, the freed correlators can be reallocated to all other satellite channels. Figure 8.4 shows a

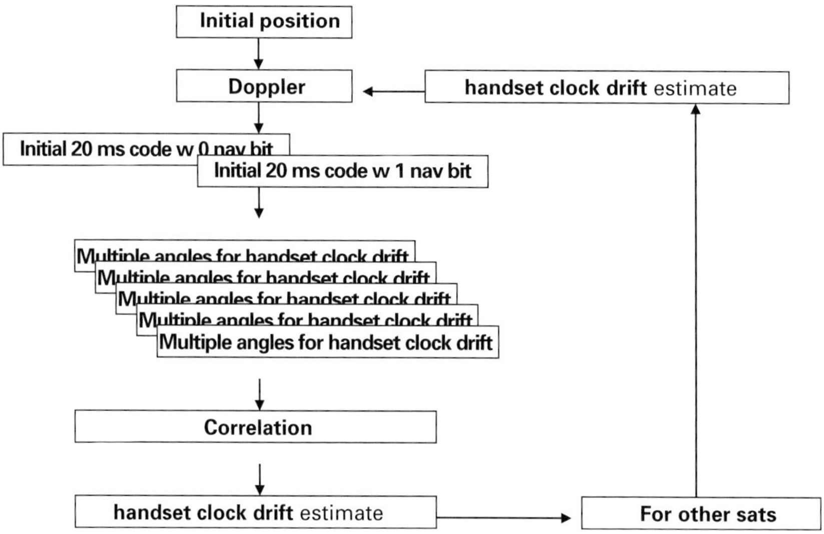

Figure 8.4 High-sensitivity acquisition algorithm flowchart.

flowchart for such an algorithm. However, this may not work if the search for other channels is conducted on sequential signal chunks, so the algorithm may require additional memory, or it may have to estimate additional parameters in a drift model.

Figure 8.5 shows the comparison between an area search with and without Doppler assistance; iPRx, whose acquisition panel is shown in Figure 8.5, used predicted ephemeris data to calculate the Doppler shift. The receiver is static, which excludes rover dynamics. Figure 8.5(a) shows that the entire possible Doppler range was

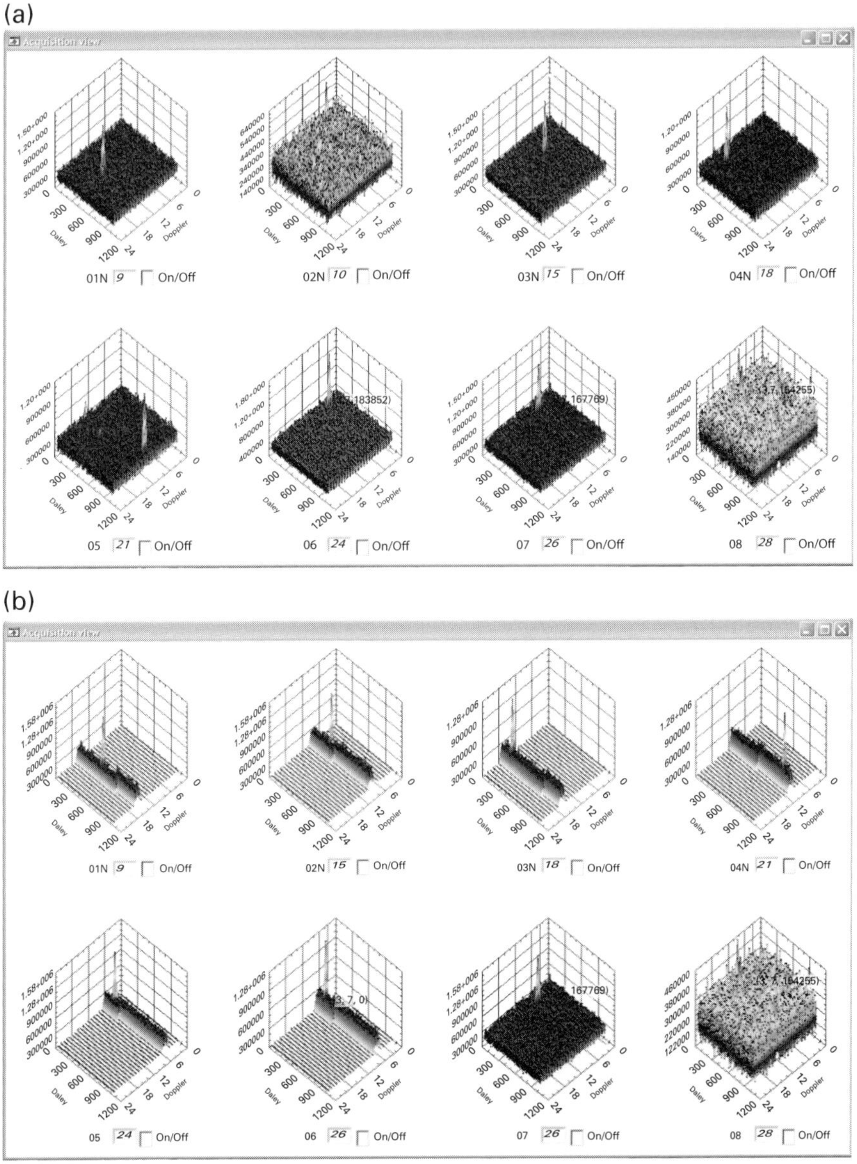

Figure 8.5 iPRx acquisition: (a) without and (b) with Doppler assistance.

searched for each satellite. The satellite PRN 26 was first acquired using a full search area and then used to exclude receiver clock drift from other channels (see Figure 8.5(b)). It is shown that the search areas for other satellites are reduced to one bin, thus reducing the acquisition time approximately by a factor of 36. That would not have been necessary if the front end had an OCXO, which would have excluded the clock drift from consideration.

8.2.2 Code phase assistance in acquisition

Code phase assistance allows us to narrow down the search along the second axis. For GPS C/A code, we have to search 1023 cases for each PRN and each Doppler frequency. For GLONASS SP, we have to search for 511 code delays, for only one PRN but 14 more frequency bands. Each frequency search area for GLONASS also can be narrowed down, for example, to three Doppler bins.

In order to calculate expected code delay we need to know the approximate time and position.

The GPS L1 C/A code sequence repeats itself every millisecond, which corresponds to a distance of 300 km. This gives us the space–time constraints for a-priori position and time. Code phase assistance cannot be used unless the a-priori position is known to an accuracy greater than 300 km and the time for the rover is known to an accuracy greater than 1 ms. It is much more difficult to provide the rover with an accurate time than with a position. For example, a cellular phone server can provide a rover with a position estimate through cell ID with an accuracy of 1–3 km.

If a receiver has managed to acquire and track at least one satellite for about 6 s, then a rather precise clock estimate can be derived from the time mark (Z-Count) in a GPS navigation message. This would require satellite ephemeris information as well in order to calculate the signal propagation time, because the time mark shows an epoch of the signal transmission by the satellite.

In this way, having a cellular phone connection can provide an estimate of position with an accuracy of about 3 km. So, by acquiring and tracking one satellite for 6 s, one can obtain a time estimate and narrow down the search areas for other satellites (Figure 8.6). This method may be useful in an obstructed environment, where at least one satellite is visible.

8.2.3 Doppler assistance in tracking

If one can provide tracking lock loop aiding, it would allow for more accuracy and stability of the lock loops, which would in turn improve the accuracy and sensitivity of the receiver.

The Doppler component from the satellite movement is again compensated for by using ephemeris or almanac information. The user dynamics may be neglected or compensated for by using INS aiding.

A receiver can use external Doppler information, which comes from a data link, and also self-assistance, which can be used after the navigation data are received. Doppler

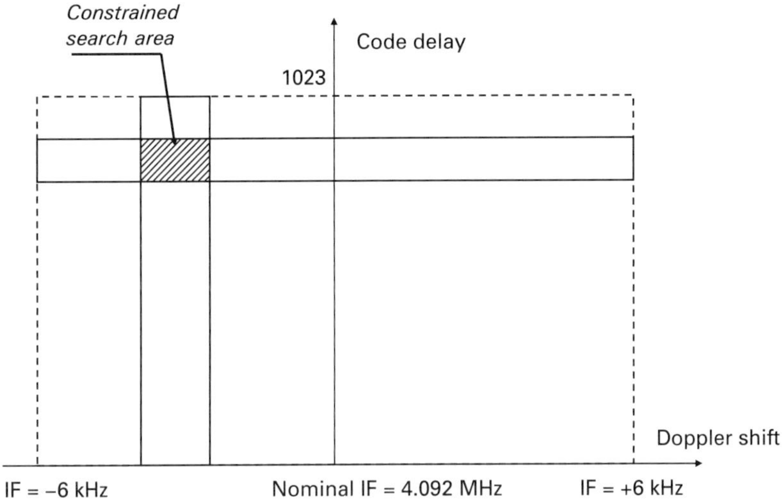

Figure 8.6 Acquisition area with Doppler and code phase constraints.

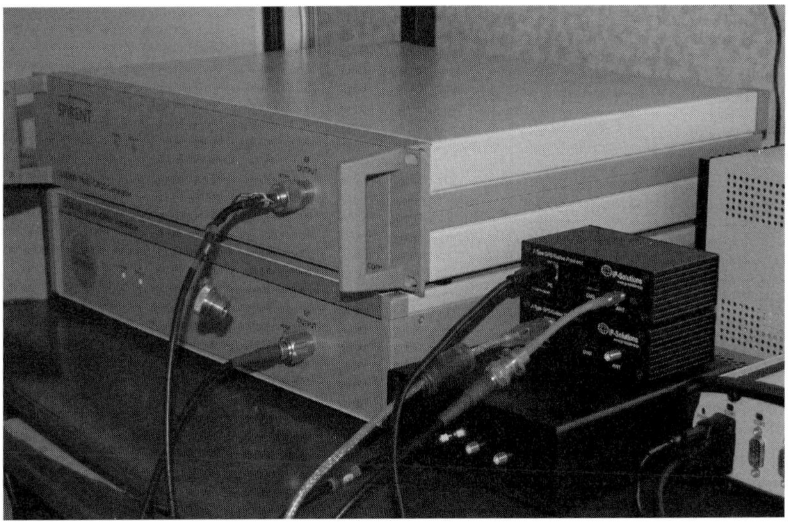

Figure 8.7 iPRx receivers under sensitivity test with Spirent simulators.

assistance is essential for coherent tracking (see Chapter 5). It is generally necessary if the coherency interval exceeds 20 ms.

If a mobile device is operating in an obstructed environment, the Doppler-assisted coherent integration can provide higher sensitivity, and the signal with the significantly lower carrier-to-noise ratio can be tracked. Figures 8.7 and 8.8 illustrate an iPRx sensitivity test with a Spirent simulator which demonstrates sensitivity improvement for coherent tracking. With coherent tracking, the sensitivity was improved by more then 20 dB and could be improved even further by extending the coherency interval.

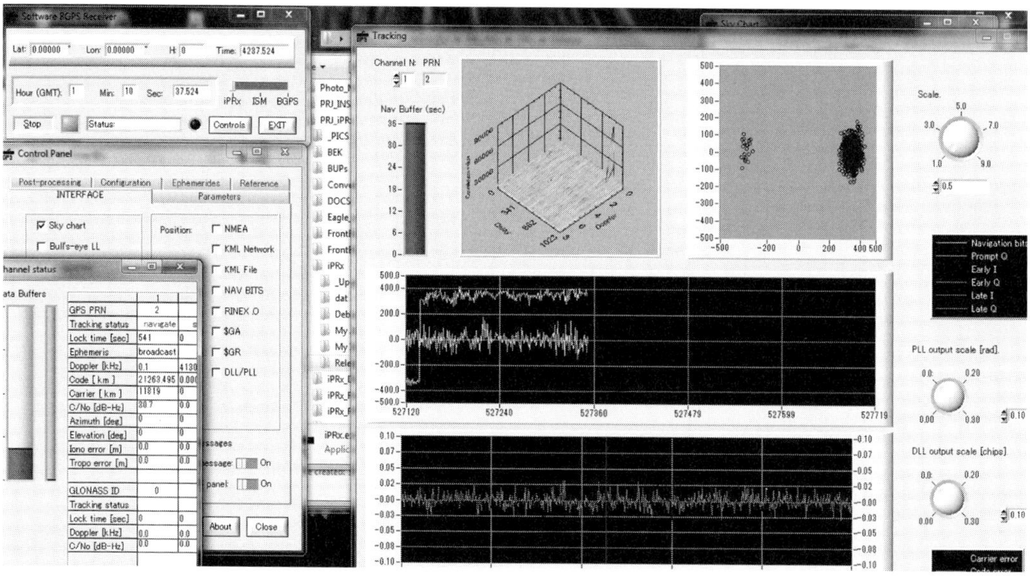

Figure 8.8 iPRx tracking using coherent tracking shows 30 dB-Hz carrier-to-noise ratio.

8.2.4 Navigation data assistance

As previously discussed, a change in the navigation bit polarity interferes with coherent acquisition and coherent tracking algorithms. If a receiver knows the navigation bit in advance, it uses this information to extend the coherency interval, which drastically increases sensitivity. Therefore, a pilot channel is adopted, which has a fixed secondary code instead of a navigation message with data.

Information about navigation data can arrive almost in real time through the network from a server at which a reference receiver is installed. The reference receiver can then obtain a clear signal from open sky, extract the navigation message, and send it to a rover located in the obstructed environment.

We can imitate these conditions using a simulator that is designed especially for AGPS tests (see Chapter 11). However, we can also achieve this by using a standard, off-the-shelf, iP-Solutions USB front end to record signals indoors and outdoors simultaneously (Figure 8.9). This simulates the situation of a reference station and a rover, with the advantage that the signals are perfectly synchronized in a recorded file because they are recorded with the same front-end clock.

This test also demonstrates the enhanced capabilities of coherent integration, i.e. the possibility of receiving the signal indoors. The outside antenna can provide the signal used to decode the navigation message. The indoor antenna provides the signal with low signal-to-noise ratio. By using provided external information about the satellite Doppler shift and the synchronized navigation bit sequence decoded from the outside antenna, the receiver can integrate the indoor signal as long as the front-end clock stability will allow, i.e. almost indefinitely. In real-life applications, the decoded navigation message can arrive as outside corrections from the reference station.

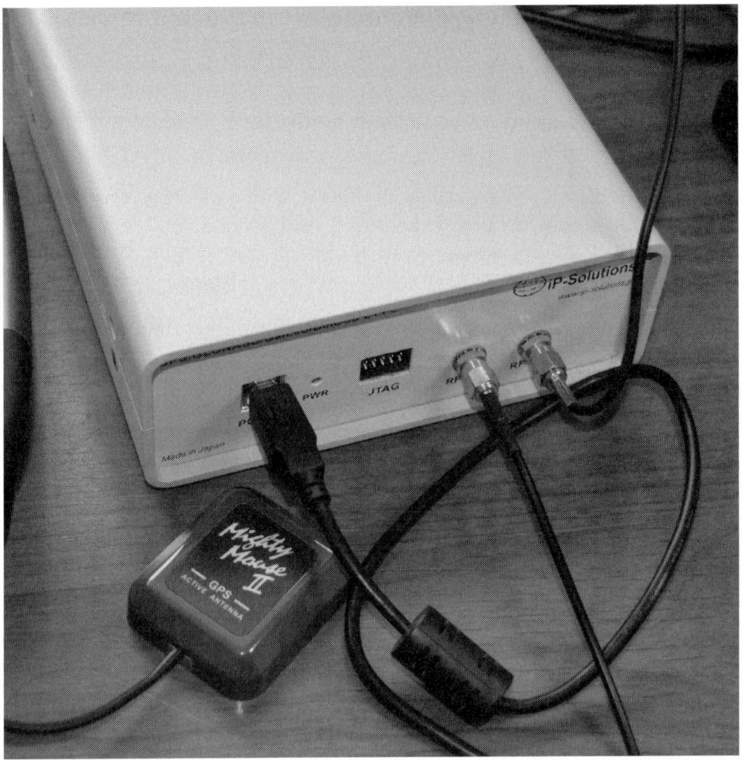

Figure 8.9 iP-Solutions off-the-shelf dual-antenna (for GPS L1 signal) front end configured for recording signals indoors and outdoors simultaneously.

This method can also be used when post-processing a recorded DIF signal. This may be required for post-processing recordings with ionospheric scintillation for geophysical research or tracking system records. Tracking systems are considered in detail in Chapter 9. In this case, the signal from a reference station is recorded and stored. When the signal from a rover is available, it can be post-processed together with the reference station signal. The main challenge is to synchronize these records. This can be easily achieved if the rover recording is commenced before the signal began to degrade, i.e. prior to scintillation or before it is moved to an environment with an obstructed satellite view.

8.3 Application of external information in the navigation processor

8.3.1 TTFF improvement: snapshot positioning

In the case of a GPS L1 C/A signal, 36 s are required to receive a complete ephemeris set for the transmitting satellite to ensure the receipt of one complete frame with duration 30 s. It is about the same time for other GNSS.

If ephemerides are available from another external source, the time to first fix (TTFF) can be reduced, because the receiver does not need to wait for ephemeris data from the broadcast navigation message.

In this case, the TTFF will be defined by the time required to obtain the time mark from the navigation message. The code sequences of any GNSS signal (with the possible exception of some types of tiered codes) repeat themselves periodically. A GPS L1 C/A signal code repeats every millisecond, i.e. every 300 km. Therefore we can use code measurements only if we resolve this ambiguity, for example by getting a time mark from the navigation message. The navigation message bits are unambiguous if we have received the message for a long enough time, so the sentence contains a time mark. For GPS L1 C/A, the time mark repeat, called the Z-Count, occurs in 6 s intervals. After 6 s, we can define exactly which bit we are receiving, and subsequently calculate the exact time, which corresponds to the receipt of a particular sample of the signal. When the receiver has measured the time of the signal propagation, either with or without the navigation message, it has in fact measured a distance to the satellite as the distance to a satellite is equal to the signal propagation time multiplied by the speed of light.

The subframe length of 6 s defines the minimum time required for a receiver to resolve code ambiguity and make a positioning without time assistance for GPS L1 C/A. For GLONASS this time is equal to 2 s, because the time mark in GLONASS is transmitted every 2 s. If, however, we can supply the receiver with a time mark from some other source, then we do not need a navigation message for the time/distance measurements from the satellites, as long as this time mark is precise enough to pinpoint the time within one code sequence, which is 1 ms. This is the main idea behind AGPS instant positioning.

Normally, however, receivers are not concerned with this process. Instead, receivers use tracking to obtain the navigation message and resolve code ambiguity. This, however, comes at a price. The receiver would need to have a few seconds of uninterrupted and uncorrupted signal reception to acquire enough of the navigation message to derive a time mark. This would be a disadvantage for many applications, in particular for positioning in urban conditions, in high-multipath environments, and when a quick positioning fix is required.

Often, it is not possible for a receiver that either is indoors or has an obstructed sky view to ensure uninterrupted tracking. In this case, a receiver may operate in snapshot mode.

We have defined snapshot positioning as positioning based on code phase measurements from an acquisition process without reading a time mark from the navigation message.

As discussed previously, to be able to make a positioning without tracking requires approximations of time and user position. This information is called assist information and is the heart, or rather the blood supply, of AGPS technology. If we do not achieve a time estimate in the receiver better than 1 ms, we cannot resolve this ambiguity without tracking and reading the navigation data.

There are two main approaches taken to obtain the time mark in order to calculate pseudorange measurements:

(1) the time mark can comprise part of the assistance data from a synchronized network;

(2) the time mark may be calculated from redundant measurements.

The implementation of assist information via cellular networks is covered by multiple patents. For example, [5] describes a method of supplying time information through a network, and [6] describes AGPS positioning using an approximate position from a cellular network.

The most common AGPS application is for cellular phones. Approximate user coordinates are roughly estimated using information about particular cellular network stations used by the host phone. The stations are located with rather high densities in many countries, especially in urban environments where the harsh conditions call for instant positioning and therefore for AGPS applications. A handset position estimate with an accuracy of a couple of kilometers is available from a cell ID. Using the information about a few of these stations allows us to use methods similar to positioning with GNSS to improve the positioning estimate even further. A rather precise time estimate can also be delivered to a user through a cellular phone network.

If this assist information is available, we can use ambiguous code estimates from the acquisition process to find a receiver's position, using measurements available just a few milliseconds after the receiver is switched on.

For snapshot positioning, however, accuracy would suffer. Figure 8.10 shows position estimates using acquisition only and using tracking. All positioning estimates are calculated in real time using broadcast ephemerides and with a mobile device front end at a cost of about $6. Tracking permits sub-meter accuracy with code phase measurements. Further improvements in accuracy may be achieved using carrier phase measurements.

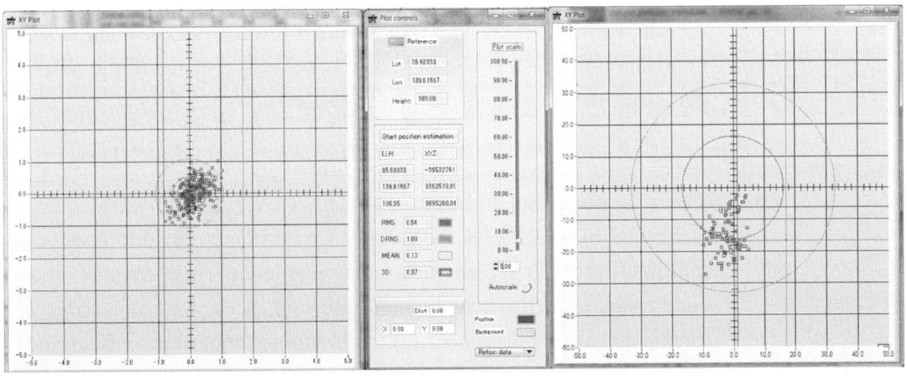

Figure 8.10 Comparison of snapshot and tracking accuracy on a scatter plot.

8.3.2 Accuracy improvement: RTK positioning

8.3.2.1 The catch: antennas

Usually, at best a mobile device incorporates GNSS in differential mode, which allows a user to calculate his or her position with an accuracy of about a meter or at most a few decimeters. This is enough for most applications. However, many applications may benefit greatly from better accuracy. Among such applications are GIS, car navigation with automatic parking and driving, and collision avoidance. For pedestrian applications, DGPS, while good enough for most people, cannot accommodate applications, for example, for users with limited visual abilities.

The main question relating to the application of RTK positioning in mobile devices is whether it is possible to achieve on a cheap mobile handset platform. At the beginning of this chapter, we referred to a test that has demonstrated RTK positioning using a mobile front-end chipset and a module from different vendors. The front ends are intended for mobile devices, and their costs meet mobile device cost requirements.

However, there is a catch. The stumbling block is the antenna. We have discussed antenna design in Chapter 7. Here we list the main factors affecting RTK positioning that is related to antennas.

The first factors that we consider are related to antenna design.

(1) Antennas must have good ground planes in order to provide strong enough signals. This is often impossible for mobile devices if the antenna is hidden in the device. Sometimes there is a trade-off between achievable signal strength and antenna unit size. Sometimes part of the device can be used as the antenna ground plane.

(2) The signal enters the antenna as an electromagnetic wave at a point called the antenna phase center. All observables relate to that point. It would not matter if the phase center were fixed at a geometrical point, but the phase center does depend on where the electromagnetic wave is coming from. In other words, the phase center depends on the satellite azimuth and elevation.

Geodetic antennas provide calibrated data to specify this dependence, so it can be compensated for during the processing [7]. The error arising from failing to account for these items can range from millimeters to centimeters. A condition for good stability of the antenna phase center is symmetry of the antenna structure, in particular using symmetric feed points.

The other group of factors affecting RTK positioning depends on environment.

(3) The most important environmental factor is multipath [4],[8]. The multipath error is created by the signal travelling to an antenna via different paths. In general, multipath error is scaled to the observables, i.e. the error is smaller for shorter code chips [4]. However, it may not help in the case of carrier phase measurements, because the code multipath would negatively affect the carrier phase ambiguity resolution process. The multipath error can be very large in urban environments. These additional paths result from signal reflections on various surfaces. A choke ring antenna is a good example of an antenna with multipath

mitigation design. It is too large and expensive for mobile applications, but this or a similar solution is essential for a reference station.

(4) Human body effects [8]: a mobile device is often hand held, and is therefore affected by the human body, which causes antenna gain degradation, asymmetrical changes in antenna gain pattern, and phase center variations.

8.3.2.2 Network RTK implementation: virtual reference station RTK system

If for some types of mobile devices, or for some positioning modes, the antenna problem can be resolved and the code and phase observables are available, the system may run into the next problem: that there must be a reference station in the vicinity of the user. The distance from the reference station when using RTK positioning should generally be no more than 10 km on average, which is significantly different from code differential GPS, where distances to reference stations can exceed several hundred kilometers. This means that the infrastructure for an RTK system must include a reference station every 10 km. A reference station is an expensive item, because it must minimize errors on its side (otherwise these errors will be added to those from the rover and effectively cut down the applicable service area or even render the service useless). The receiver should provide high-quality measurements. The reference station should have good antennas, for example a choke ring antenna, which we mentioned in Section 8.3.2.1. Also, the antenna must be installed in a low-multipath environment with open sky. The multiple site allocation for the network may also present a large problem. The only way to overcome these problems related to setting up a dense reference station network, without sacrificing accuracy or time for initialization, is to apply an approach similar to that adopted for code differential GPS SBAS (space-based augmentation systems, for example WAAS), where the corrections are averaged over the coverage area. This would allow a service provider to decrease the number of reference stations drastically. For carrier differential, this concept was developed as a network RTK system network, which was discussed in Chapter 4.

A network RTK system permits the extension of the service area with a limited number of reference stations. The reference stations are linked to a control center, which calculates area corrections for the largest part of the GNSS error budget.

The largest errors that reference station corrections intend to compensate for decorrelate with distance. Among these errors, the most significant are ionospheric errors. Fortunately, these errors mostly decorrelate linearly with distance under normal operating conditions. Other errors that decorrelate with distance are tropospheric and orbital effects. Orbital errors are usually insignificant over a medium-length baseline, and tropospheric error distribution usually fluctuates on a medium-length baseline, and in some cases may be very significant. Using the distance-weighted errors at the known locations of the real reference stations, one can find the approximate magnitude of the most probable error between them. In order to find this averaging error near the user position, we can use different criteria to optimize this estimate, using various algorithms. The difference between the results of these algorithms is not generally significant. A natural extension of the network concept is that network RTK increases the integrity and reliability of the service when compared with a single-baseline solution.

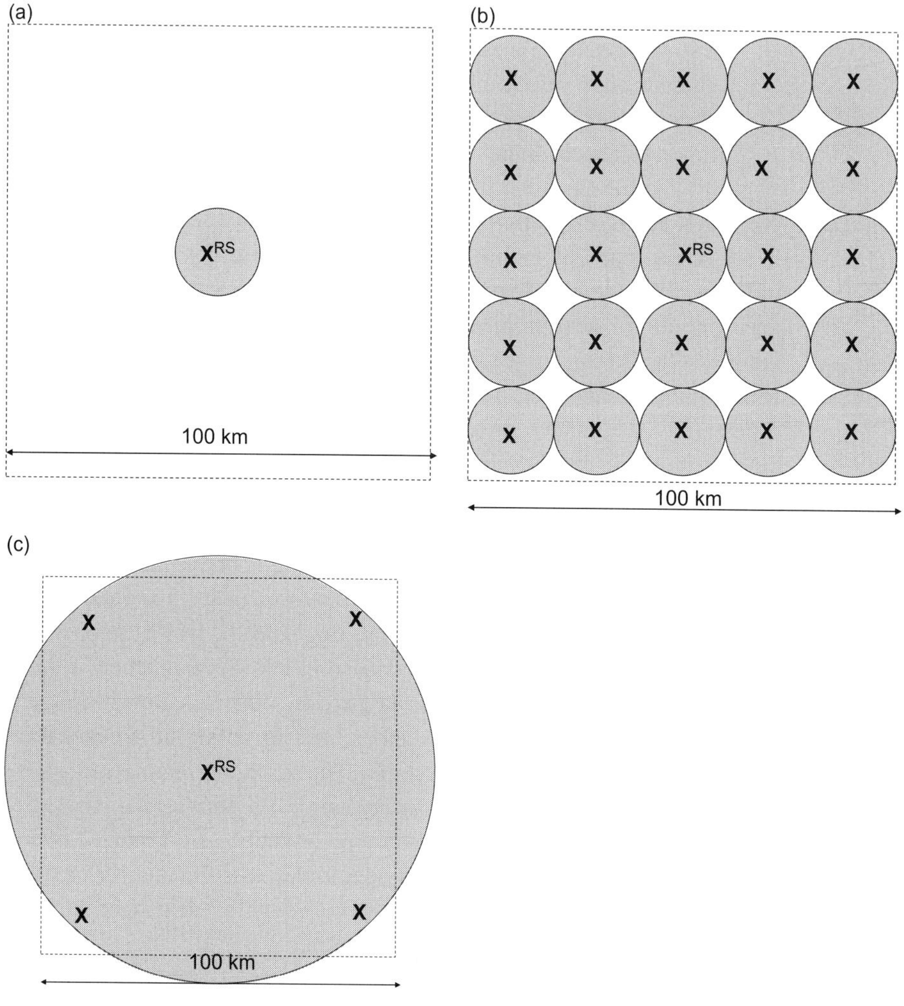

Figure 8.11 (a) Single reference station; (b) RTK network; (c) VRS coverage. After [10].

This makes more efficient use of the reference stations. By using network RTK, the inter-reference-station distance can be increased by a factor of 3, and the number of reference stations required in area can be decreased by a factor of 5 [9]. Figure 8.11 demonstrates the concept [10],[11]. Figure 8.11(a) shows a single reference station, which can provide RTK within a 10 km distance. Figure 8.11(b) shows a number of reference stations, which can cover an area of 100 km. Figure 8.11(c) shows how five times fewer stations provide RTK service within the same area using the virtual reference station (VRS) concept.

The required distance between reference stations can be from 30 to 50 km for the network RTK system. A server can send to a rover either VRS corrections specifically calculated for that rover or corrections together with area correction parameters with an update rate of 1–2 Hz. Figure 8.12 shows how the VRS concept can benefit in terms

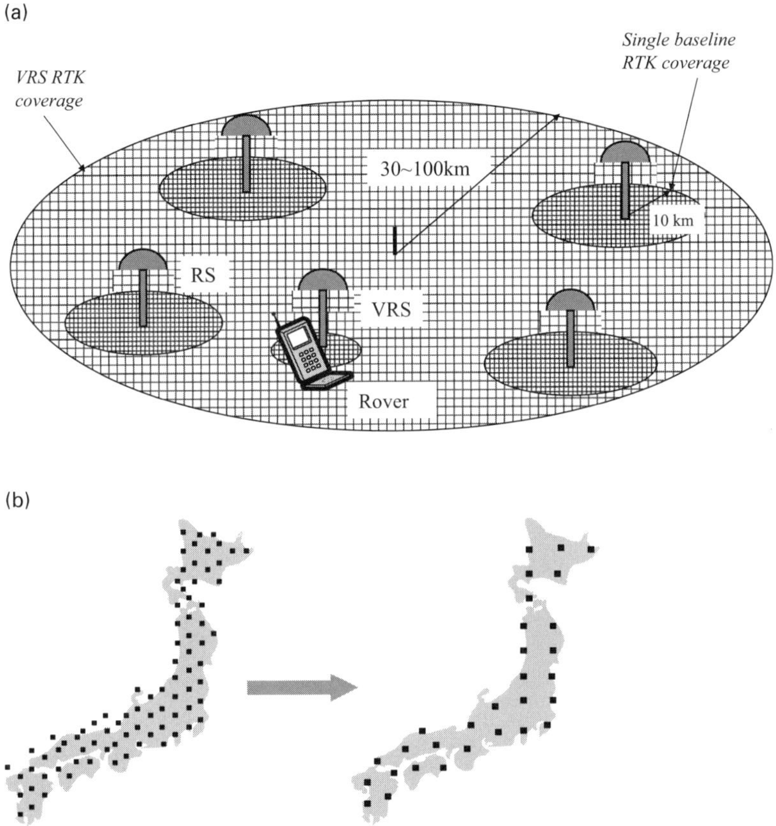

Figure 8.12 (a) VRS concept (not to scale). (b) Possible RTK network reduction due to VRS implementation in Japan. After [10].

of the number of required reference stations for the same area. The integrity and reliability of the service will be increased, with the same level of accuracy and significantly fewer stations.

If the server transmits a set of area corrections, a rover calculates the appropriate corrections for its position. If a server sends VRS corrections, the network appears to the rover receiver as a single reference station. Then the server feeds the data directly into the GNSS receiver. When the rover sends its information to the server, the server can perform all the calculations based on the grid information, and then send network corrections, which can be seen as the usual RTK corrections, to the rover. These corrections are calculated for a specific rover. In this case, an RTK-capable rover can be used without additional modifications for VRS. In practice, the rover receiver tries to take into account the known baseline length, which is undesirable because the RTK network corrections emulate a short, or zero-baseline, case. The position of the VRS can be varied, and therefore the baseline length from the VRS to the rover receiver can be controlled. The VRS corrections can be calculated at the rover location. Figure 8.13 shows a flowchart of one of the VRS system implementations.

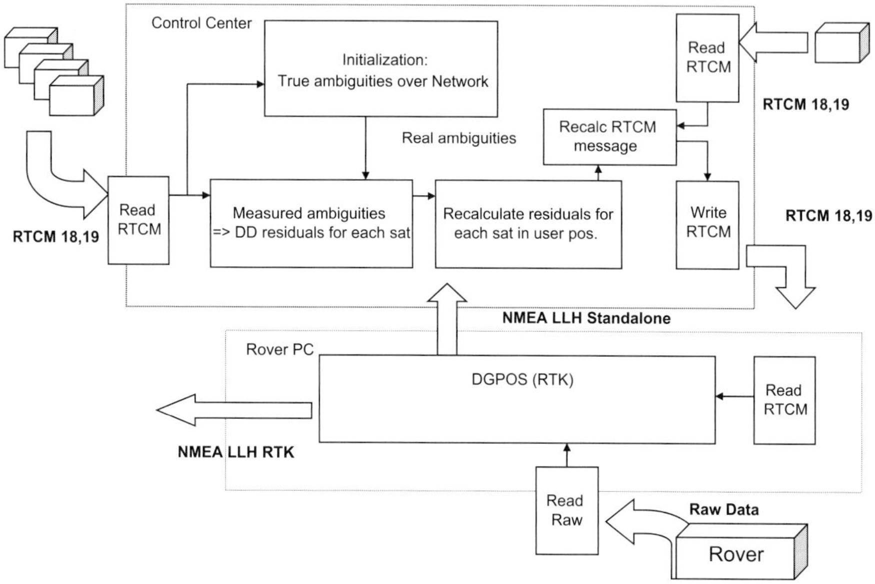

Figure 8.13 VRS software block scheme. After [10].

Consider the case when VRS corrections are calculated on a rover. An important characteristic of VRS and, in general, RTK corrections is latency. If the carrier phase corrections are delayed, they simply cannot be used. This important factor, on the scale of seconds, makes a huge difference to the code phase corrections, which in comparison are basically not sensitive to latency at that scale. Therefore, we can design a network RTK which consists of two layers: (i) a primary reference station that supplies the user with the main correction stream, and (ii) all the other reference stations that function to provide information on error decorrelation. These secondary reference stations supply users with less urgent information related to a relative correction distribution around the network, in relation to the primary reference station. A latency of up to a few minutes is tolerable for the secondary reference station corrections because these corrections are due to the ionosphere, which is generally not very dynamic. The primary reference station, in contrast, supplies a user with absolute corrections, which are valid in its vicinity. Corrections from the primary station are critical and should be supplied with minimum latency. A latency period of 1 s in the primary station corrections may result in approximately 1 cm error in the user position [12],[13].

RTK corrections can be supplied by a server either as raw measurements or as corrections. These two formats are not equivalent. Townsend *et al.* [14] provide the following considerations for using corrections or raw measurements.

(1) The dynamics of the corrections and the raw measurements are different. The corrections change slowly, as they are affected only by error sources. The measurements include satellite dynamics as well. Therefore a rover must account for any differences in satellite position between epochs of reference and rover measurements.

(2) In the case of corrections, the rover must use the same ephemeris and clock parameters that were used in the server for the correction calculations. Additionally, as the calculations on the rover side may be carried out by different algorithms from different vendors, approximation errors may differ from those in server, which in turn may become a major error source for the RTK algorithm. A changeover of ephemerides may also cause different ephemerides to be used at the rover and the server, which would make RTK positioning impossible.

An implementation of a network-based VRS service has been developed and tested in the Tokyo area using an Internet-based reference station network and a TV sound multiplexed subcarrier data link [15],[16]. The successful test was conducted for a region of approximately 100 km radius. The corrections data were distributed through an audio sub-channel (ASC). The Asahi TV ASC broadcast system has been officially adopted by the Japanese Ministry of Post and Telecommunications for broadcasting DGPS and RTK corrections. The ASC provided two data channels: one for transmitting RTK, DGPS, and differential GLONASS corrections, and another one for weather and traffic information. The VRS RTK coverage via the ASC broadcast was from 40 to 70 km radius, depending on antenna type, in comparison with the RTK coverage of 10 km radius.

8.4 External information content

External data for a mobile device may include the following components.

8.4.1 Group 1: assistance data

Originally, the transmission of assistance data was patented by Qualcomm. Assistance data were developed for AGPS and include the following.

(1) Satellite orbit information, which can be packed into various formats. The format may be different from that of the broadcast navigation message. The ephemeris data may be in Keplerian parameters or tabular form. The format may also be optimized for a specific period of validity.

(2) Satellite clock corrections. The main difference with satellite orbits is that satellite clocks are less stable and generally must be supplied in real time or at least updated more often than ephemerides.

(3) Current GPS time, which is required for code ambiguity resolution and estimation of satellite position at the time of transmission. The estimation of satellite position at the time of transmission can be much less accurate, because the satellite is moving with a speed of a few kilometers per second.
Current GPS time may often not be accurate enough for ambiguity resolution, but in any case it can be used to narrow the search area.
For a synchronized cellular network, the uncertainty can be within tenths of milliseconds. For a non-synchronized cellular network, it can be within seconds.

If the estimate is very rough, like in the case of a non-synchronized cellular network, its effect on calculation of satellite position must also be taken into account, while conducting code ambiguity resolution.

(4) Frequency: a cellular network may supply a rover with frequency assistance, which can be used to compensate for the drift of the front-end clock. It can be used, for example, to narrow the search area in acquisition.
(5) Doppler frequency shift for visible satellites, used for narrowing the search area during the satellite acquisition.
(6) Doppler first derivative, which can be used for code ambiguity resolution and to predict the Doppler shift.
(7) Elevation and azimuth for visible satellites. This information can be used to calculate ionospheric and tropospheric errors. Calculation of the ionospheric error requires transmission of the additional parameters.
(8) Using an almanac is an alternative to the transmission of the elevation, the azimuth, and the Doppler shift. All these parameters can be calculated in the handset from the almanac.
(9) The initial position estimate can be deduced from cellular ID information.
(10) Code phase estimate.
(11) Current navigation bit.

These data provide a handset with information, which allows

- rapid signal acquisition, and therefore improved TTFF,
- receiver work in snapshot mode,
- smaller search area and increased sensitivity.

For examples of specific AGNSS protocols, see [2] and [3]. These protocols are subject to change. The amount of data would depend first of all on the channel bandwidth and cost of transmission. Handset capability is secondary, because the handset algorithms and hardware can be developed according to requirements. The main obstacle from the handset point of view is the antenna, which limits the level of accuracy the handset can achieve.

There are also some disadvantages related to the assistance data.

(1) The transmission of assistance data has been patented by Qualcomm, though this is sometimes challenged or ignored.
(2) The data transmission requires a *permanent* connection to the network for time assistance. We look at the alternatives in Chapter 9.
(3) Data are localized to the network and cannot be used regardless of user location. This in particular is related to some privacy issues.

8.4.2 Group 2: additional parameters

These include various parameters used to assist code positioning.

(1) Ionospheric model parameters, for example Klobuchar model parameters.
(2) UTC to GPS shift.

(3) Time shifts between various GNSS and GPS.
(4) Integrity, an alarm that provides a warning that a satellite cannot be used for positioning within a certain period (usually a few seconds) after satellite failure.

8.4.3 Group 3: differential corrections

(1) Code phase differential corrections.
(2) Carrier phase differential corrections.
(3) VRS corrections [16].

This group requires a larger bandwidth than previous corrections. However, the bandwidth available for mobile devices is drastically increasing in order to support streaming video.

There are a few RTK correction message formats available. The most common are the RTCM SC-104 (Radio Technical Commission for Maritime Services, Special Committee 104) and the Trimble CMR (Compact Measurement Record) protocols. The RTCM RTK messages are implemented either as corrections to the code and carrier phases (Types 20 and 21) or as raw measurement information (Types 18 and 19).

In the RTCM case, the server uses Type 1, Type 18/19, Type 3, Type 20/21, and Type 59 for area correction parameters. The baud rate should be limited to a range of 2400–4800 baud. Latency effects should be minimized by correction prediction.

Usually, either one of the pair of RTK messages in the RTCM SC-104 standard is used. For example, the message types 18 and 19 bear raw carrier phase and code phase measurement information. Message types 20 and 21 bear correction information for the same measurements. An arbitrary constant integer number of cycles is chosen at the start of tracking to keep the size of the correction small.

It is preferable to use type 20 and 21 correction messages rather than the type 18 and 19 uncorrected measurement messages, for the reasons we have listed in Section 8.3.2.2.

The corrections are summed up in Table 8.1.

8.5 Pseudolites

8.5.1 Pseudolite applications

Pseudolites, or pseudo-satellites, are ground-based transmitters of GNSS-like signals [17]. Pseudolites were used for positioning even before GPS satellites. In the early 1970s, at the beginning of the GPS era, the U.S. Department of Defense used pseudolites at the Yuma Proving Ground to test the first GPS receivers. (It also gave its name to the YUMA format of the almanac data.)

One of the main applications of pseudolites today is to test GNSS receivers without actually having a full constellation in view, for example the German Galileo test and development environment (GATE) [18]. The main purpose of these pseudolites is to simulate GNSS, and the whole system operates pretty much like a huge GNSS simulator. The pseudolites have been also used for Japan's QZSS simulation [19],[20]. There

Table 8.1. External information content

Content	Current category	Purpose	Affected specification parameter
		Group I	
Ephemeris	AGPS	snapshot positioning	TTFF
Time		snapshot positioning	TTFF
Frequency		high-sensitivity positioning	sensitivity
Doppler		high-sensitivity positioning	sensitivity
Doppler first derivative		snapshot positioning	sensitivity, TTFF
Visible satellite list		snapshot positioning	TTFF
Satellite azimuth and elevation data		snapshot positioning	TTFF
Almanac		snapshot positioning	TTFF
Approximate position		snapshot positioning	TTFF
Code phase estimate		snapshot positioning	TTFF
Current navigation bit		high-sensitivity positioning	sensitivity
		Group II	
Ionospheric error model parameters	various	any positioning	accuracy
UTC to GPS shift		various	TTFF
Time shifts between various GNSS and GPS		positioning in obstructed environment	minimum number of satellites
Integrity message		reliability	integrity
		Group III	
Code phase differential corrections	differential corrections	differential and RTK positioning	accuracy
Carrier phase differential corrections		RTK positioning	accuracy
Code phase measurements		differential and RTK positioning	accuracy
Carrier phase measurements		RTK positioning	accuracy
VRS corrections		network RTK positioning	accuracy

are many more applications for pseudolites, but they are not growing in number, mostly because of possible in-band interference with GNSS.

We included pseudolites in this chapter for the following reason. The use of pseudolites essentially requires a data link to a rover, which would provide a user with immediate pseudolite clock error information. This is inevitable if the rover uses pseudolites for positioning in the same sense as navigation satellites. For comparison, the Japan indoor messaging system (IMES) concept [21] is completely different and

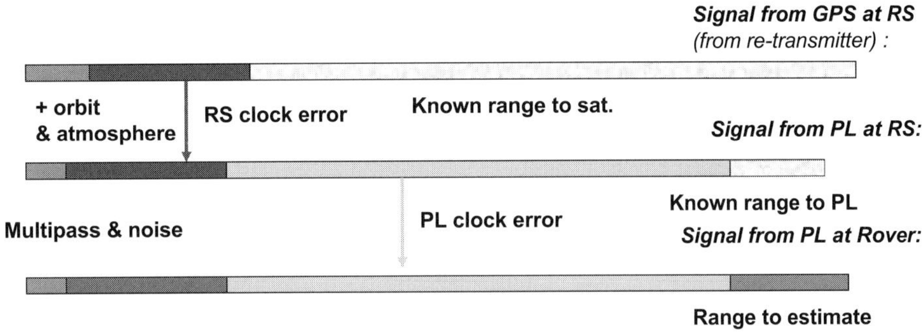

Figure 8.14 Pseudolite-related error budget and its compensation in the user receiver.

related to RF markers [22] rather than to GNSS. In the case of IMES, the rover just receives the transmitter's own coordinates encoded into the signal navigation message.

Compensation for clock errors can be achieved with reference station or master clock synchronization. Figure 8.14 depicts a method of compensation for clock errors using a reference station. The reference station clock error can be estimated using the satellite signal. The pseudolite clock error then can be estimated at the reference station and transmitted to a receiver.

The next most important pseudolite application was for aircraft approach and landing [23]. These pseudolite systems use the RTK method. The system was successfully developed by Stanford University and then by their spin-off company IntegriNautics (later became Novariant), the company which made the first commercial pseudolite. The pseudolite theory is given in Cobb's comprehensive work on the subject of pseudolite theory in his thesis [24]. In this section we also use some information from the author's works [25]–[29]. Figure 8.15 shows the author and Dr. S. Cobb working on a pseudolite system installation in Tcukuba, Japan. Further Novariant development has concentrated on GNSS agricultural applications and off-band pseudolites [30].

After the first commercial pseudolite system became available, many universities and companies started to work on pseudolite systems. Most of their research concentrated on synchronized pseudolites for indoor applications. The main distinguishing feature of such systems is that they do not require a data link to the user's rover receiver to supply it with pseudolite clock corrections. For example, all pseudolites can be synchronized by using the same master clock. There remain to overcome a number of problems for the majority of such applications. One of these problems is related to multipath. Indoor premises always feature a high-multipath environment. Sometimes multipath on this scale makes positioning with code phase data almost impossible.

Pseudolites are very attractive as augmentations to GPS satellites, both for code and carrier phase positioning. Pseudolites add more "GPS satellites" to a constellation, and they are very useful in places where satellite visibility is limited, such as urban canyons, mining pits, etc. Also, pseudolites could even provide a user with a means of seamless navigation indoors, such as in parking lots and tunnels. Moreover, pseudolites improve the geometry of the new constellation. For carrier phase positioning, it is

Figure 8.15 The author and Dr. S. Cobb working on a pseudolite system installation in Tcukuba, Japan. Taken from [26].

important that this geometry changes more rapidly for a dynamic user than for the usual satellites, because if a user passes on the signal sources, which are ground based pseudolites in this case, the ambiguity resolution occurs more quickly. Also, in comparison with a satellite signal, a pseudolite error budget does not contain atmospheric and orbital errors.

There is a variety of pseudolite system designs and concepts, including pseudolites combined with DGPS or monitored by a master station, single-site installation vs. network installation, GPS synchronization vs. external clock synchronization, etc.

The modification of positioning software so it works with pseudolites presents yet another challenge. The navigation equations produce two solutions which are difficult to choose between because of the proximity of a receiver to the signal sources; this does not cause problems for GPS satellites, because these solutions are very distant from each other.

Further, the software has to provide a solution for a classic system of non-linear equations. Figure 8.16 demonstrates the positioning algorithm's sensitivity to initial position error. In the case of closely located pseudolites, the error in the directional cosines matrix due to initial position error is much more significant than for distant GPS satellites. Therefore, the directional cosines matrix varies substantially as a function of the initial position. A few meters in the case of pseudolites will cause a far more significant error in the matrix than thousands of kilometers for GPS.

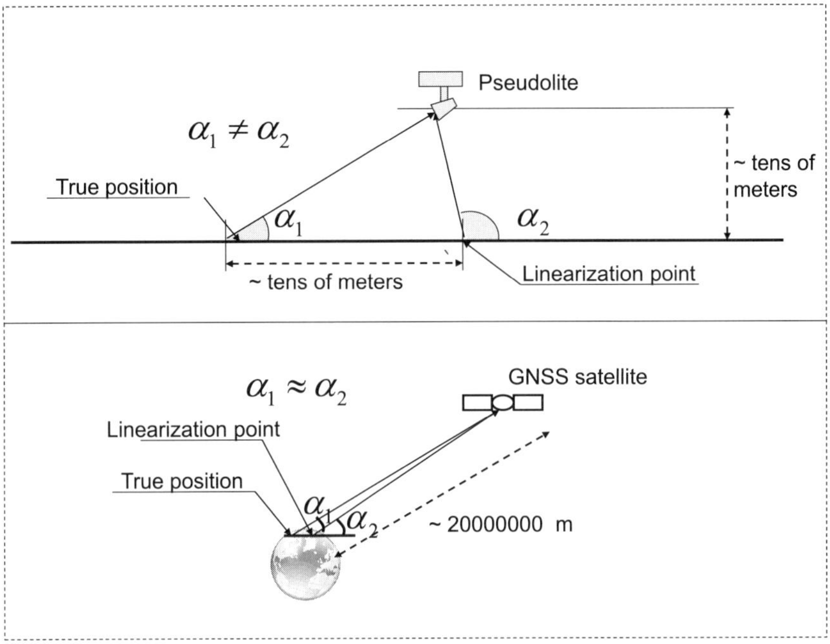

Figure 8.16 Positioning algorithm sensitivity to initial position error: (a) pseudolite; (b) GNSS satellite.

The conventional positioning method is critical in relation to an initial position error, because non-linear pseudorange equations require linearization in the true position vicinity:

$$\delta \vec{X} = ([A]^T \cdot [A]^{-1}) \cdot [A]^T \cdot \delta \vec{r}\ , \tag{8.4}$$

where $\delta \vec{X}$ is an error in user position and time, $[A]$ is the directional cosines matrix, and $\delta \vec{r}$ is the offset of the pseudoranges. The directional cosines $[A]$ are not critical to this error in the case of GPS satellites located far away. For closely set pseudolites, the error in the directional cosines due to initial position error is much more significant, and matrix $[A]$ varies significantly as a function of the initial position (see Figure 8.16). The easiest way to handle this problem is to start from a known position, which is not always possible for real-life applications. Therefore, positioning algorithms less sensitive to initial position error are required.

The most general problem for pseudolite implementation is interference with GNSS, which for pseudolites manifests itself as a near–far problem.

Different types of pseudolite can cause different types of interference, and there are a number of ways to cope with this. It may be necessary to set up a number of tests to ensure the absence of interference on non-participant receivers from different manufacturers.

The following measures can reduce the interference:

(1) power attenuation;
(2) frequency shift;
(3) pulsing scheme with a variable duty cycle.

The problem of interference can be resolved to some extent by using signal pulsing. However, it is not confirmed how massive pseudolite usage affects a noise floor and therefore the so-called *non-participant* receiver behavior. However, the GNSS frequency band remains attractive for potential users due to the simplicity of the solution in terms of the receiver design. Indeed, using in-band pseudolites one can use existing front-end solutions and a baseband processor. Basically, a pseudolite receiver reduces to a standard off-the-shelf receiver and a customized navigation processor or attached PC with application software, which can log raw data from the receiver. Only minor modifications in firmware may be required on the part of the receiver in order to construct a receiver to search for additional PRNs.

None of the interference mitigation methods listed above can guarantee the complete absence of interference for all types of receivers. Even with the pulsing technique, a noise floor will be affected, especially in the case of widespread use. Therefore to implement them on a wide scale, for example in the social infrastructure, pseudolites should be moved into an unshared frequency band.

8.5.2 Indoor positioning with carrier phase

Indoor pseudolites can be used in the GNSS frequency band, providing that premises are certified to make sure that there is no signal leakage to the outside. Positioning indoors with GPS-like signals can be realized through high-sensitivity AGPS. However, AGPS is not available everywhere. There are lots of places where the outside GPS signal is not strong enough, even for high-sensitivity AGPS, or where assist information is not available. At the time of writing, AGPS is capable only of code positioning, which limits its accuracy to the pseudorange level.

Indoor positioning with carrier phase data encounters other obstacles. The problem is that pseudolites are fixed in space, which does not permit implementation of a conventional carrier phase ambiguity resolution technique. It works in the case of a landing system because the ambiguities are resolved with the assistance of GNSS satellites, and because the aircraft moves rapidly past the pseudolites. The way to implement a conventional technique for most other applications, including those indoors, is to start from a known location.

Indoor positioning software can be represented by two main subroutines, one of which is for initial position estimation, whilst the other is for tracking. A code positioning algorithm does not use knowledge of the initial position. Tracking can be implemented using the carrier phase. To resolve the ambiguities to provide an initial position estimation for the carrier phase algorithm, one must start from a known location. It is impossible to resolve carrier phase ambiguities indoors when the rover receiver is static. This in particular limits the carrier phase positioning to a reckoning. If, during the rover movement carrier cycle, slip has occurred, the coordinates become erroneous.

A standard algorithm of dead reckoning for GPS usually makes use of the fact that GPS satellites are very far away and the user can linearize the positioning equations. Thus, algorithms of dead reckoning indoors become non-linear. As an example, we can look at such an algorithm. One pseudolite is chosen as the key pseudolite, in the same manner as for forming double differences with the satellites. We form the carrier phase double difference $\nabla\Delta C$ for satellite i and *key* pseudolites for two sequential epochs as follows:

$$\nabla\Delta C_{ikey}(k-1) = [d_{i\,RS}(k-1) - d_i(k-1)] - [d_{key\,RS}(k-1) - d_{key}(k-1)], \quad (8.5)$$

$$\nabla\Delta C_{ikey}(k) = [d_{i\,RS}(k) - d_i(k)] - [d_{key\,RS}(k) - d_{key}(k)]. \quad (8.6)$$

Then deduct one from the other to obtain

$$\nabla\Delta C_{ikey}(k) - \nabla\Delta C_{ikey}(k-1) = [d_{key}(k) - d_i(k)] - [d_{key}(k-1) - d_i(k-1)]. \quad (8.7)$$

Denote

$$\begin{aligned}&[d_{key}(k-1) - d_i(k-1)]\\ &= \sqrt{[x(k-1) - x_{key}]^2 + [y(k-1) - y_{key}]^2 + [z(k-1) - z_{key}]^2}\\ &\quad - \sqrt{[x(k-1) - x_i]^2 + [y(k-1) - y_i]^2 + [z(k-1) - z_i]^2} = d_{i\_init}, \end{aligned} \quad (8.8)$$

to obtain

$$\begin{aligned}&\nabla\Delta C_{ikey}(k) - \nabla\Delta C_{ikey}(k-1)\\ &= \sqrt{[x(k) - x_{key}]^2 + [y(k) - y_{key}]^2 + [z(k) - z_{key}]^2}\\ &\quad - \sqrt{[x(k) - x_i]^2 + [y(k) - y_i]^2 + [z(k) - z_i]^2} - d_{i\_init}, \quad i = 1, 2, 3. \end{aligned} \quad (8.9)$$

Finally, denote

$$\nabla\Delta C_{ikey}(k) - \nabla\Delta C_{ikey}(k-1) + d_{i\_init} \triangleq m_i \quad (8.10)$$

to obtain the following set of non-linear equations:

$$\begin{aligned}&\sqrt{[x(k) - x_{key}]^2 + [y(k) - y_{key}]^2 + [z(k) - z_{key}]^2}\\ &\quad - \sqrt{[x(k) - x_i]^2 + [y(k) - y_i]^2 + [z(k) - z_i]^2} = m_i, \quad i = 1, 2, 3. \end{aligned} \quad (8.11)$$

We now have three equations with three unknowns, which can be solved using globally convergent methods. References [27] and [28] demonstrate that the accuracy of the dead reckoning algorithm depends on multipath and can reach the sub-centimeter level. Thus, an indoor pseudolite system can achieve sub-centimeter-level relative accuracy and about meter-level absolute accuracy from code initialization. *Relative accuracy* is a measure of the accuracy of the current rover position relative to its previous position; *absolute accuracy* is the accuracy of the current rover position relative to pseudolites.

8.5.3 Repeaters

A GNSS repeater is a device which consists of an antenna (located outside) which acquires GNSS signals, an amplifier, a cable, and an indoor transmitting antenna. One can use a repeater as a signal source instead of a pseudolite.

A receiver located indoors can use only a single GNSS satellite signal out of all the signals received at the roof antenna. If the receiver uses all the satellites, it calculates

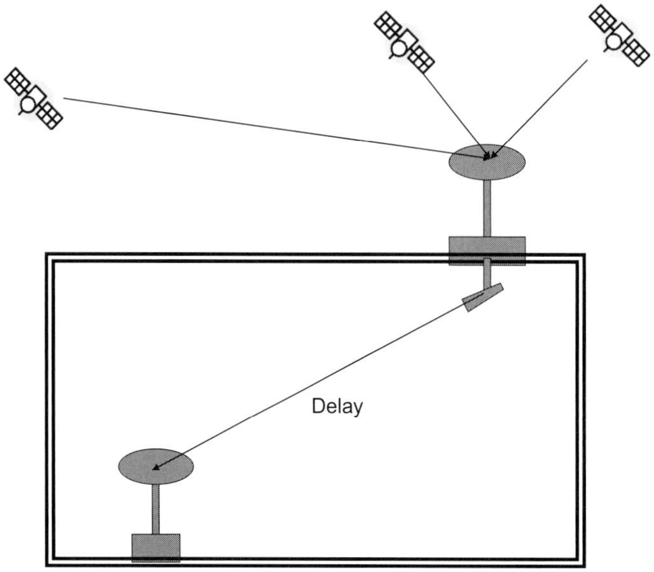

Figure 8.17 GNSS signal repeater.

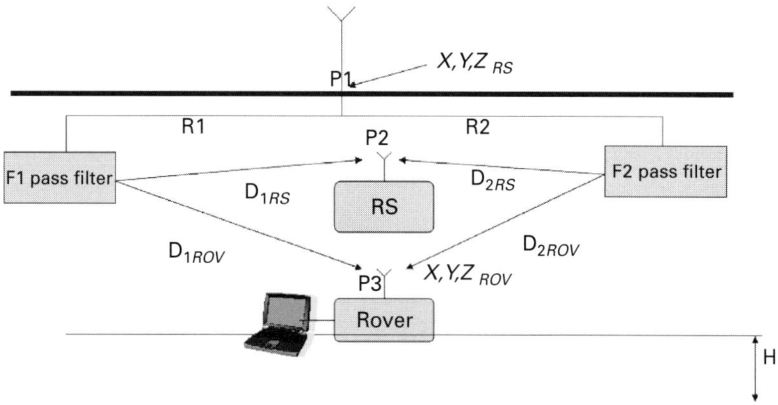

Figure 8.18 Indoor positioning with frequency antenna separation.

the position of the roof antenna. This position will be the same as that calculated by the receiver if it is connected directly to the rooftop antenna, with the difference caused by errors added during indoor signal propagation (Figure 8.17). The delay caused by this propagation is completely negligible. The main error, which dominates the indoor positioning error budget, is multipath. Thus, this technique, namely the comparison of position from an indoor located receiver with one from the same receiver connected directly to an outdoor antenna, can give us a good estimation of the multipath at the indoor antenna position. Another function of repeaters is the synchronization of pseudolites.

It may be possible to construct an indoor positioning system based on multiple GPS repeaters using either multiple antennas with an obstructed view or the FDMA technique.

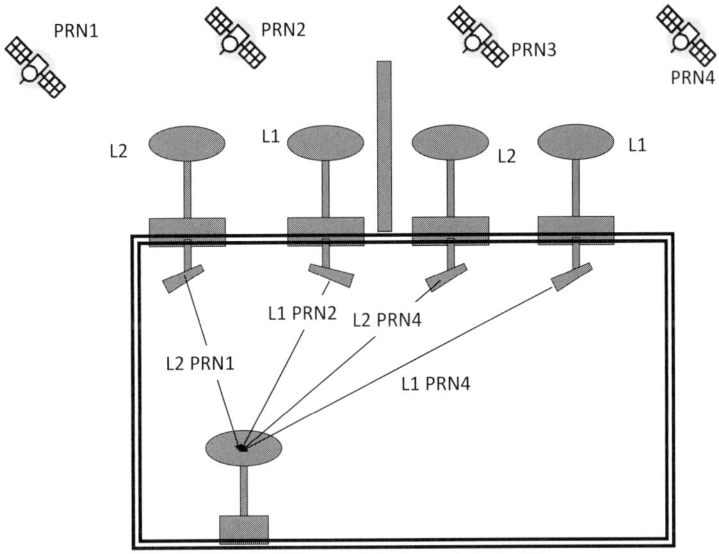

Figure 8.19 Indoor positioning with spatial antenna separation.

In order to use more than one repeater for indoor positioning, we need to make sure that each repeater will transmit a different signal. This can be achieved by frequency separation (see Figure 8.18). Suppose that a GPS signal from an outdoor antenna is divided by a splitter and then that the L1 and L2 signals are filtered out before they reach the corresponding transmitting antennas. Then, using a dual-frequency GPS receiver, we can use the same satellite signal from two repeaters, one of which is on L1, with the other on L2. The other signals can come, for example, from GLONASS or BeiDou. Three civilian frequencies (such as for GPS and GLONASS) provide enough signals such that repeaters can be used for 2D indoor positioning with a receiver that will require only firmware modifications.

In comparison, a less elegant, but much less expensive, solution is to use spatial separation only (see Figure 8.19). This "poor man's" solution can be easily realized, but can suffer from extra multipath and outages. It also requires some kind of mission planning to let the receiver know which satellite signal to expect from each repeater.

References

[1] I. Petrovski, T. Tsujii, and H. Hojo, First AGPS – now BGPS. Instantaneous precise positioning anywhere, *GPS World*, Nov. 2008.

[2] F. van Diggelen, *A-GPS: Assisted GPS, GNSS, and SBAS*. Boston, MA: Artech House, 2009.

[3] N. Harper, *Server-side GPS and Assisted-GPS in Java*. Boston, MA: Artech House, 2009.

[4] I. Petrovski and T. Tsujii, *Digital Satellite Navigation and Geophysics: A Practical Guide with GNSS Signal Simulator and Receiver Laboratory*. Cambridge: Cambridge University Press, 2012.

[5] N. F. Krasner, Method and apparatus for determining time for GPS receivers, US Patent 5,945,944, Aug. 31,1999.

[6] F. van Diggelen, Method and apparatus for time-free processing of GPS signals, US Patent 6,417,801, July 9, 2002.

[7] M. Rothacher, W. Gurtner, S. Schaer, R. Weber, and H. Hase, Azimuth- and elevation-dependent phase center corrections for geodetic GPS antennas estimated from GPS calibration campaigns, in *IAG Symposium No.115*, W. Torge, Ed. Berlin: Springer-Verlag, 1996, pp. 335–339.

[8] X. Chen, C. Parini, B. Collins, Y. Yao, and M. Rehman, *Antennas for Global Navigation Satellite Systems*. Chichester: John Wiley & Sons Ltd., 2012.

[9] G. Lachapelle, P. Alves, L.P. Fortes, M.E. Cannon, and B. Townsend, DGPS RTK using a reference network, in *Proc.13th Int. Tech. Meeting of the Satellite Division of The Institute of Navigation (GPS 2000)*, Salt Lake City, UT, Sept. 20–22, 2000.

[10] I. Petrovski, E. Cannon, G. Lachapelle, *et al.*, The issues of practical implementation of the commercial RTK Network Service, in *Proc. 14th Int. Tech. Meeting of the Satellite Division of The Institute of Navigation (ION GPS 2001)*, Salt Lake City, UT, Sept. 2001, pp. 2654–2664.

[11] I. Petrovski, B. Townsend, S. Hatsumoto, S. Kawaguchi, H. Torimoto, and K. Fuji, An impact of high ionospheric activity on multiref RTK network performance in Japan, *Proc. 15th Int. Tech. Meeting of the Satellite Division of The Institute of Navigation (ION GPS 2002)*, Portland, OR, Sept. 2002, pp. 2247–2255.

[12] B. Townsend, I. Petrovski, S. Kawaguchi, M. Kelly, M. Ishii, and H. Torimoto, Concept for a regional satellite-based carrier-phase correction service, in *Proc. 16th Int. Tech. Meeting of the Satellite Division of The Institute of Navigation (ION GPS/GNSS 2003)*, Portland, OR, Sept. 2003, pp. 2631–2636.

[13] I. Petrovski, K. Fujii, K. Sasano, *et al.*, Practical issues of virtual reference station implementation for nationwide RTK network, in *Proc. GNSS 2001*, Seville, Spain, May 2001.

[14] B. Townsend, K. VanDierendonck, J. Neumann, I. Petrovski, S. Kawaguchi, and H. Torimoto, A proposal for standardized network RTK messages, in *Proc.13th Int. Tech. Meeting of the Satellite Division of The Institute of Navigation (ION GPS 2000)*, Salt Lake City, UT, Sept. 19–22, 2000.

[15] I. Petrovski, K. Sasano, B. Townsend, M. Ishii, and H. Torimoto, TV broadcast utilization for code and carrier phase differential GPS, *Proc. GNSS 2001*, Seville, Spain, May 2001.

[16] K. Sasano, I. Petrovski, M. Ishii, H. Torimoto, and B. Townsend, Method of using a TV sound multiplexed subcarrier data link for a DGPS/RTK service, in *Proc. 13th Int. Tech. Meeting of the Satellite Division of The Institute of Navigation (ION GPS 2000)*, Salt Lake City, UT, Sept. 19–22, 2000, pp. 2418–2423.

[17] B. Elrod and A. J. Van Dierendonck, Pseudolites, in *Global Positioning System: Theory and Applications, Vol. II*, B. W. Parkinson and J. J. Spilker, Eds. Washington, D.C.: American Institute of Aeronautics and Astronautics Inc., 1996.

[18] G. Heinrichs, E. Loehnert, E. Wittmann, and R. Kaniuth, First outdoor positioning results with real Galileo signals by using the German Galileo test and development environment, in *Proc. 20th Int. Tech. Meeting of the Satellite Division of The Institute of Navigation (ION-GNSS) 2007*, Fort Worth, TX, Sept. 25–28, 2007, pp. 1576–1587.

[19] T. Tsujii, M. Harigae, K. Okano, and I. Petrovski, Test results of augmented GPS by the pseudolite installed on a helicopter, in *Proc. Int. Symp. GPS/GNSS*, Tokyo, Japan, Nov. 15–18, 2003.

[20] T. Tsujii, H. Tomita, Y. Okuno, *et al.*, Development of a BOC/CA pseudo QZS and multipath analysis using an airborne platform, *Proc. 2007 Natl. Tech. Meeting of The Institute of Navigation*, San Diego, CA, Jan. 2007, pp. 446–451.

[21] D. Manandhar, K. Okano, M. Ishii, *et al.*, Development of ultimate seamless positioning system based on QZSS IMES, *Proc. 21st Int. Tech. Meeting of the Satellite Division of The Institute of Navigation (ION-GNSS 2008)*, Savannah, GA, Sept. 16–19, 2008, pp. 1698–1705.

[22] I. Petrovski, M. Ishii, H. Torimoto, and T. Hasegawa, Development of highway ITS and pedestrian ITS based on RTK network, pseudolites and PN coded magnetic markers, in *Proc. 13th Int. Tech. Meeting of the Satellite Division of The Institute of Navigation (ION GPS 2000)*, Salt Lake City, UT, Sept. 19–22, 2001.

[23] C. Cohen, B. Pervan, S. Cobb, D. Lawrence, D. Powell, and B. Parkinson, Precision landing of aircraft using integrity beacons, in *Global Positioning Systems: Theory and Applications, Vol.2*, B.W. Parkinson and J. Spilker, Eds. Washington, D.C.: American Institute of Aeronautics and Astronautics, Inc., 1996.

[24] H. S. Cobb, GPS pseudolites: theory, design, and applications. Unpublished Ph.D. dissertation, Stanford University, CA, 1997.

[25] I. Petrovski, K. Okano, M. Ishii, H. Torimoto, Y. Konishe, and R. Shibasaki, Pseudolite implementation for social infrastructure and seamless indoor/outdoor positioning, in *Proc. ION GPS2002*, Portland, OR, Sept. 2002.

[26] I. Petrovski, K. Okano, M. Ishii, H. Torimoto, Y. Konishe, and R. Shibasaki, Pedestrian ITS in Japan, *GPS World*, vol. 14, no. 3, pp. 33–37, Mar. 2003.

[27] I. Petrovski, K. Okano, K. Suzuki, *et al.*, Indoor code and carrier phase positioning with pseudolites and multiple GPS repeaters, in *Proc. ION GPS 2003*, Portland, OR, Sept. 2003.

[28] I. Petrovski, K. Okano, K. Suzuki, *et al.*, Precise indoor positioning with pseudolites, *2003 Int. Symp. on GPS/GNSS*, Tokyo, Japan, Nov. 15–18, 2003.

[29] S. Sugano, K. Fujii, I. Petrovski, *et al.*, It's a robot life, *GPS World*, Sept. 2007, pp. 48–55.

[30] D. Bevly and S. Cobb, *GNSS for Vehicle Control*. Boston, MA: Artech House, 2010.

9 Positioning without data link: from BGPS to PPP

This chapter is not about standalone positioning as such, but rather about all the enhancements to standalone positioning, which we have discussed in the previous chapter, but achieved without data link. Here we use the term BGPS (BGNSS) to describe technology that achieves results similar to AGPS (AGNSS), but without immediate corrections data.

9.1 Advantages of positioning without a data link

In a 1996 article in *GPS World* magazine [1], the author stated that GPS is calmly but rapidly penetrating mass markets, and finds itself in cellular phones, cars, watches, cameras, and golf carts. By 2013, this process is almost complete. The GNSS is the essential part in all these devices. This has increased the number of GPS users drastically. With the many navigation satellites in the sky, and a handful of GNSS-enabled gadgets, a user would expect to experience a seamless positioning service – instant positioning at any time and at any place. And that is without becoming a specialist in satellite navigation. Technology in general is trying to meet this new requirement, sometimes quite drastically, as in the Windows 8 interface.

In relation to positioning, and especially for such applications as the emergency services, a user should be able to achieve instant positioning at any location, at any time, at the stroke of a key, and it should not depend on assistance data provided through a communication channel. BGPS was developed as a method to enhance GPS availability in a similar fashion to AGPS, but without immediate assistance information and, consequently, a communication link.

As we have discussed in Chapter 8, external information can be used to enhance receiver performance in terms of accuracy, speed of operation, sensitivity, and other parameters. However, in order to enjoy this external information, a receiver had to have a data link to a server at the time of positioning. This requirement is inconvenient and constrains a user. Let's consider just a few of these inconveniences.

(1) Additional equipment is required in order to facilitate a data link. As we have discussed in Chapter 8, the data link could be a radio, the Internet, a cellular phone network, a satellite signal, and so on.

(2) An additional fee may be required for each particular data link to function and/or for a specific service. For example, to get a positioning fix with a cellular phone requires a general service from the operator, and a fee may be payable for a service specific to positioning. Some other services have dedicated data links used only for these correction services, for example corrections from private satellite systems.
(3) Most services (almost all besides global), require the rover to be within a certain proximity to a server. For example, the RTK method, which we have discussed in Chapter 8, would only work within a 10 km radius of a reference station. Global differential services, such as satellite systems, are not limited in terms of location, but are rather expensive and provide only the information required to enhance accuracy.
(4) Another important issue is privacy. In many cellular phone systems, a mobile device sends its raw data, or chunks of code phase measurements, to a server. The server calculates the rover position and sends it back; thus, the rover knows its exact location. Even when the rover calculates its own position, the server needs to provide the rover with an approximate position, thereby pinpointing its location with a certain accuracy.
(5) Even if privacy is not an issue, many places may not have cellular networks at all. Such places may include mountainous areas, forests, oceans, and so on. People who are traveling through these areas may still require all or some of these enhancements to a standard GNSS receiver spec.
(6) It is also desirable to be able to make an instant positioning without a network because a network can malfunction or even be destroyed during natural disasters such as earthquakes. It is still very important in such cases to be able to make instant and snapshot positioning.
(7) When using the GPS function in cellular phone applications, the cellular phone cannot be used at the same time as GPS because of interference. This means that a user has to wait until all the data from the navigation message are acquired.

For many mobile devices, a user may not foresee the need for a data link. For example, when using a camera to take a photo, one does not expect that a cellular phone network would be required. On the other hand, it would be desirable to have an instant service, for example a time and location stamp on a photo at the moment it was taken. One would not appreciate waiting for 36 seconds either prior to or after taking a shot.

High sensitivity also may be very useful in areas obstructed with vegetation. A user can benefit from features similar to high sensitivity of cellular phones. In many cases, it is still essential to be able to determine coordinates with a receiver. Figure 9.1 demonstrates the problem conceptually. Vegetation may interfere with signal tracking, which makes the navigation message inaccessible. Methods described in Chapter 8 require a data link at time of positioning in order for the receiver to operate.

One more important issue to consider concerns the changes in the market place for GNSS devices. A few decades ago, the user of a GNSS device had to have special education and training. This user estimated and anticipated system limitations. For example, consider the then popular mission-planning software which was necessary

Figure 9.1 "Oh no, that won't work. We have to get out of the woods first, but we just don't know which way to go." First appeared in the online version of [1]. Based on artwork created by N. Petrovskaia.

to ensure performance at the time of the test. The user knew all the system requirements relating to unobscured skies, uninterrupted signal reception for a certain period of time, and the tolerable multipath environment. The user did not expect something from the system that it could not deliver due to its limitations.

Today, however, millions of users use mobile devices without knowing much about GNSS. They expect the same availability and ease of use as, for example, for a radio. Although it was not the case at the time of radio's inception, for many decades it has been possible simply to just turn a switch and get music in a car, at home, in the street, or inside a shopping center. It does not matter where one is, or what station provides the service. All the user needs to do to receive the information is to turn on a switch.

In this chapter, we discuss the methods that allow a user to enjoy most if not all of the enhancements described in Chapter 8, but without having a data link at the time of positioning. External information is still required, but it can be compiled days or even weeks in advance of the time of positioning. Therefore, a data link is not required at the time of positioning. Moreover, the required information can be compiled using free services on the Internet. Therefore there are no additional fees.

Additionally, because these methods do not require a server, this information can be used without compromising privacy anywhere on the globe, on the high seas, in rain forests, or anywhere else.

9.2 BGPS: instant positioning without network

9.2.1 Advantages of BGPS

We use here the term BGPS, which was introduced in [2] in a description of a general approach to positioning without a time mark to emphasize the difference from AGPS, in which the reception time estimate is delivered via a network.

Instant positioning without network, which we call BGPS, can provide a user with the same advantages as AGPS.

9.2.1.1 Instant positioning

Instant positioning with BGPS can be achieved by finding the time of reception without reading a time mark from the navigation message. The algorithms for this have been known for some time, but the challenge was to make these algorithms operate quickly enough to support real-time positioning. Our iPRx software receiver, which was modified to test these algorithms, makes a positioning within parts of a second on a general processor. The positioning can be achieved within milliseconds using FPGA.

9.2.1.2 Power savings

Power savings are possible because a receiver may operate for just one-hundredth of a second in comparison with the conventionally required time of 36 s. BGPS can provide a user with all the advantages of AGPS without the necessity of being connected to a network at the time of positioning. This feature makes such technology an excellent candidate for animal- and bird-tracking devices. This is an additional market outside that of AGPS.

In comparison with a normal GPS position logging device, for the same period of time BGPS must stay powered on over much shorter intervals of time: for example about $36/0.006 = 6000$ times less. For example, positioning fixes for tracking an animal are required every 10 min (600 s) for a period of 1 yr. A conventional GPS receiver would have to be powered on for about 36 s over every 5 min, leading to a total of 36 s $\times$ 365 days $\times$ 24 hr $\times$ 12, which is about 44 days of continuous work. A BGPS receiver, however, is powered on to make a snapshot position for only 0.006 s for each positioning fix, which would require the battery to work for only 10 min in total for the same period of 1 yr. This means that in general, the battery for a device that is required to work for years, will be required to support operation for hours only. This results in the capability of reducing the size and mass of such devices such that they can be used in tracking small animals and birds.

9.2.1.3 Less interruption during cellular operation

In a cellular phone, the phone functionality and GNSS positioning functionality cannot be realized at the same time due to interference and processing power, if the positioning is using a host processor. Instant positioning minimizes the time required for positioning to the level of milliseconds, without accessing a server.

9.2.1.4 High sensitivity

An application such as a GPS-enhanced camera would require an instant position fix without network assistance as it is unlikely that the camera would always be used in places where a network is available. Another application would be for an aircraft, which is continuously maneuvering in the skies. The capability of using any snapshots of data will benefit its navigation in two respects. First, no data are lost to a navigation system, even if a satellite was in view for a few milliseconds. Second, such data have much smaller latency, which is extremely important in the case of high dynamic vehicles such as aircrafts. Network assistance would probably be unavailable for such application.

An immediate position fix using BGPS can rely for a few weeks on the additional information existing in the handset without connection to a network or any other data link.

9.2.2 History of the approach

The idea behind all instant positioning methods was introduced for the first time in work on a totally different subject by Wu *et al.* [3]. Due to the importance of this article for instant positioning, we briefly describe some essential information from that point of view.

The LEO satellite onboard receiver described in the article was a front end called a "bit grabber," which produced DIF signal data and sent them to a memory for later processing. The front end consisted of an oscillator, a down-converter, and an analog-to-digital convertor.

Because of this structure, it was also the first GPS software receiver. The receiver required very low power because it worked for only a small percentage of time, recording short snapshots of GPS signal, most of its time being in sleep mode. We describe possible applications that exploit this particular benefit later in the chapter.

The receiver provided ambiguous code phase measurements and unambiguous Doppler observables. The signal duration for one snapshot was 2 ms. It was required to take only a few samples per orbit.

The resolution of pseudorange ambiguities had been performed in two steps. An approximate orbit with an accuracy within 50 km was calculated using Doppler measurements. Then this orbit was used to construct unambiguous pseudoranges, because 50 km orbit accuracy permits the resolution of 300 km ambiguities (see Figure 9.2).

The GPS ephemerides available from a worldwide reference station network were used in the processing.

In the same year, a 1997 US Patent for determining the time for GPS receivers was filed which describes a method of supplying time information through a cellular network [4].

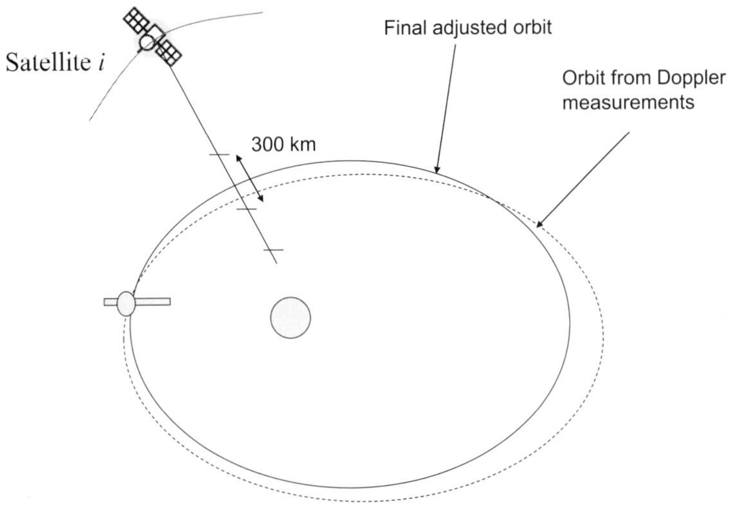

Figure 9.2 LEO satellite code phase ambiguity resolution with Doppler measurements.

Now we take a look at the details of the BGPS approach. It is a continuation of an approach, presented in [3], which was designed to make a positioning without reading a navigation message and without any time assistance data. As we have learned, receiver clocks are significantly less accurate than satellite clocks, and they are handled in a completely different manner in comparison with satellite clocks. Whereas satellite clocks are measured and algorithmically compensated for in the system outside of the user receiver, receiver clocks are estimated in the user receiver outside the system. User receiver clocks are treated as an extra unknown in the positioning task. We recall that the number of unknowns defines the number of equations in the system, and that the number of equations in turn should be more or equal to the number of satellites. Therefore, as described earlier, if the user needs to define three coordinates, it requires four satellites, because one unknown, the receiver clock error, has to be added to the three unknown user coordinates. The additional fifth satellite in this example can be used to find an additional unknown, which is the time of signal reception.

This approach has been already realized in many practical solutions by various receiver manufacturers. The ideas behind this approach have been continuously developed, in particular by Syrjärinne [5] and Sirola [6]. The approach is described for AGPS in general in [7]. Alternative methods may include analytical solutions [8]. However, the methods and algorithms required for the practical implementation of this approach have not yet been fully described.

9.2.3 BGPS in a nutshell

Let's suppose that our rover receiver has measured GPS signals from a few satellites for a short interval of time. Figure 9.3 shows these measurements, P_i^m. When there are short data chunks, the receiver cannot receive a time mark from the navigation message t_t and

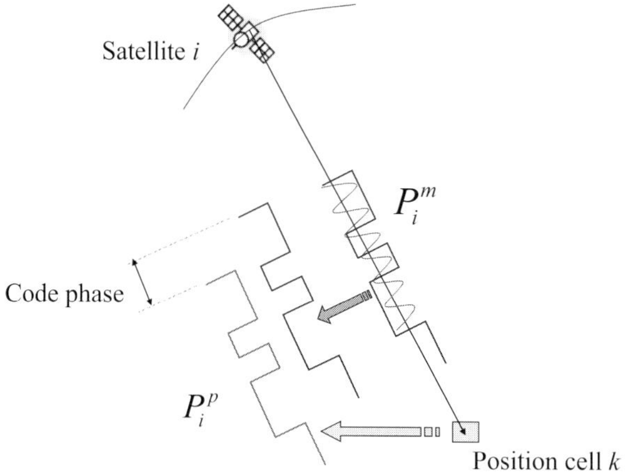

Figure 9.3 Concept of positioning with short signal chunks.

cannot find the time of the signal reception t_r. The time marks actually give us the time of transmission, but the receiver needs to change it to the time of reception anyway, because the time of transmission is different for all satellites.

We can assume the receiver's approximate position and time of reception. From these assumed values we can calculate the predicted value of the code phase, P_i^p, which we should be receiving as measurements if the receiver indeed was at the assumed position and time.

We can then calculate the difference between the predicted, P_i^p, and measured, P_i^m, code phases (Figure 9.3). The pseudorange residuals should be at a minimum if our assumptions of position and time are correct. If the residuals look unreasonably large, we should move to another cell and assume the next position or time.

When we assume a certain epoch for the signal reception, the satellite position for the corresponding transmission epoch should be calculated. The code phase residuals can be calculated roughly as follows:

$$dP = \sum_m \sum_k \left\{ \left[X_i^{SAT}(t_j(t_{ROV})) - X_i^{ROV} \right] - \left[\Delta P_i + N_i \cdot \mathrm{CL} \right] \right\} \rightarrow min, \tag{9.1}$$

where the pseudorange ambiguity number is given by

$$N_k = floor\left(\frac{\left[X_i^{SAT}(t_j) - X_i^{ROV} \right]}{CL} \right), \tag{9.2}$$

where CL is the code length, ΔP_i is the measured code phase, and $X_i^{SAT}(t_j(t_{ROV}))$ is the satellite position calculated based on assumed time t_{ROV}.

An initial time estimate can come from the device internal clock. From the initial time uncertainty, for example 10 s, we can estimate an uncertainty in satellite position, which would be in the range of 40 km. Then we can assume a user position uncertainty of 100 km. Altogether these parameters roughly define a search area in space-time dimensions (Figure 9.4). The position should be searched along available vertical and

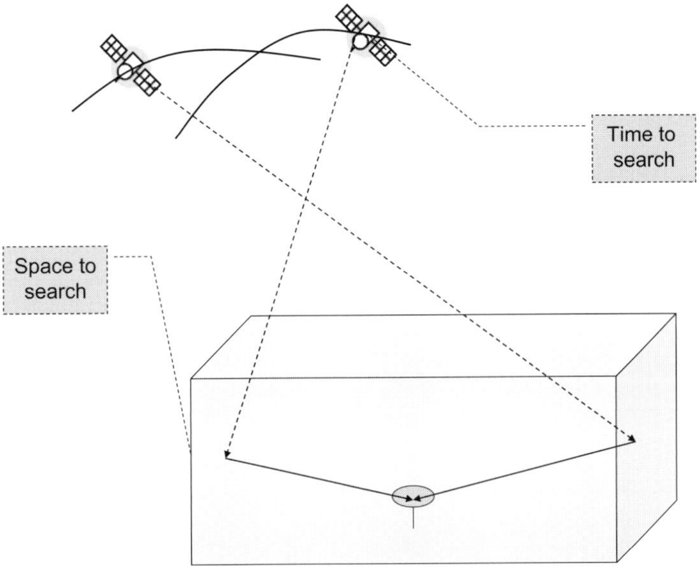

Figure 9.4 Search area in space and time continuum.

horizontal cells for the receiver. The time should be searched among available satellite positions for the possible time interval. The predicted measurements are then calculated for assumed receiver–satellite positions.

As soon as we find a position with reasonable accuracy, we can make further improvements by using familiar methods, because we have resolved the code phase ambiguities.

Figure 9.5 illustrates the brute-force search method. As an example, we consider a search area of length 100 km with 1 km cells. This would yield 10 000 cells to search for a particular epoch. A search range along the time axis of 10 s would yield 40 km of satellite uncertainty, so we need to search for satellite position cells as well.

9.2.4 Formalization

The time estimate required for resolving ambiguities in the code phase can be either derived from assist information, as for AGPS, or calculated using an additional satellite, as for BGPS. If we do not have time delivered from a time mark on the navigation data or via an external data link, an additional unknown should be added to the equations.

We will consider a solution for our original non-linear GNSS equations (3.10) for code phase observables in a matrix form,

$$\overline{Z} = [A(\overline{X}')], \tag{9.3}$$

where $\overline{X}'$ is now an extended state vector

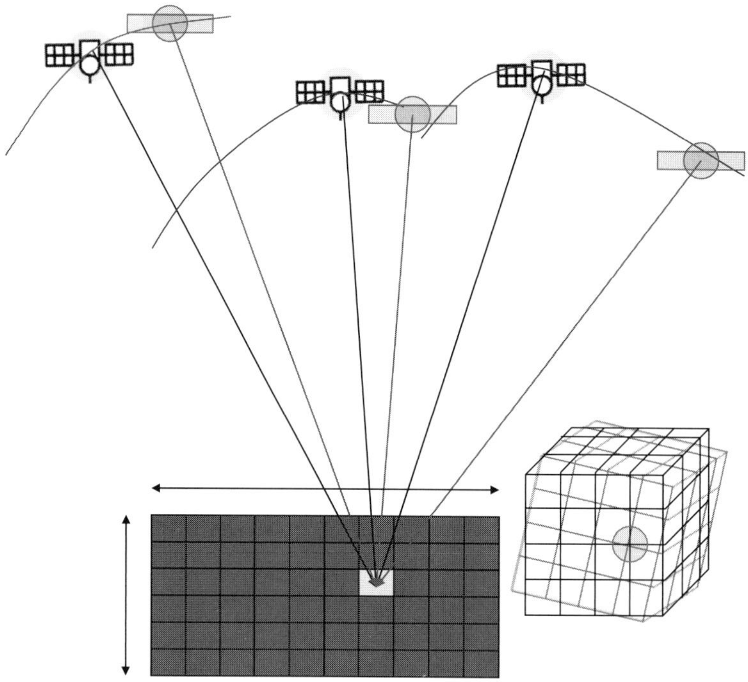

Figure 9.5 Brute-force searching concept.

$$\overline{X}' = \begin{bmatrix} x_r \\ y_r \\ z_r \\ \delta t_r \\ t_r \end{bmatrix}, \tag{9.4}$$

where t_r is the time of signal reception.

The difference between this and previous equations is that now we have introduced an additional variable, which is the time of signal reception, t_r, into the state vector. In the equations in the preceding chapters the time of signal reception was assumed to be known as it was derived from the navigation message with a certain accuracy.

Let us also note that $\overline{X}'$ is now a vector of antenna position coordinates rather than that of an error in its estimate as in the linear equations.

In general, we need to find a vector of unknowns, which would satisfy a certain criterion. For linear equations, the solution for a given criterion of the minimum of squared residuals between the calculated and measured values of the vector is well known and given by, for example, the LSE algorithm, as described in Chapter 3. In the general case of non-linear equations, the solution can be found using, for example, a brute-force method, which we shall consider first.

In the search for a solution one can define an area for the search. It can be better defined in local horizontal coordinates, because it will impose constraints on the third coordinate, which becomes the altitude. The extended state vector can be rewritten as follows:

$$\overline{X}'_{LOC} = \begin{bmatrix} E_r \\ N_r \\ H_r \\ \delta t_r \\ t_r \end{bmatrix}, \tag{9.5}$$

where E_r, N_r, and H_r are the east, north, and height components in local horizontal frame, respectively.

To complicate matters further, if we do not have this time estimate, the satellite positions cannot be treated as known.

We can assume that the sought position is in a certain area defined by the maximum and minimum values of the state vector components. The area, which is located in a five-dimensional space-time continuum with coordinates defined by state vector components, will be covered by a grid of possible $\overline{X}'_{LOC}$ solutions. In the brute-force method, the search will be conducted by changing one component of the state vector at a time by a step value:

$$\overline{X}'_{LOC}(i) = \begin{bmatrix} E_{min} + i_E \cdot \Delta E \\ N_{min} + i_N \cdot \Delta N \\ H_{min} + i_H \cdot \Delta H \\ \delta t_{min} + i_\delta \cdot \Delta \delta t \\ t_{min} + i_t \cdot \Delta t \end{bmatrix}, \tag{9.6}$$

where i_E, i_N, i_H, i_δ, and i_t define the point on the grid. The maximum number of iterations is in general different for each dimension.

Each resulting state vector should be calculated by evaluating a cost function. At each point we can recalculate the vector of measurements as

$$\tilde{\overline{Z}}(i) = \left[A(\hat{\overline{X}}'_{LOC}(i))\right]. \tag{9.7}$$

The criterion can be derived, for example, as follows:

$$C = (\overline{Z} - \tilde{\overline{Z}})^T \times (\overline{Z} - \tilde{\overline{Z}}). \tag{9.8}$$

The global minimum of the cost function C gives the sought position. It is very important that the cost function has multiple local minima, so standard optimization algorithms are not applicable for this task.

In general, a solution to a non-linear equation of this type can be found by finding a minimum of a given criterion. Syrjärinne [5] and Sirola [6] describe a method of finding position using the criterion similar to that in (9.8) using squared residuals, which are based on the code measurements to specific satellites.

In practice, criterion formation is more suitable for mobile devices if it is based on forming residuals in relation to a key or master satellite. In general, we can present such a criterion, or cost function, as follows:

$$C_1 = \sum_{i=1}^{N-1} \left(\left\{ f\left(P_i^m - P_j^m\right) - f\left(P_i^p - P_j^p\right) \right\} \right), \tag{9.9}$$

where P_i^m is the code phase measurement for the ith satellite, P_i^p is the predicted code phase for the ith satellite, P_j^m is the code phase measurement for a chosen master satellite, P_j^p is the predicted code phase for the chosen master satellite, and f is a function to be applied to the residuals (in particular, the time derivative).

For example, we consider the squared residual; then the equation can be rewritten in the following form:

$$C_1 = \sum_{i=1}^{N-1} \left(\left(P_i^m - P_j^m\right)^2 - \left(P_i^p - P_j^p\right)^2 \right). \tag{9.10}$$

Alternatively, the cost function can use the moduli of residuals,

$$C_1 = \sum_{i=1}^{N-1} \left(\left|P_i^m - P_j^m\right| - \left|P_i^p - P_j^p\right| \right). \tag{9.11}$$

If we are looking for a position in a two-dimensional plane, with fixed altitude, the search space can be significantly reduced by removing one dimension. This brute-force method currently does not allow real-time implementation, even with added constraints.

In AGPS, the cost function is required in the vicinity of a known position in a four-dimensional space-time continuum. Such a function is unambiguous. Figure 9.6 depicts a screenshot of the iPRx software receiver with two cross-sections of four-dimensional cost function for an initial position error of 10 km and a time error of 10 s. The corresponding pseudorange errors are on the order of a couple of kilometers. These two plots represent the AGPS case.

If assisted information about an approximate initial time and position is unavailable, the cost function has to be considered over a large area in four-dimensional space. The complexity and ambiguity of such a function make it impossible to use brute-force methods for real-time operation. The cost function for unknown initial time and position is depicted in Figures 9.7 and 9.8.

Figures 9.7 and 9.8 show the cost function for an uncertainty in initial position of 1000 km and a maximum time inaccuracy of 10 min. The left-hand plot in both figures shows the pseudorange residuals as a function of distance in meters from the central cell to the east and north directions. The right-hand plot in both figures shows the pseudorange residuals as a function of horizontal distance from the central cell and time offset in seconds. Figure 9.8 shows a two-dimensional projection of the plots in Figure 9.7(a).

We can see that in the case of small position and time uncertainty, as for AGPS, we can employ a downhill optimization method to find a solution for the non-linear equation (9.7). However, if the initial approximation from a network is not provided, we have the situation that the global minimum is masked by multiple local minima, although it is distinguishable. Figure 9.9 shows a blow-up of the plots similar to those

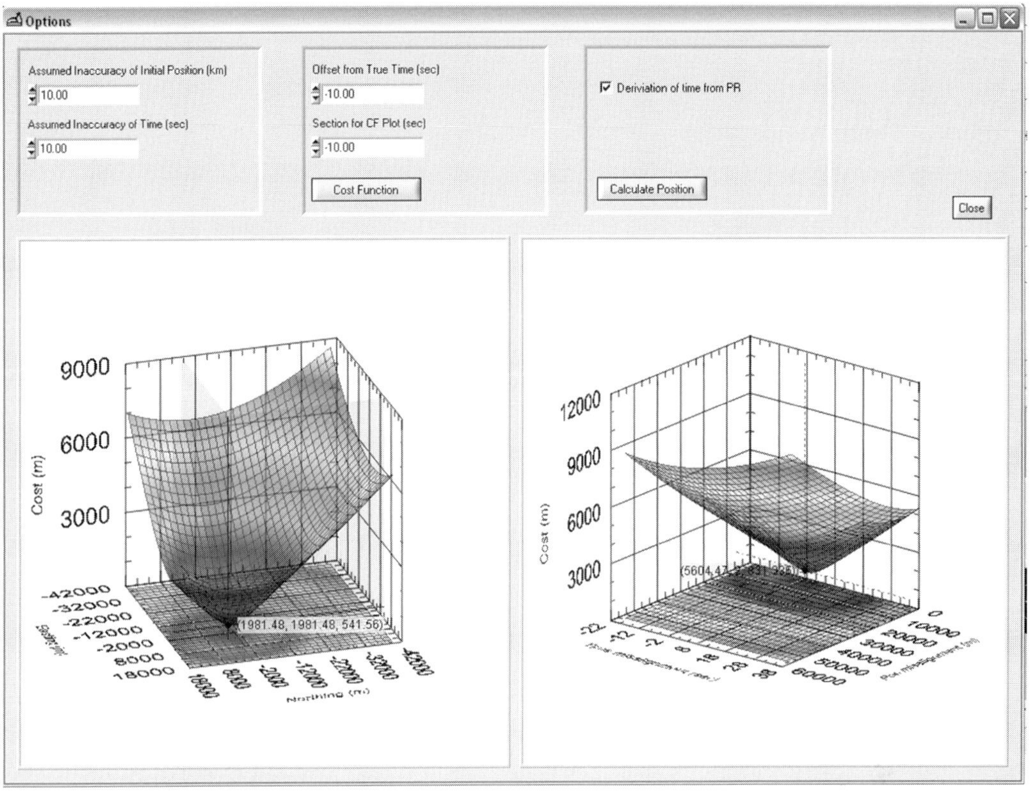

Figure 9.6 Screenshot of iPRx software receiver, with two cross-sections of four-dimensional cost function for AGPS.

in Figure 9.8 for a different search area. Figure 9.9(a) shows a two-dimensional projection of the pseudorange residuals as a function of distance in meters from the central cell for position uncertainty of 10 km and a time uncertainty of 10 min. Figure 9.9(b) also depicts a position uncertainty of 10 km, and the time uncertainty in this case is 1 min.

Using a brute-force method to calculate position without assistance information and an a-priori position may take rather a long time and require substantial resources. As a result, it may negate all the advantages of not reading the navigation message. In order to make real-time application feasible, it would be necessary to use an initial position with an accuracy of a few kilometers and to receive the time from a synchronized cellular phone network.

The methods and algorithms presented in the following allow us to overcome these limitations and provide us with an instant position fix without prior knowledge of user location and time of positioning. This permits the calculation of a position more than 30 times quicker than with conventional GPS algorithms and methods, i.e. within a few milliseconds, instead of 30 s using conventional GPS algorithms.

(a)

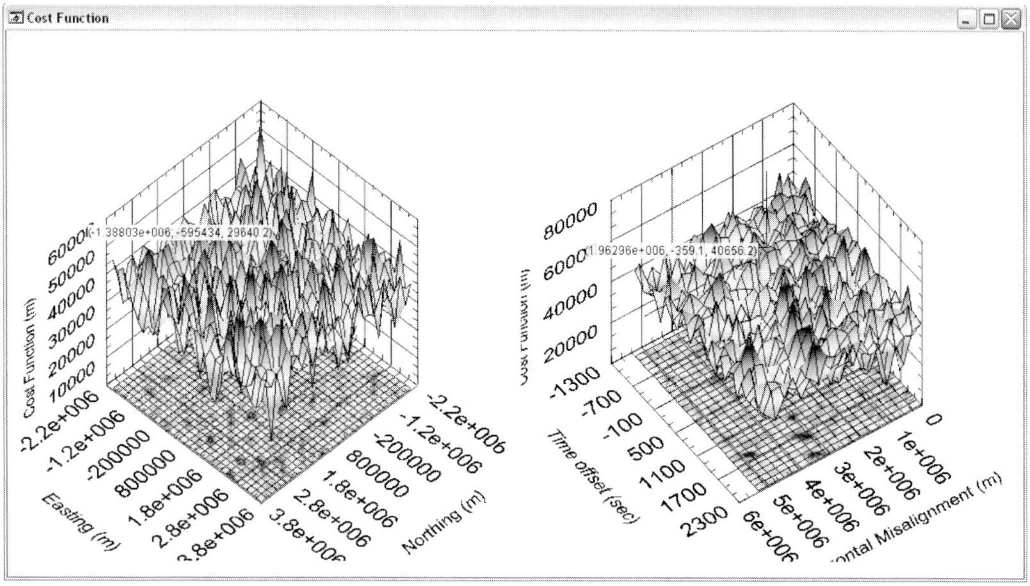

(b)

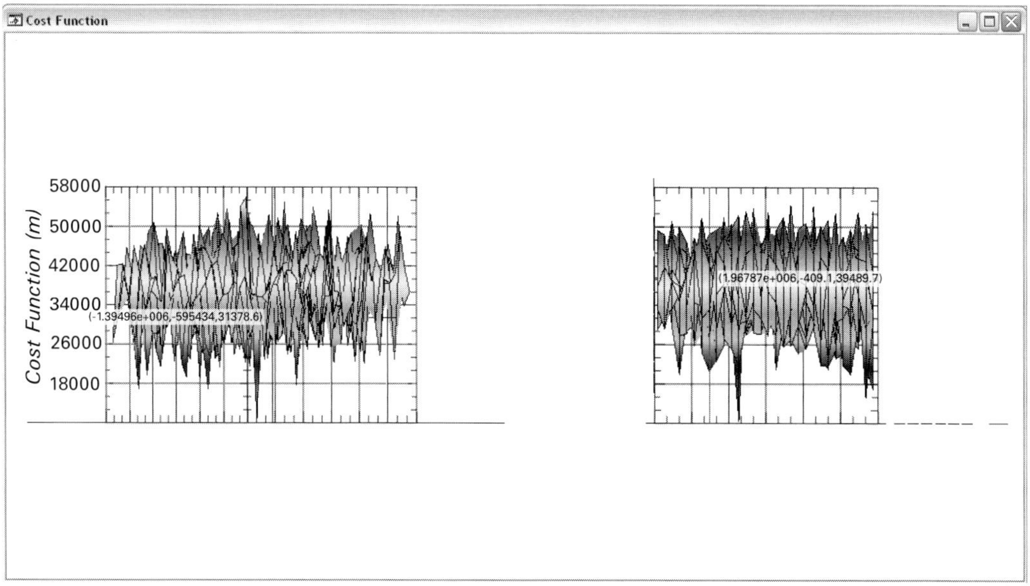

Figure 9.7 (a) Two cross-sections of four-dimensional cost function for unknown initial time and position, as for BGPS. (b) Projection of cost function from (a).

9.2.5 Algorithm criteria

We can now form a criterion that the algorithms need to minimize in order to get a solution. The criterion can be formed as a cost function, consisting of functions of differences in magnitude of the predicted code measurement functions and the

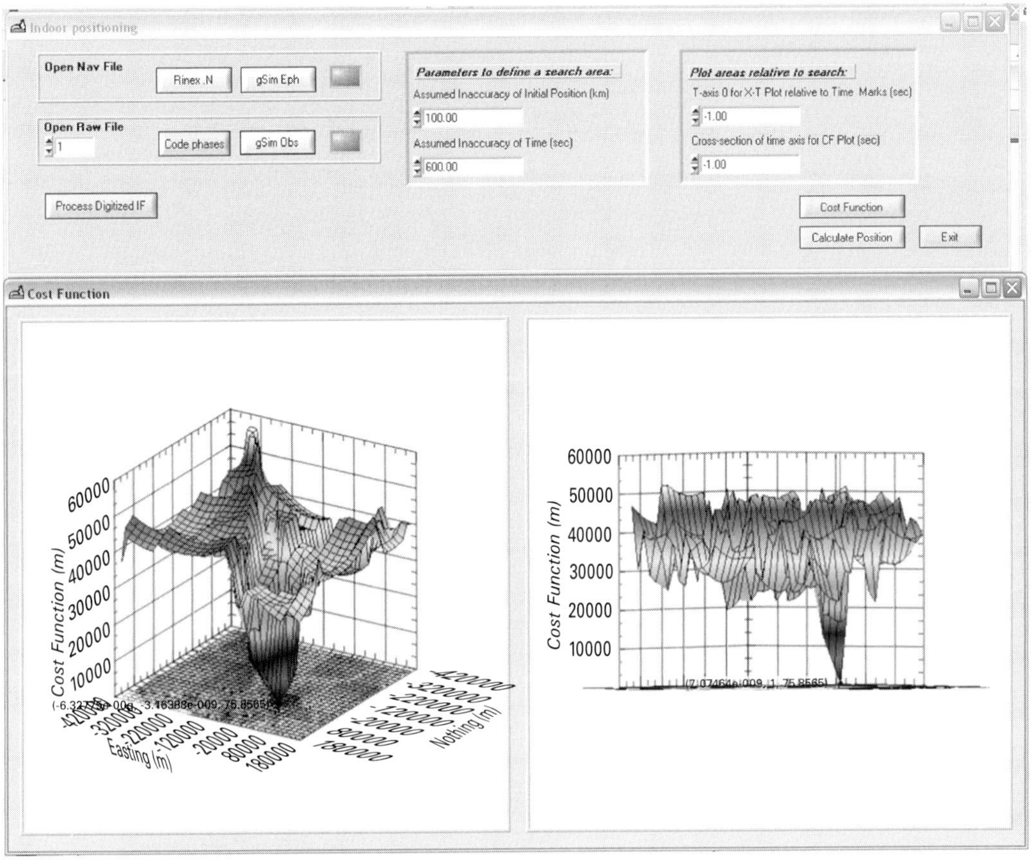

Figure 9.8 Cost function for a position uncertainty of 100 km and time uncertainty of 10 min.

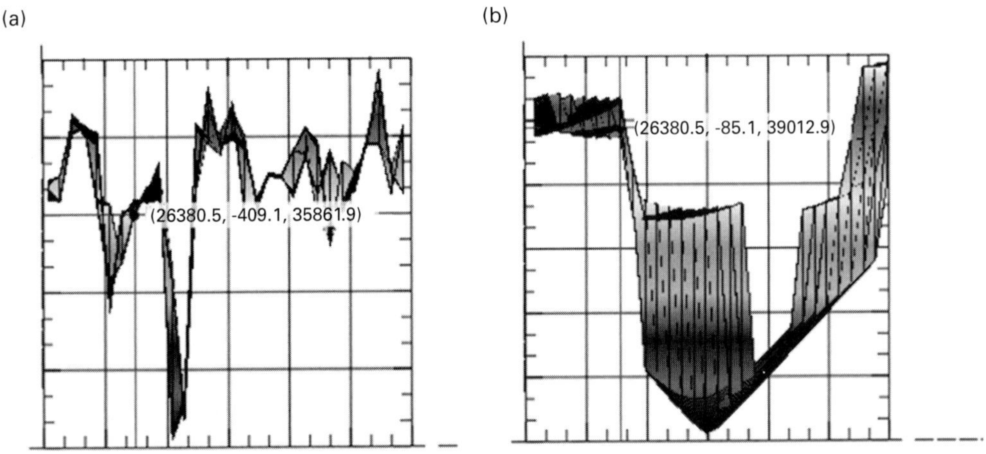

Figure 9.9 Cost function for position uncertainty of 10 km and time uncertainty of (a) 10 min and (b) 1 min.

code measurement functions similar to those in (9.9), but in a more general form. So, we have

$$C_1 = \sum_{i=1}^{N-1} \left(w_i \cdot g\left\{ f\left(P_i^m - P_j^m\right) - f\left(P_i^p - P_j^p\right) \right\} \right), \tag{9.12}$$

where P_i^m is the code measurement of the ith satellite, P_i^p is the predicted code of the ith satellite, P_j^m is the code measurement of a chosen master satellite, P_j^p is the predicted code measurement of a chosen master satellite, and w_i is a weight function based on the signal-to-noise ratio for the ith satellite signal.

To put constraints on the search area for the cost function, the receiver finds an approximate position estimate, using another cost function, which consists of functions of differences in the magnitude of Doppler measurements and predicted Doppler measurements:

$$C_2 = \sum_{i=1}^{N-1} \left(w_i \cdot g\left\{ f\left(dP_i^m - dP_j^m\right) - f\left(dP_i^p - dP_j^p\right) \right\} \right), \tag{9.13}$$

where dP_i^m is the Doppler measurement of the ith satellite, dP_i^p is the predicted Doppler measurement of the ith satellite, dP_j^m is the Doppler measurement of a chosen master satellite, dP_j^p is the predicted Doppler measurement of a chosen master satellite, and w_i is a weight function based on the signal-to-noise ratio for the ith satellite signal.

In order to mitigate the predicted satellite clock errors and facilitate usage of ephemerides with longer validity, the receiver can use another cost function, which is based on the functions of differences in magnitude of the differences over time of the prediction code measurement functions and the differences over time of code measurement functions in order to use orbit satellite data only for the positioning calculation as follows:

$$C_3 = \sum_{i=1}^{N-1} \left(w_i \cdot \left(g\left\{ f\left(P_i^m[t_{k+1}] - P_j^m[t_{k+1}]\right) - f\left(P_i^p[t_{k+1}] - P_j^p[t_{k+1}]\right) \right\} - g\left\{ f\left(P_i^m[t_k] - P_j^m[t_k]\right) - f\left(P_i^p[t_k] - P_j^p[t_k]\right) \right\} \right) \right), \tag{9.14}$$

where $P_i^m[t_{k+1}]$ is the code phase measurement of the ith satellite for epoch t_{k+1}, $P_i^m[t_k]$ is the code phase measurement of the ith satellite for epoch t_k, $P_i^p[t_{k+1}]$ is the predicted code phase measurement of the ith satellite for epoch t_{k+1}, $P_i^p[t_k]$ is the predicted code phase measurement of the ith satellite for epoch t_k, $P_j^m[t_{k+1}]$ is the code phase measurement of a chosen master satellite for epoch t_{k+1}, $P_j^p[t_{k+1}]$ is the predicted code phase measurement of a chosen master satellite for epoch t_{k+1}, $P_j^m[t_k]$ is the code phase measurement of a chosen master satellite for epoch t_k, $P_j^p[t_k]$ is the predicted code phase measurement of a chosen master satellite for epoch t_k, and w_i is a weight function based on the signal-to-noise ratio for the ith satellite signal.

9.2.6 Required a-priori information

The only assisting information required by BGPS is the predicted ephemerides, which are freely available from IGS through the Internet. The IGS provides a number of various services, including ephemerides predicted for a five-day period. The ten-day ephemerides can be calculated using only raw data from IGS stations. This means that

BGPS can provide an instant position fix using a connection to an outside source of information only once in ten days. Moreover, the information can be theoretically updated for a short period using broadcast ephemerides.

Orbit information can be valid for a month or even longer for most applications. The ephemeris information that is prone to aging is related to the satellite clock. With emerging satellite systems, clock information will last much longer. Therefore it will be possible to make an instant positioning with BGPS without any update information from a server within one month, or even a few months.

BGPS technology can use international geomatic services data to collect ephemeris information. There are no restrictions on the use of these data. Alternatively, the predicted data useful over even longer times can be calculated by the user or service provider. There are two options here. One is to use freely available IGS raw data and the other is to have a special network of reference stations. A proprietary network cannot compete with the IGS network in terms of the number and distribution of stations, but it can provide integrity and quality assurance.

Mitigating clock errors for predicted ephemerides using redundant measurements and a majority vote algorithm is required for practical implementations.

9.2.7 Time resolution in real time

9.2.7.1 Task example

We can assume that the internal reference device clock (for example of a GPS receiver, or cellular phone) can provide us with an initial time estimate with an accuracy, for example, of 5 s.

From this initial time of 5 s we can estimate the satellite position with an uncertainty of 20 km along the satellite track. We can assume that the user knows their position with some uncertainty, for example within 100 km. These parameters will define the search area. If the uncertainties in user position and/or current time are larger, the search area will also increase.

We assume a certain accuracy with which we can estimate our position. This mostly depends on multipath. For example, 50 m is the accuracy normally defined as standard for indoor positioning. The number of cells to search will be defined as the initial uncertainty in the user position divided by the available accuracy. In our example, the number of cells to search in two dimensions is 10 000 km^2/10 000 m^2 = 1 000 000 cells.

The number of satellite positions will depend on time uncertainty. In our example, 20 000 m/100 m = 200 positions. This is the same for all satellites. It means that for one epoch we have to check 2×10^8 solutions. For ten epochs we have to check 2×10^9 solutions. The set of solutions with time and position assumptions must be checked for solution consistency. The brute-force method can be combined with the downhill method, which should take over after the solution is in the vicinity of the global minima.

We designate a cost function to be the difference between the solution based on current measurements and the solution for the error-free reference station at the same position simulated in the rover processor. When the minimum of the cost function is found, the corresponding time–position pair will constitute the solution. The size of each cell will define the accuracy of the solution; in our example this is 100 m. From the solution, we find the time accuracy will be on the same scale as the multipath. In our example, we assume the rover position accuracy is 50 m, which leads to 150 ns for the time accuracy. If we consider less accurate initial assumptions of time and/or user positions, the process can be done in a few iterations.

The minimum of the cost function can be found by brute force. The brute-force method used to calculate position without assistance information usually does not permit any practical application, taking minutes, for example, to calculate position [6] even for small search areas. The method loses any advantage it may have over a conventional GPS algorithm, which takes about 6 s to retrieve a time mark from a navigation message. This section describes a method that allows the resolution of pseudorange ambiguity and calculates a position in less than 1 s, instead of the 6 s for conventional GPS algorithms. We describe methods that permit positioning in real time, which is essential for mobile applications.

9.2.7.2 Heuristic approach to search strategy

The optimization methods that allow us to reduce the load on a processor and shorten the time required for calculations include genetic algorithms and simulated annealing methods for finding global minima of multiple-minima functions. The methods are described in many references; see, for example, [9]–[12].

In solving equations (9.3), we need to find a value for the state vector (9.4) that minimizes the chosen criterion. This can be achieved via various methods.

(1) Brute-force method. This method is not applicable for real-time applications, because there are too many cells in the search space.

(2) Standard gradient-based methods. These methods can only find local minima. They stick with the first minimum they encounter. The search space for time resolution has multiple minima (see Figure 9.7).

(3) Random search methods. In a way, this method is similar to the brute-force algorithm, because optimum results can be expected only when all the search space is covered. It can run into the global minima by chance, but the result cannot be used practically before the complete area is covered.

(4) Heuristic algorithms. One of the working algorithms well suited for the particular application is a *genetic algorithm*. The genetic algorithm has been implemented in the iPRx software receiver. The test results (see Figure 9.10) show that positioning was available within 1 s after the receiver was turned on. The position output with rate 1 Hz was based on independent position fixes. In a real device, sequential position fixes do not require code ambiguity resolution.

The scatter plot shown in Figure 9.10 demonstrates a rather poor accuracy, with about 30 m RMS. This accuracy was limited by the sampling rate, because very short (a few milliseconds) chunks of data were used. Though it is possible to improve accuracy without tracking, this was not the purpose of the implementation at this stage.

9.2.8 Preliminary position estimation methods

BGPS does not require immediate information from a server. However, any constraints in terms of initial position or time may help to reduce both the time required for calculations and the processor load.

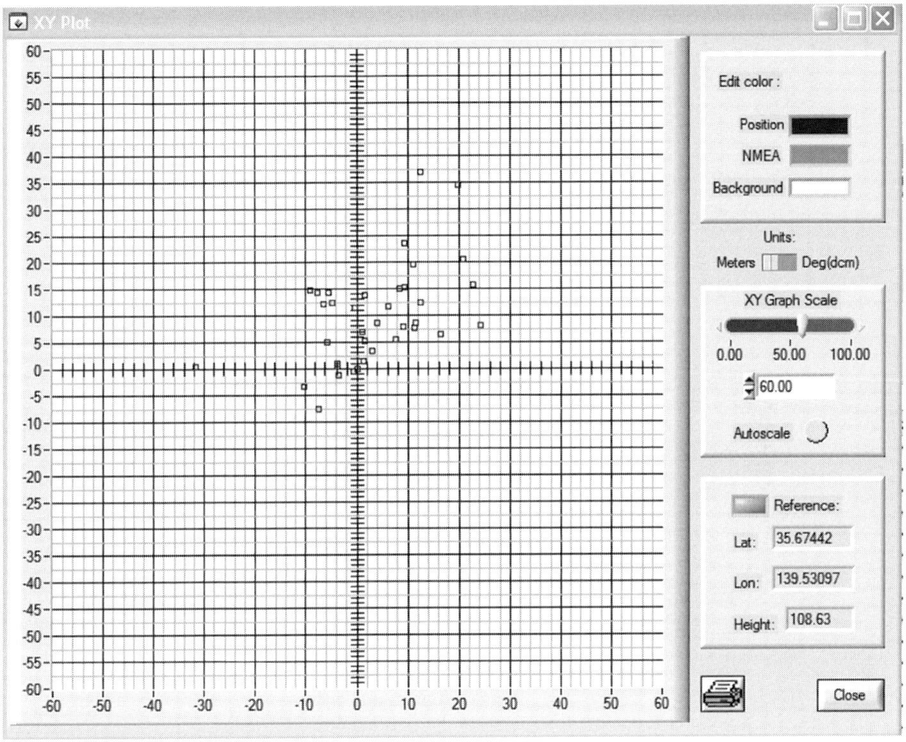

Figure 9.10 Test results for iPRx receiver in BGPS mode.

Initial approximations can be obtained, for example, by introducing a country name into a receiver. Another constraint can be introduced from the PRN (pseudo random noise) numbers of visible satellites. This limits the area because certain satellites can be visible only from certain areas at a given time.

The most significant contribution to narrowing the initial search area comes from Doppler measurements. The use of Doppler measurements for pseudorange ambiguity resolution was proposed in a JPL report in 1997 [3]. It is not useful to use Doppler measurements alone for positioning, especially without tracking loops, because the accuracy of such positioning will be very poor. However, it is good enough to put constraints on the initial search area to find global minima of multiple-minima functions. The search area can be easily minimized to a few hundred kilometers with Doppler measurements.

9.2.9 Instant positioning implementation in a device

BGPS implementation requires an acquisition module, correlators, a position calculation module, memory, a time reference module, and a clock.

The device implementation is depicted in Figure 9.11(a). It is using its own time source. It does not use a wireless communication transceiver or any other

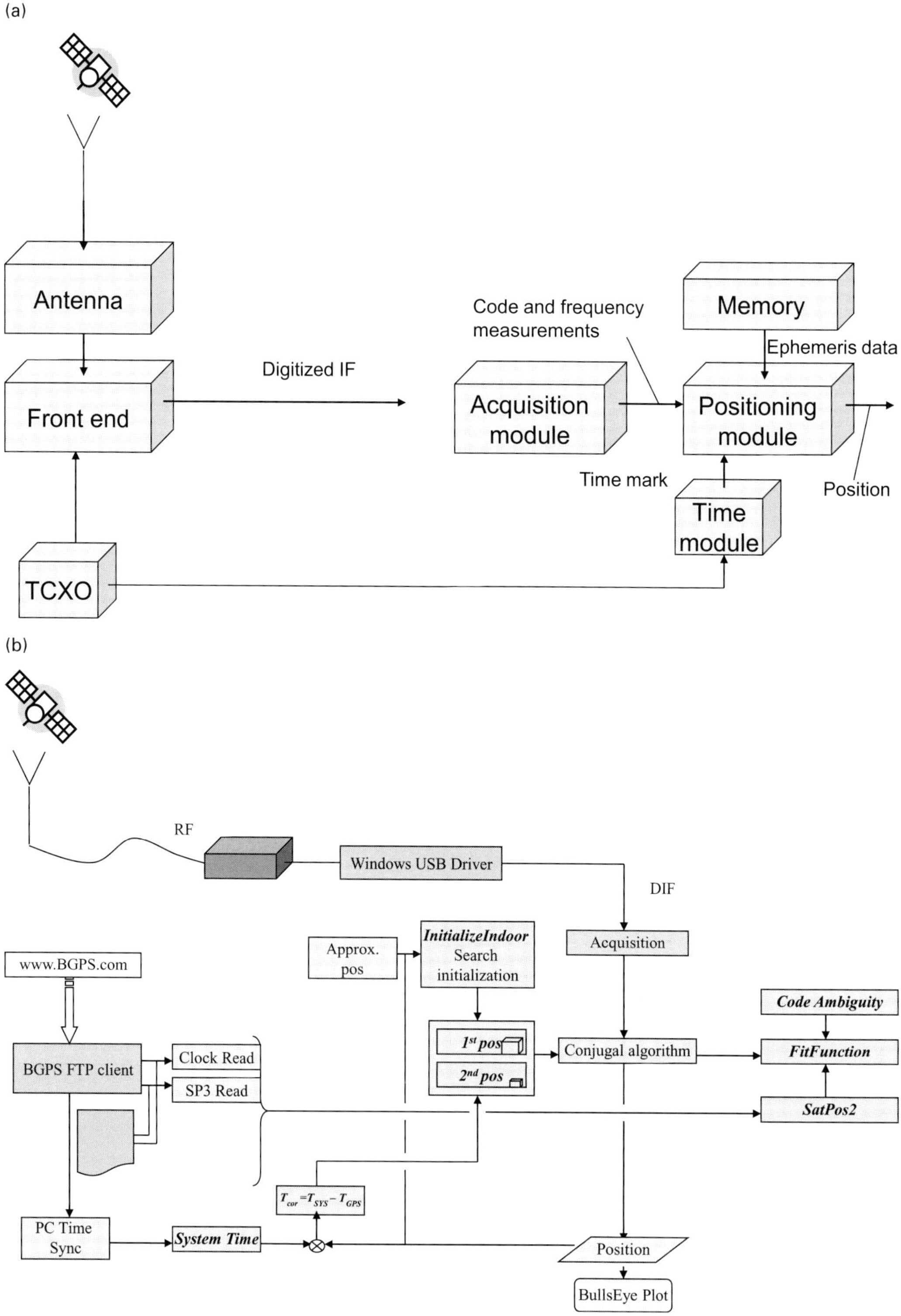

Figure 9.11 BGPS implementation in a mobile device.

data link to supply the assistance data. Its distinguishing feature is that the receiver does not require a data link to a network at the time of positioning. Therefore it can make an instant positioning outside of AGPS networks and can be used for a much wider range of applications. Figure 9.11(b) shows the details of the implementation of the positioning methods and their interface with infrastructure, required for preparing information for a prototype on a personal computer.

The cost function has multiple local minima. The position calculation module uses a genetic or simulated annealing algorithm, which we mentioned in Section 9.2.7.2, to provide a realization of the search cost function global minima.

Prior to positioning, predicted satellite ephemerides can be downloaded into a memory module; we consider sources for this information later in this chapter.

The acquisition module (after acquiring signals) outputs the PRN, the number of acquired satellites, Doppler measurements, and code phase measurements.

In addition, an algorithm used to locate a particular area can use predicted satellite ephemerides and system clock to find a region in which satellites with the same PRN numbers as those of acquired satellites are visible for given time (host device time) and satellite orbits. The region defines the initial search area, which is supplied to a downhill search algorithm. This uses, for example, a downhill simplex method [13] or similar to find the minimum of a cost function with only one minimum. The cost function is formed by a Doppler-based cost function forming algorithm, using GPS Doppler measurements and predicted Doppler measurements. Predicted Doppler measurements are calculated using predicted satellite positions in the time search area and antenna positions in the coordinate search area. The time of reception and the antenna position form the four-dimensional search area. The receiver clock error is excluded from these measurements by forming single differences between satellites and a chosen key satellite, typically with the highest elevation angle.

The BGPS family methods use predicted ephemerides. The predicted orbit data can be used over very long times without updates, and still provide good positioning accuracy. Satellite clock error prediction data degrade much more quickly. It is important for such methods to be able to use only orbit data, because this increases the validity period for predicted ephemerides; thus, mitigating clock errors for predicted ephemerides is an important component of the algorithms.

BGPS technology is applicable to any GNSS signals, such as GPS, GLONASS, Galileo, and BeiDou, the only requirement being that GNSS signals are acquired from enough satellites. The data from the acquisition module can be immediately processed to find a position. As a result, such technology can successfully replace AGPS in many applications, especially those which cannot easily gain access to a network or in which the network is not synchronized. BGPS uses global data that are not localized. A user can obtain data in one country, then go to another and switch the receiver on almost a week after the data were downloaded and get a position fix in under 1 s.

9.3 Precise positioning without reference station

9.3.1 From a network to the global network

9.3.1.1 Global correction information for mobile devices

As we have discussed in Chapter 8, in order to achieve enhancements in accuracy, we need corrections data from a reference receiver. The coordinates of the reference receiver antenna should be known precisely.

If we want to set up a high-accuracy receiver as a reference station, we need to survey the reference antenna position in advance. Before processing of the measurements is possible, special attention is required to ensure that the antenna is positioned in a low-multipath environment. The receiver raw data should be collected from the receiver within one day, better still over three days in a row. The measurements can be processed using data from the global IGS network, which are available free of charge for the benefit of the global GNSS community [14]. These data can be processed with GNSS software, such as Bernese GPS from the Astronomical Institute of the University of Berne, GAMIT from MIT, or Gipsy from JPL.

What result from the processing are the accurate antenna coordinates. Now the receiver measurements can be used as corrections for rover receivers in the same vicinity.

The measurements from the reference receiver are used to compensate for major parts of the error budget in the rover receiver. As the distance to the reference station increases, the correlation between the reference receiver and the rover receiver errors decrease. The rule of thumb is that there is a limit of 10 km over which the reference receiver carrier phase observables may not be used for rover RTK positioning. The code differential positioning will work at any distance, but the accuracy will be degraded due mostly to ionospheric error. In order to compensate for this error, we can use a few distant reference receivers instead of just one. We have discussed referenced positioning with a network in Chapter 8.

Surveying the reference station positions is usually carried out using the global International GNSS Service (IGS) network as a source of correction data for the surveyed receivers. When the exact coordinates of these receivers are available, those receivers are used in turn as the reference receivers for a rover receiver. Why do we not use these global IGS corrections directly as a reference network for a mobile device, essentially skipping an intermediate step to make the reference corrections global?

Although this would be possible, it would require a lot of data to be transferred to a mobile device and a lot of processing power to operate on these data. However, when we process IGS data, we do not use reference receiver measurements. The IGS processing center uses measurements from the global network; in this way, a huge amount of corrections data is transferred into the final products, which are precise ephemerides and ionospheric maps. These data actually accumulate all the correction data. Figure 9.12 demonstrates this approach graphically.

When we survey the coordinates of our receiver with these geodetic software, we do not use the raw data from the global network, but rather the final IGS products that accumulate these data. In a similar way, we need only the IGS final products to be

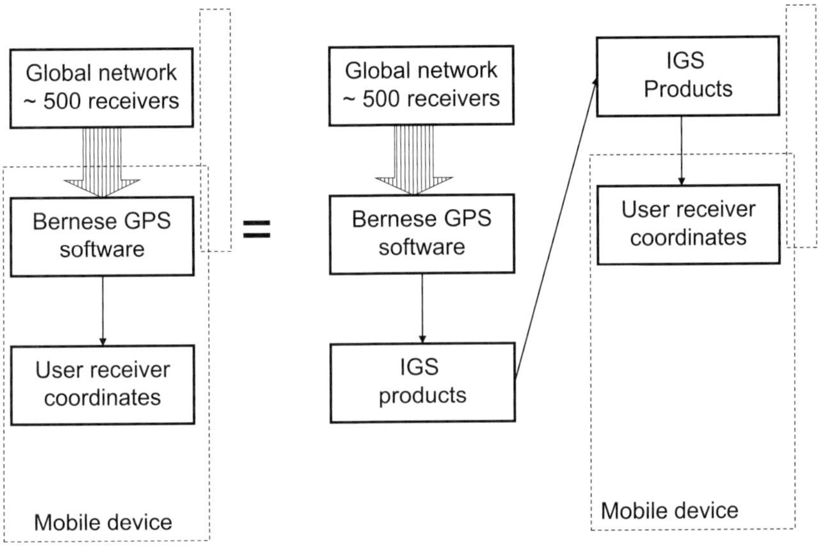

Figure 9.12 Receiver positioning using global network vs. IGS final products. Bernese GPS software is used as an example only.

presented in the rover receiver. The size of these data is small, and processing does not differ much from the algorithms we have considered so far.

9.3.1.2 Free global corrections

A global reference network, such as IGS, is an example of an extreme case of network positioning, with the important benefit that IGS products are of high quality and have been freely available to the worldwide GNSS community via the Internet since 1994. These products can always be used instead of a regional or local network. In reference to a network RTK, the method can actually be used without ambiguity fixing to integers or with float ambiguities.

One issue that arose in the past was that precise ephemerides were provided with a three-day delay. The only other available data comprised predicted ephemerides. This situation has now changed, and IGS provides even more services suitable for mobile applications. The IGS real-time service (RTS) allows real-time access to IGS products, providing orbit and clock corrections based on the IGS global infrastructure of network stations. GLONASS data are also provided. Other GNSS constellations will be added as they become available.

9.3.1.3 Orbit prediction

BGPS requires predicted ephemerides to be stored in a mobile device in advance. There are a few ways to obtain these predicted orbits.

(1) Five-day prediction orbits may be downloaded via the Internet from IGS.
(2) Raw observation data in a RINEX format can be downloaded from IGS for the receivers that constitute the global network. The predicted orbits can be

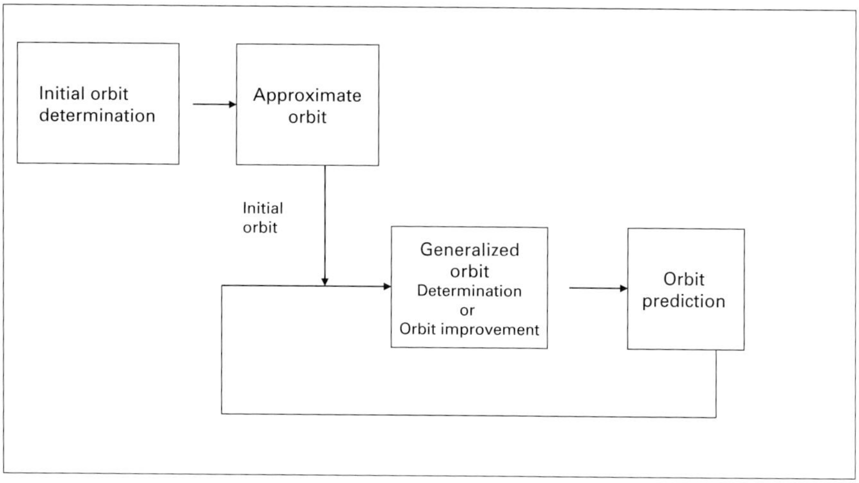

Figure 9.13 Orbit prediction flowchart.

calculated using, for example, Bernese GPS software, which is used to calculate the orbits that are part of the CODE products. This allows calculation of orbits over a longer prediction interval. The author has calculated predicted orbits over a period of ten days and successfully used them for positioning.

(3) One can establish one's own global or regional reference network. The observation data from this network can be used in a similar manner to those in (2). Such a network cannot compete with the IGS in terms of accuracy, but it can add some flexibility, and possibly integrity, into the service.

In this section, following [15] and [16], we briefly consider how predicted orbits can be calculated. This is important if we require orbit data to be available in a handset for a period longer than five days. The accuracy degradation in the predicted orbits after five days, and up to two weeks, is still insignificant for many applications.

In the first step (Figure 9.13), an a-priori orbit has to be determined. IGS or broadcast orbits can be used to gain initial information on a-priori satellite orbits. The a-priori orbit can be specified to be known to an accuracy of a few meters along each axis. The accuracy of an a-priori orbit, however, is not important; if it exceeded 20 m, a second iteration step in the orbit improvement would be required. The second step is orbit determination. In order to describe satellite orbits with enough precision for positioning, the Keplerian model must be refined. A satellite orbit is not perfectly elliptical, due to numerous forces that affect the satellite movement. The equation for satellite movement can be written in a general form as follows:

$$\frac{d^2\vec{r}}{dt^2} = -GM\frac{\vec{r}}{r^3} + \vec{f}\left(t, \vec{r}, \frac{d\vec{r}}{dt}, p_0, p_1, \ldots\right), \tag{9.15}$$

where the non-linear function $\vec{f}(t, \ldots)$ includes all the forces that are functions of the satellite position at a particular epoch, which can be sufficiently well modeled, and the

Table 9.1. Contributions from various forces to satellite movement[a]

Force	Satellite acceleration caused by force (m/s^2)	Error in satellite position after one day, if the force is not accounted for (m)
Oblateness of the Earth	5×10^{-5}	10 000
Lunar gravitation	5×10^{-6}	3000
Sun gravitation	2×10^{-6}	800
Other harmonics of the Earth's gravitational field	3×10^{-7}	200
Sun radiation pressure	9×10^{-8}	200
Y-bias	5×10^{-10}	2
Solid Earth tides	1×10^{-9}	0.3

[a] After [17].

unknown parameters p_i are associated with radiation pressure. These parameters should be estimated along with the satellite position $\vec{r}$ if included in the consideration. All these forces, including other harmonics of Earth's gravitational field, have less, but a still significant, impact on satellite movement that is periodic in nature. Therefore, all Keplerian parameters, semi-major axis, eccentricity, inclination, ascending node, and argument of perigee are oscillating. Consequently, the orbits cannot be precisely described as ellipses; the real orbits can be presented as an envelope of a set of arcs with specific osculating Keplerian orbits, each corresponding to positions of the satellite within a specific interval.

Satellite orbits are calculated based on precise models of the forces on the satellites in (9.15) and multiple observation sets from network receivers for periods from three to five days. Table 9.1 summarizes the contributions to satellite movement from various forces by combining results from [15], [16], [18], [19], and [20].

Up-to-date data are required to model these forces. Even when the predicted orbits are calculated using a proprietary network, the following data from the IGS are required:

- data files with Earth rotation parameters; this information originally comes from the International Earth Rotation and Reference Frames Service (IERS) Bulletin A or B;
- ocean loading data files.

Information about the satellites can be accessed through IGS or satellite system information. It includes the following:

- information about satellite problems and maneuvers;
- data on satellite parameters, such as antenna and biases.

Finally, the data can come either from a proprietary network or from IGS. These data should include the following:

- observations (code and carrier phases, Doppler, and carrier-to-noise ratios); observations from IGS are usually in RINEX format;
- antenna phase center information;
- antenna height information.

Orbit estimation then can be seen as an orbit improvement achieved by fitting the solution of (9.15) to observations. Orbit prediction is carried out by an extrapolation procedure. In this case, it is desirable to use more observation data from the preceding days.

SP3 format

The National Geodetic Survey, National Ocean Service, NOAA SP1 (where SP is an abbreviation for standard product) format was introduced by B. Remondi in 1985 [21]. A further enhancement of this format, in particular concerning satellite clock errors, was introduced later in [22] and [23], which defined the SP3 format as it is used today. The SP3 format becomes de facto standard for writing precise ephemerides, especially those widely used in geodetic applications, which as we shall see later, became closely related to mobile applications. The SP3 format can support all GNSS. An example of the SP3 format is given in Figure 9.14.

```
#cP2013  5 19  0  0  0.00000000      96 ORBIT IGb08 HLM  IGS
## 1741      0.00000000   900.00000000 56431 0.0000000000000
+   31   G01G02G03G04G05G06G07G08G09G10G11G12G13G14G15G16G17
+        G18G19G20G21G22G23G24G25G26G28G29G30G31G32  0  0  0
+          0  0  0  0  0  0  0  0  0  0  0  0  0  0  0  0  0
+          0  0  0  0  0  0  0  0  0  0  0  0  0  0  0  0  0
+          0  0  0  0  0  0  0  0  0  0  0  0  0  0  0  0  0
++         2  2  2  2  2  2  2  2  2  2  2  2  2  2  2  2  2
++         2  2  2  2  2  2  2  2  2  2  2  0  2  2  0  0  0
++         0  0  0  0  0  0  0  0  0  0  0  0  0  0  0  0  0
++         0  0  0  0  0  0  0  0  0  0  0  0  0  0  0  0  0
++         0  0  0  0  0  0  0  0  0  0  0  0  0  0  0  0  0
%c G  cc GPS ccc cccc cccc cccc cccc ccccc ccccc ccccc ccccc
%c cc cc ccc ccc cccc cccc cccc cccc ccccc ccccc ccccc ccccc
%f  1.2500000  1.025000000  0.00000000000  0.000000000000000
%f  0.0000000  0.000000000  0.00000000000  0.000000000000000
%i    0    0    0    0      0      0      0      0         0
%i    0    0    0    0      0      0      0      0         0
/* RAPID ORBIT COMBINATION FROM WEIGHTED AVERAGE OF:
/* cod emr esa gfz jpl ngs sio usn whu
/* REFERENCED TO IGS TIME (IGST) AND TO WEIGHTED MEAN POLE:
/* PCV:IGS08_1740 OL/AL:FES2004  NONE     Y  ORB:CMB CLK:CMB
*  2013  5 19  0  0  0.00000000
PG01  13644.158073 -18836.920110  12728.487730     27.576991  7  6  9 123
PG02 -14646.775754  -5795.571279 -21205.997146    432.802493  9 10  4 119
PG03  23203.451690   6460.584762  10919.476968    208.641314  6  9  8 115
PG04  -9841.714128 -16788.391583 -18237.562966    221.670784  9  8  6 102
PG05 -25001.316073    690.076154  -9172.344697   -394.956191 10 11  9 150
PG06  22777.861570  12117.111198   7159.773573    226.295917  4  8  9  88
PG07   5618.173073 -25751.366866   1473.946443    215.298466 10  8  8 126
PG08    103.033277 -23396.302673  12078.032378      6.203766 11  7  8 137
PG09  -9982.684001 -14242.540171  19486.192325    241.803793  8  8  8 123
PG10 -13566.380994 -11385.021428 -20198.910203    -80.330631 10  9  8 113
PG11  13915.248930 -12849.804161  18087.855686   -383.183693  8  7  7 110
PG12 -22972.883302  11009.794980  -7148.884114    134.521561 10  7  7 144
PG13   2318.060313 -17584.594381 -19870.244310    117.954903 10  6  9 110
PG14  14220.773910  21701.290210   6239.765484    220.857593  5  8  8  90
PG15 -17222.106199   4040.915648  19854.503780   -119.674253  8 10  8 113
PG16  24864.801874    595.156854  -9813.141245   -252.022340  4 10 10 131
PG17 -14610.469266 -21405.086604   6568.804624     18.095692  5  8 11 133
PG18  -2638.059090  16927.389068  20631.020608    254.361829 11  4  7 119
PG19  18800.111585   1534.478016  18777.762492   -374.015443  7  9  8 117
```

Figure 9.14 IGS ephemeris data in tabular SP3 format.

Continued

Continued

Lines from 1 to 24 represent the file header, which describes

Line 1 details the type of ephemeris (FIT (fitted), EXT (extrapolated or predicted), and BCT (broadcast)), and name of the computing agency.
Line 2 shows the start epoch at the beginning of the orbit and the interval.
Lines 3–7 give the number of satellites and their respective identifiers.
Lines 8–12 specify orbit accuracy.
Lines 13–22 are reserved for further use, additional parameters, or modifications.
Line 23 is the epoch header date and time.
Line 24 is the satellite ephemeris parameters; "P" in the first column indicates that the parameters are coordinates; "V" indicates that the parameters are velocities. Coordinates are given in kilometers with an accuracy down to 1 mm. The velocity components are given in decimeters per second, with an accuracy down to 10^{-4} mm/s. The last column represents clock errors in microseconds, with an accuracy down to 1 ps, in the case of the coordinate line, and the rate of change of clock correction given in 10^{-4} μs/s in the case of the velocity line, with accuracy 10^{-10} μs/s.

Bad or absent coordinate values are set to 0.000000. Bad or absent clock values are to be set to _999999.9999.

9.3.2 Embedded algorithms

9.3.2.1 Satellite ephemeris interpolation procedure inside mobile device

Broadcast GNSS ephemerides are estimated by ground segments which belong to the corresponding country authorities. The international geodetic community also calculates ephemerides for some constellations, and these are freely available via the Internet from the IGS. The common format for ephemerides is the SP3 format (see Box). The ephemerides are free for all applications, including commercial ones [14]. Available ephemeris products are summarized in Table 9.2.

These global ephemerides may be used in a tabular format for navigation; this avoids having to wait until the navigation message is decoded by a receiver. It also allows a receiver to calculate a position using short chunks of GNSS signal, as may be required in snapshot positioning, in particular for indoor applications. In order to use these ephemerides for navigation solutions, we may use geometrical interpolation within the points of tabular orbits without applying numerical integration or considering force models.

There are a number of various types of interpolation suitable for this purpose. An analysis of comparison and accuracy for both polynomial and trigonometric interpolations is given in [24]. The first such interpolation method was devised by Isaac Newton. In the following, we consider an interpolation employing a Fourier series.

Table 9.2. Some of the available GNSS ephemerides update

Ephemerides	Orbit accuracy (cm)	Conditions	Availability
Broadcast	<100	none	real-time
Real-time	~10	free	real-time
Predicted 24 hours	<10	free	predicted
Predicted five days	<20	free	predicted
Final orbits	<5	free	after five days

We use a sidereal day as our base period for the interpolation, as osculating Keplerian parameters have these harmonics [25]. Therefore the trigonometric series can be expressed as follows:

$$x_i = A_{i0} + A_{i1} \sin\left(\frac{2\pi}{T_{SD}}t\right) + B_{i1} \cos\left(\frac{2\pi}{T_{SD}}t\right) + A_{i2} \sin\left(\frac{4\pi}{T_{SD}}t\right) \ldots + B_{in} \sin\left(\frac{2n\pi}{T_{SD}}t\right), \tag{9.16}$$

where x_i are the orbit coordinates and $T_{SD} = 0.99726956634$ is one sidereal day.

Given $2n + 1$ measurements, which are the coordinate values in the tabular orbit table, we can estimate coefficients A_{ij} and B_{ij}.

If we are not sure about what the harmonics in these series are, we can estimate them from the data.

It is shown that implementation of five terms in the interpolation series gives a standard deviation error of less than 0.4 cm and a maximum error of less than 70 cm. We can assume that three terms will give us approximately the same accuracy as broadcast ephemerides. The optimum number of seven members gives a standard deviation error of less than 0.1 cm and a maximum error of less than 8.2 cm. This provides limits on the orbit accuracy achievable by pure geometric approximation of precise ephemerides. Using precise ephemerides with force models allows us to achieve millimeter-level accuracy. In that case, numerical integration with force models should be used instead of geometrical interpolation methods.

As we can see, there are trigonometric series also in broadcast GPS ephemerides representations. Accuracy analysis also implies that broadcast ephemerides are limited in accuracy due to the limited number of members in their series, rather than to anything else.

9.3.2.2 Precise error models

As we have established in Section 9.3.2.1, the real-time, and even the predicted, ephemerides are available with centimeter-level accuracy. We cannot use carrier phase measurements to determine an absolute position without a reference station. In order to achieve centimeter-level positioning, we need code phase measurements in a mobile device with the same level of accuracy. As we know from Chapter 3, code phase measurements are corrupted by various errors on a much larger scale.

The method of precise point positioning (PPP) is employed to remove all possible errors in GNSS observables down to the centimeter level and so use accurate ephemerides for positioning.

An essential part of PPP therefore is the application of precise models for the GNSS error budget. The PPP algorithms should include extended models besides those basic models considered in Chapter 3 when we discussed the GNSS error budget.

The geodetic community have developed extremely precise, sophisticated models which are able to account for basically all the errors in the GNSS signal. Resulting from this was the capability to define coordinates with a unprecedented millimeter-level accuracy.

These models may include the Earth geopotential model, such as JGM3, models of the gravitational attraction of the Sun, Moon, Jupiter, Mars, and Venus, using DE200 JPL ephemerides, elastic Earth tides, ocean tides, general relativistic corrections, and effects of the polar tide. A mobile device should be able to store these additional data sets related to these models, which should be used as the input for PPP algorithms. The data are similar to those required for orbit determination, and may include Earth orientation information, ocean loading tables, and JPL ephemerides. For single frequencies, IGS ionospheric maps can be used for some periods of time. For dual-frequency devices, ionospheric error can be compensated.

9.3.2.3 Filtering

As described in the preceding sections, IGS ephemerides and atmospheric model products can be treated as corrections from the extended ground network. The only limitations are related to the fact that the integer character of the carrier phase ambiguities is lost in these corrections. If we treat the ambiguities as float ambiguities, we should still be able to achieve an accuracy similar to that in RTK using only IGS products. Therefore, PPP uses corrections, but these corrections are encapsulated into the IGS product rather than being used in their implicit form. The advantage is that we can use predicted IGS products and thus be able to achieve high accuracy in real-time modes without a data link, which should provide real-time corrections.

The last essential part of PPP is filtering. Code phase measurements are essential for PPP. The noise associated with code phase measurements is relatively high; for example, for GPS L1 C/A it is around 30 cm. The required accuracy for PPP is on the centimeter scale, and therefore the filter should be applied to reduce code phase noise. The pseudorange observables can be combined with carrier phase observables using various filters. We consider here a simple filter for combining pseudorange and carrier observables as it was proposed by R. Hatch [26]. Smoothed pseudorange observables for the ith satellite, P_{s_i}, can be calculated as a weighted sum of code and carrier phase increment as follows:

$$P_{s_i}(t_k) = W_p \cdot P_{s_i}(t_k) + W_\phi \cdot [P_{s_i}(t_{k-1}) + \Phi(t_k) - \Phi(t_{k-1})]. \qquad (9.17)$$

The weight coefficients W_ρ and W_ϕ are defined as follows:

$$\begin{cases} W_\rho = 1 - \dfrac{k-1}{N_S}, W_\phi = \dfrac{k-1}{N_S} \quad (k = 1, 2, \ldots N_S), \\ W_\rho = \dfrac{1}{N_S}, W_\phi = 1 - \dfrac{1}{N_S} \quad (k > N_S), N_S = \tau_S/T_S, \end{cases} \tag{9.18}$$

where τ_S is the smoothing time constant and T_S is the sampling interval. For example, if the time constant is 100 s and the sampling rate is 1 Hz, then $N_S = 100$. After 100 s, the weight coefficient for code phases (W_ρ) is 0.01 and the weight coefficient for carrier phases (W_ϕ) is 0.99.

If the multipath error and the ionospheric error can be considered to be small for the particular application, the pseudorange noise is effectively reduced and more precise range information can be obtained. However, if there is a significant multipath error, the corresponding range error remains for a long time since the multipath error of the pseudorange is much larger than that of the carrier phase, and they never cancel.

Ionospheric delay affects the smoothing process because the signs of the delay for the pseudorange and for the carrier are opposite. Therefore, some methods have been proposed that reset smoothing filters after a specified interval [27]. The code–carrier divergence becomes evident after about 30 min of filtering depending on receiver noise and multipath errors.

We can say therefore that the PPP method has three essential components, which are as follows.

(1) Global corrections. Usually PPP refers to the situation for which global corrections are available in real time, for example from a satellite service. The difference for reference positioning is that, because corrections are global, they do not allow for carrier phase ambiguity resolution. However, as we have seen, global corrections can be considered to be encapsulated in IGS products. Because the corresponding products can be predicted, the global corrections can also be used offline, which allows for PPP to be made without immediate data link.
(2) Precise models.
(3) Noise filtering.

9.3.2.4 The catch

So, what is the catch? In fact, there are two.

(1) Low integrity. Predicted ephemerides have a low level of integrity in comparison with broadcast data.
(2) Poor long-period satellite clock predictability. Satellite clocks are predicted with much less accuracy and low integrity, although it needs to be said that the broadcast satellite clock information is also predicted. The satellites use atomic clocks, which are very precise, but they still can create very large position errors. The satellite clocks are measured and estimated by ground stations, and the estimations of satellite clock errors are made by ground facilities for the

> following 24 hours or more. Then the estimates are uploaded to the satellites as parameters of the clock error model to be transmitted as part of the navigation message, so users will be able to compensate for satellite clock errors in their receivers. In the case of referenced positioning satellites, clock errors are almost completely canceled on short and medium baselines.

Estimates of the satellite clock errors form part of the satellite navigation message. For GPS, they are estimated in the form of a short second-order polynomial. They comprise a constant component, a component which is a function of time, and a component which is a function of time squared.

In the case of IGS, the clock errors are included in the precise ephemeris files, produced by the IGS on a day-by-day basis. These estimations of clock errors are presented as error magnitude values in seconds at the particular epoch. Usually, the precise ephemerides are calculated with a 15 min interval. To obtain data within this interval, one needs to interpolate them to the specific epoch of interest within the interval. Depending on the required accuracy, the interpolation can be performed using a system of equations that describes the satellite movement, taking into account various models or just a number of neighboring points in the ephemeris file. Figure 9.14 shows clock errors measured by IGS in the SP3 format.

Exercise

Estimate the currently achievable accuracy in clock and orbit prediction by comparing five-day predicted IGS ephemerides with IGS final products for all satellites. Compare how the orbit and clock errors differ for various satellites. Estimate the possible effects of these errors on positioning.

9.4 Applications

BGPS can be implemented in a mobile device for a conventional receiver, a tracking system, or an asset monitoring (fleet management) system. The BGPS-based tracking system is different from those based on conventional receivers in one major respect: only the front end and memory are needed on the rover side. The first application of such an approach to a JPL low-cost orbit determination system was given by Wu [3]. Applying this approach to mobile devices can significantly improve user experience for existing applications, especially in mobile devices. BGPS can substitute for AGPS in almost all applications. All privacy concerns raised by AGPS are resolved with BGPS. Applications such as GPS enhanced cameras would require an instant position fix without network assistance. Many mobile devices, which could benefit from AGPS features, are often used in places where a network is not available or cannot be used (such as in high-dynamics aircraft, racing cars, etc.). The BGPS method can be used for such applications.

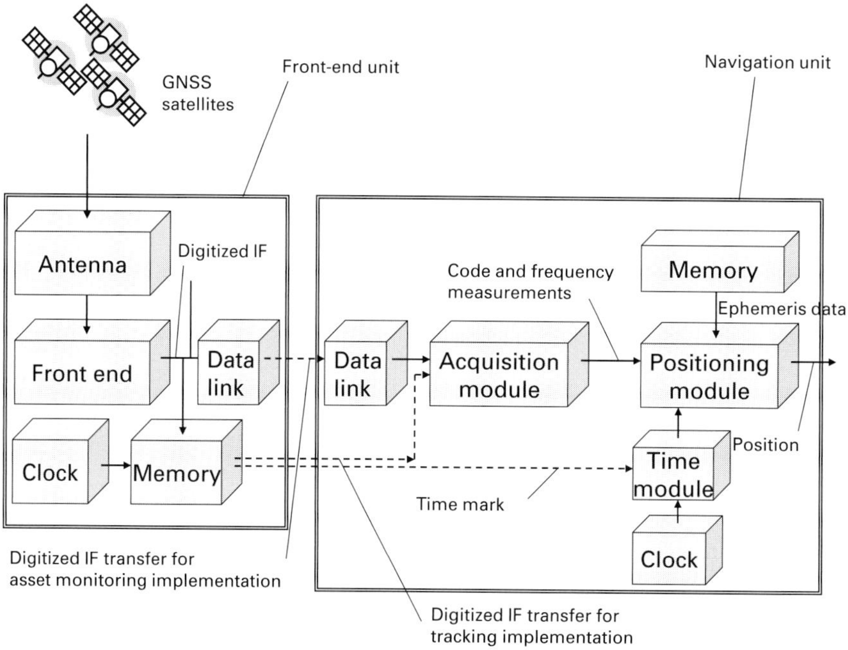

Figure 9.15 Flowchart of BGPS device for tracking or fleet management.

9.4.1 Fleet management

Figure 9.15 shows an implementation of a device for tracking or fleet management. It is similar to the device depicted in Figure 9.11, but the digitized intermediate frequency (DIF) is recorded to a memory instead of being processed in the mobile device. Therefore the mobile device has only the front end and memory.

Alternatively, the rover side may have only a front end and a communication device. In this case, the communication device is used not to receive corrections, but to transmit recorded chunks of the signal. Such a system may be useful also for wildlife tracking.

Conventional tracking and asset monitoring systems require a complete receiver on the rover side which includes, besides the front end, a baseband processor and a navigation processor. This approach allows us to create tracking systems, which can be, by definition, smaller, lighter, cheaper, and have lower power consumption on the rover side than conventional systems. This is possible because each position calculation is based on only a very short (as small as one code period, which is 1 ms for a GPS L1 C/A signal) snapshot of GNSS data. These data can be kept in memory or sent to a control center for processing. The control center can calculate a rover position using theoretically only 1 ms of data (normally we use 3 ms minimum). The size of these data can be as small as a few hundred bytes. Conventional GPS processing will require 30 000 times more data. It is impractical to make a tracking system that would require such a large amount of data to be transmitted or kept in memory for each waypoint. Therefore it is necessary to process all data on the rover side. This implies that the rover side must include a complete receiver or AGPS. In contrast, the BGPS approach allows use of only the front end instead of the complete receiver.

9.4.2 Bird tracking

This section describes an ultra-light tracking system [28]. The system was designed to track large birds, such as bar-headed geese (*Anser indicus*), for a period of about one year.

The bird-tracking application is very important. Since the industrial revolution, humans have had an adverse impact on Earth's ecology. It is possible to control this impact through knowledge of ecological changes. The migratory pathways and habitats selected by avian species are sensitive, among other things, to anthropogenic climate change and contamination of the environment. Tracking data for birds can be collected and synchronized with data from other sensors, such as cardiac activity and body temperature, in order to provide more information and understanding of migratory flights. Tracking information over long periods leads to the collection of important information about foraging habits, diurnal activity, flight physiology, and migratory routes of wild birds. Similar information can be collected for various wild animals, and can be extremely important in preserving endangered species.

In the case of bird migration, birds tend to return to the same place that they started from. Therefore it is possible to deploy recoverable systems that do not send data via, for example, satellite links, which is a high-power operation. The tracking systems should be as light as possible in order not to endanger the subject or interfere with its natural behavior. The system should comprise less than 2% of the bird's body mass. This is based on previous studies, where it was learned that short flights are barely influenced if birds are loaded with a mass of up to 5% of their body weight. A common GNSS device used to collect data over a long period of time, such as one year, requires a lot of power, and consequently heavy batteries.

BGPS technology can be adapted for the proposed system in order to process very short chunks (tens of milliseconds) of recorded GPS signal and locate the corresponding antenna positions. BGPS is an approach to global positioning that presents ultimate advantages over alternative methods when it is required that a position fix should be made after a few milliseconds of GPS signal reception. BGPS also meets real-time requirements. Providing an initial estimate of time within seconds, or even tens of seconds, is available, as the raw signals from a constellation of GPS satellites need only be sampled for a few milliseconds in order that antenna position can be obtained in real time or in post-processing. The presented system requires only the post-processing mode, which eases some of the restrictions for initial time constraints. This processing can be performed using a fast desktop PC whose power consumption is unimportant, while the raw data are collected and stored very efficiently in bursts by the system in the field. Unlike other approaches common in battery-powered GPS systems, there is no need for a lengthy warm-up interval of the order of 30–120 s while the satellite navigation message is obtained. Such a device was developed by scientists at Bangor University. It is depicted in Figure 9.16.

Figure 9.16 Bird tracking device, developed by Bangor University.

Using processing software in such devices has an advantage over operation in post-processing mode. The software applies three steps to provide a position fix: orbit preparation, signal acquisition, and raw data processing. The orbit preparation step uses IGS products, which are tabular orbits in SP3 format. The orbits are pre-processed by the software, which creates a continuous interpolated orbit representation using optimized trigonometric interpolation.

The signal acquisition step acquires the signal using FFT algorithms. The acquisition algorithms implement coherent acquisition to improve sensitivity and allow shorter records. The acquisition step outputs raw data, which are measured code phases.

The accuracy of the code phase measurements is limited by the sampling rate, which is chosen to minimize the memory requirements for the records. The essential difference between acquisition and tracking in terms of accuracy is that acquisition accuracy is usually limited by the distance between the samples, whereas the tracking accuracy is defined by the signal bandwidth, and is therefore usually much higher. If the signal chunks are large enough, it is possible to apply further processing which refines the pseudorange accuracy. The accuracy achievable without further enhancements is on the order of 25 m RMS.

The main consideration is the choice of antenna. High-gain active antennas allow the duration of a GPS snapshot to be decreased, but there may be a trade-off with size and weight constraints. An alternative solution may be the addition of an on-board amplifier.

9.4.3 Positioning with pilot signals

In the case of new and modernized GNSS signals, BGPS methods can be used to make a positioning using the pilot signals only.

Pilot signals are signals with a secondary code instead of a navigation message. Examples of pilot channels that are or will be available on civil signals are given in Table 9.3.

Some civil signals, such as GPS L1C, L5, GLONASS L3, and Galileo E1, have pilot channels. For indoor positioning, signal power is attenuated to a level that a

Table 9.3. Pilot channels

Signal	GPS L1C$_P$		GLONASS L3$_P$		Galileo E1-C	
Code	Primary	Secondary and tiered	Primary	Secondary and tiered	Primary	Secondary and tiered
Code length	10 ms	18 s	1 ms	10 ms	4 ms	100 ms
Code rate	1.023 Mbps	100 bps	10.23 Mbps	1.023 Mbps	1.023 Mbps	250 bps
Code ambiguity	~3000 km	unambiguous	~ 300 km	~3000 km	~1200 km	unambiguous

conventional receiver cannot acquire. High-sensitivity receivers use long coherent integration to acquire the low-power signal. The coherency interval describes an interval during which results from accumulators are added to each other either coherently or as they are without manipulating their sign (see details in Chapter 8). For GPS L1 C/A, an accumulator output can be summed for 20 ms. If the accumulator results are summed over a longer period, they may run into a navigation bit change, which may occur every 20 ms. This would negate the accumulation, because after the bit changes the result of accumulation will decrease with every millisecond, even if the signal is aligned with the replica. It is possible to extend this period to 40 ms by a trial-and-error method; however, an extension for long intervals, which is required for indoor positioning, would be problematic.

Pilot signals have a sequence known in advance instead of a navigation message. Therefore they can be used indoors. In their case, the coherent integration is limited only by the receiver clock stability.

The pilot channel can be used to assist in acquisition of the data channel or can be used for making a positioning, which can be defined in three categories as follows.

(1) Positioning with all pilot channels within a network. This is an obvious urban indoor application. It works in a similar way to current high-sensitivity GPS. A receiver uses a time mark and ephemerides from the network to make a fix. External time information can be used to assist an acquisition.
(2) Positioning with all pilot channels outside of a network. This can be used, for example, for positioning in woods, in a situation similar to that illustrated in Figure 9.1. In this case, the receiver can use predicted ephemerides and the time resolution methods of BGPS technology to make a fix. If the length of the code is sufficient, the time information can be derived directly from the acquired pilot signal.
(3) Positioning with mixed pilot and data channels. When one or a few data channels are available, the time mark can come from the available data channel. Therefore only ephemerides for pilot channels are needed. They can be either predicted or from a data link.

There are other differences between pilot channels and data channels in respect to the positioning, besides an absence of data, which allows for a longer coherent integration.

If the pilot channel features tiered code, then either the code ambiguity can be resolved more easily or the time can be calculated directly from the tiered code.

Code ambiguity is a function of the code length. A code length of 1 ms is equal to 300 km. Time can be found directly from the code without external time information or time resolution methods if the length of code is large enough.

References

[1] I. Petrovski, Expert advice – everywhere, without waiting, *GPS World*, Oct. 2006. Available at http://www.nxtbook.com/nxtbooks/questex/gps1006/index.php?startid=12 (last accessed May 2013).

[2] I. Petrovski, T. Tsujii, and H. Hojo, First AGPS – now BGP: instantaneous precise positioning anywhere, *GPS World*, vol. 19, no. 11, pp. 42–48, Nov. 2008.

[3] S. C. Wu, W. I. Bertiger, D. Kuang *et al.*, MicroGPS for low-cost orbit determination, TDA Progress Report 42–131, Jet Propulsion Laboratory, Pasadena, CA, Nov. 15, 1997.

[4] N. F. Krasner, Method and apparatus for determining time for GPS receivers, US Patent 5,945,944, Aug. 31,1999.

[5] J. Syrjärinne, Time recovery through fusion of inaccurate network timing assistance with GPS measurements, in *Proc. 3rd Int. Conf. on Information Fusion, Vol. II*, July 2000, pp. WeD5-3–WeD5-10.

[6] N. Sirola, Exhaustive global grid search in computing receiver position from modular satellite range measurements, *J. Phys.: Conf. Series*, vol. 52, pp. 73–82, 2006.

[7] N. Harper, *Server-side GPS and Assisted-GPS in Java*. Boston, MA: Artech House, 2009.

[8] J. Awange and E. Grafarend, *Solving Algebraic Computational Problems in Geodesy and Geoinformatic*. Berlin: Springer-Verlag, 2005.

[9] D. L Chambers, Ed., *Handbook of Genetic Algorithms. Vol. 1, Applications*. Boca Raton, FL: CRC Press, 1995.

[10] S. N. Sivanandam and S. N. Deepa, *Introduction to Genetic Algorithms*. Berlin: Springer-Verlag, 2008.

[11] D. L. Chambers,Ed., *Handbook of Genetic Algorithms. Vol. 2, New Frontiers*. Boca Raton, FL: CRC Press, 1995.

[12] D. L. Chambers, Ed., *Handbook of Genetic Algorithms. Vol. 3, Complex Coding Systems*. Boca Raton, FL: CRC Press, 1999.

[13] W. H. Press, S. A. Teukolsky, and W. T. Vetterling, *Numerical Recipes*, 3rd edn. Cambridge: Cambridge University Press, 2007.

[14] J. M. Dow, R. E. Neilan, and G. Gendt, The International GPS Service (IGS): celebrating the 10th anniversary and looking to the next decade, *Adv. Space Res.*, vol. 36, no. 3, pp. 320–326, 2005.

[15] G. Beutler, *Methods of Celestial Mechanics. Vol. I: Physical, Mathematical, and Numerical Principles*. Berlin Heidelberg: Springer-Verlag, 2005.

[16] G. Beutler, *Methods of Celestial Mechanics. Vol. II: Application to Planetary System, Geodynamics and Satellite Geodesy*. Berlin Heidelberg: Springer-Verlag, 2005.

[17] I. Petrovski and T. Tsujii, *Digital Satellite Navigation and Geophysics: A Practical Guide with GNSS Signal Simulator and Receiver Laboratory*. Cambridge: Cambridge University Press, New York, 2012.

[18] W. M. Kaula, *Theory of Satellite Geodesy – Applications of Satellites to Geodesy*, Waltham, MA: Blaisdell Publishing Company, 1966.
[19] O. Montenbruck and E. Gill, *Satellite Orbits, Models, Methods, Applications*. Berlin, Heidelberg: Springer-Verlag, 2000.
[20] M. Capderou, *Satellites, Orbits and Missions*. Paris: Springer-Verlag, 2005.
[21] B. W. Remondi, Distribution of Global Positioning System ephemerides by the National Geodetic Survey, in *Proc. 1st Conf. Civil Applications of GPS – Institute of Navigation*, Sept. 12, 1985.
[22] B. W. Remondi, Extending the National Geodetic Survey geodetic orbit formats, NOAA Tech. Rep. 133 NGS 46, 1989.
[23] B. W. Remondi, NGS second generation ASCII and binary orbit formats and associated interpolated studies, in *Proc. 20th General Assembly, Int. Union of Geodesy and Geophysics*, Vienna, Austria, Aug. 11–24, 1991, 28 pp.
[24] M. Schenewerk, A brief review of basic GPS orbit interpolation strategies, *GPS Solutions*, vol. 6, pp. 265–267, 2003.
[25] J. J. Spilker Jr., Satellite constellation and geometric dilution of precision, in *Global Positioning System: Theory and Applications, Vol. I*, B. W. Parkinson and J. J. Spilker, Eds. Washington, D.C.: American Institute of Aeronautics and Astronautics Inc., 1996.
[26] R. Hatch, Instantaneous ambiguity resolution, in *Proc. IAG Int. Symp.107 on Kinematic Systems in Geodesy, Surveying and Remote Sensing*. New York: Springer Verlag, pp. 299–308, 1991.
[27] X. Jin, *Theory of Carrier Adjusted DGPS Positioning Approach and Some Experimental Results*. Delft: Delft University Press, 1996.
[28] I. Petrovski, C. Bishop, and R. Spivey, Ultra-light bird tracking system based on BGPS™, in *Proc. GPS/GNSS Symp.*, Tokyo, Oct. 2011, pp.177–183.

10 Trends, opportunities, and prospects

10.1 From Cold War competition to a business model

The development of global satellite navigation was sparked by the Cold War competition between the USA and the USSR, which led to the birth of GPS and GLONASS.

The concept of satellite navigation was created in the 1960s. In 1963, Aerospace Corporation in the USA began a study on a space-based satellite navigation system. At that time, the concept of GPS had already been developed [1]. By the mid 1960s, a similar research program had started in the Soviet Union.

In a few years, both programs resulted in first satellite launches. In May 1967, the USA launched the first navigation satellite for the US Navy satellite system Timation. This satellite system provided a foundation for GPS. In November of the same year, the USSR launched the first Soviet navigation satellite, Kosmos-192, which transmitted navigation signals on 150 and 400 MHz.

In 1973, the NAVSTAR GPS program began, and in the following year the first NAVSTAR satellite was launched. The first GLONASS satellite, Kosmos-1413, was launched nine years later, in 1982. GLONASS started to be used by the international geodetic community in 1987.

The start of GPS business development can probably be marked by Trimble Navigation, the biggest manufacturer of commercial high-end GPS receivers founded in 1978, becoming a publicly traded company in 1990. At the beginning of the 1990s, a few events took place that sparked global interest and investment in research development of GPS user equipment and applications. These events have stimulated business success unprecedented in the history of technology, with GPS creating millions of businesses worldwide and billions of dollars in revenue.

In 1991, the United States announced that, starting from 1993, the GPS standard positioning service (SPS) would be available to the international community on a continuous, worldwide basis, with no direct user charges for a minimum of ten years. In 1992, this offer was extended by offering SPS to the world for the foreseeable future, and, subject to the availability of funds, to provide a minimum of six years' advance notice of termination of GPS operations or elimination of the SPS.

In 1993, the US Secretary of Defense formally declared the initial operational capability of GPS. GPS had 24 satellites in orbit at that time, and was capable of sustaining 100 m accuracy and continuous availability worldwide.

GLONASS started formally in the same year, but achieved a full constellation of 24 satellites only in 1995. Though also free for the international community, GLONASS was not successful in creating business. There are a few reasons why GLONASS failed to create a worldwide business at that time [2], and we discuss two of these in the following.

First of all, most of the GLONASS satellites had to be replaced at approximately the same time once they had reached the end of their lifespan. This time coincided with a difficult period – the Soviet Union had collapsed and its socialistic economy had been replaced. Consequently, the budget for space programs suffered. At that time, not only GLONASS, but also almost all other space programs were frozen. Many companies that had started to produce combined GPS/GLONASS receivers worldwide stopped production at that time.

The other reason is that GLONASS uses FDMA on top of CDMA. The FDMA was probably initially introduced into GLONASS to make the system more immune to interference, because it is more difficult to jam a wider band. Unfortunately for the business component of GLONASS, FDMA also resulted in larger user equipment, higher power consumption, and increased requirements for processing power. As discussed in the preceding chapters, GLONASS's wider bandwidth requires a higher sampling rate, which in turn creates a trade-off between processing power and accuracy. One either has to decimate the digitized signal below the Nyquist sampling rate and lose accuracy, or carry out much more processing.

Today, however, the situation has changed. The commercialization of GLONASS has become one of its important directions, strongly encouraged by the Russian government. It is most likely now that the GLONASS business will be successful.

First, the modernized GLONASS incorporates CDMA signals without FDMA, along with legacy signals; the legacy FDMA signals will be kept "until the last legacy signal user exists" [3].

Second, and even more important, GNSS user equipment is currently multi-frequency, often using various combinations of GPS, GLONASS, BeiDou, and Galileo on L1, L2, and L5 frequencies. This is boosting the technology, resulting in the miniaturization of multi-frequency and wide-band RF components, which are consequently becoming smaller, cheaper, and less power hungry. On the other hand, the processing power of mobile devices increases tremendously. Therefore, the FDMA feature of legacy GLONASS signals no longer creates problems in that respect. For example, as we saw in the preceding chapters, the RF front-end chipset for mobile devices is now equally suitable for any GNSS in the L1 band.

All these considerations indicate exceptional opportunities for GLONASS-related businesses, and GLONASS is now successfully creating its own business niche. But what about other GNSS; would we expect all of them to feature in the same mobile device?

10.2 Would you go for a "multi-mighty" receiver?

The mass market is driven by cost versus benefits criteria. A specification with the smallest size, minimum power consumption, *and* the best price, which meets the

Table 10.1. Satellite orbit inclinations

	GPS	GLONASS	Galileo	BeiDou
MEO inclination	~55°	~65°	~55°	55°

Figure 10.1 Soyuz VS03 launch from Europe's Spaceport in French Guiana on its mission to place the second pair of Galileo In-Orbit Validation satellites into orbit. Copyright ESA: S. Corvaja, 2012. Printed with permission.

accuracy, TTFF, and sensitivity requirements, is a winner. There are also, of course, country regulations and redundancy to be considered.

New satellite systems from Europe and China have appeared. By 1999, the final concept for Galileo had been decided. The first stage of the Galileo program started in 2003, led by the European Union and the European Space Agency. In October 2011, the first two Galileo satellites were launched to start system validation. Figure 10.1 shows the Soyuz VS03 launch from Europe's Spaceport in French Guiana on its mission to place the second pair of Galileo In-Orbit Validation satellites into orbit.

The first BeiDou satellite, BeiDou-1A, was launched in October 2000. In 2006, China announced that from 2008 BeiDou would offer an open service with an accuracy of 10 m.

From a business point of view, new GNSS may appear in similar situations, mostly with large, but still regional, markets.

There are many advantages of using GPS and GLONASS together in a receiver. Some of the advantages are specific to GLONASS. For example, GPS orbit inclination provides better coverage in southern latitudes, whereas GLONASS, due to its inclination and because it was designed for use in the territory of the former Soviet Union and Europe, provides better coverage in northern latitudes. A combination of the two systems allows better coverage over the globe. A user, however, will not benefit any further in that respect from adding other systems. Table 10.1 gives the inclinations for all GNSS.

More systems mean a more reliable service, and healthy competition can only benefit users. Compatibility of systems has improved, and will be improving further. More systems will provide higher accuracy and higher integrity, especially in urban environments. In this respect, more systems are better.

There are other benefits of using more than one system, such as availability in an obstructed environment. In addition, availability of RTK and BGPS requires at least one extra satellite for ambiguity resolution. Also, one additional satellite is required for each additional satellite system in use, if the time shift between satellite systems from the navigation message cannot be used.

A GPS L1 C/A or BeiDou MEO user receives a time mark within 6 s of navigation message, which is the length of one frame. A GLONASS user can resolve time with a time mark within 2 s of navigation message, provided that the receiver keeps its time track accurate within 2 s. The preamble in GLONASS is called the time mark and is located at the end of each subframe (string). This, in general, allows better TTFF if the receiver is using GLONASS. If the receiver time is accurate within 2 s, this time mark (preamble) allows the resolve code ambiguity to be resolved directly. The GLONASS time marks are ambiguous, because they are in fact preambles. If the receiver time is accurate within 4 s, two solutions can be compared to resolve the code ambiguity, and so on. If the receiver does not keep the time with sufficient accuracy, the whole frame must be received, which bears the explicit unambiguous time information. In that case, for GLONASS, a receiver needs to receive 16 subframes (strings) (15 subframes in a frame + 1 subframe) to ensure reception of the first subframe with time information. Each subframe takes 2 s, so altogether this takes 32 (i.e. 16×2) s. For GPS, the same process requires 1 subframe, which accounts for 6 s. This is because not only do we we use a preamble in GPS, but also each subframe bears explicit unambiguous time information.

Would adding a third or even fourth system provide enough benefits to a mobile device user to outweigh the shortcomings?

The international GLONASS market has increased due to the fact that countries that do not have their own satellite navigation system can include some redundancy in their infrastructure if they implement GNSS from different owners. This is especially important today, when a satellite navigation system has the capability to deny or encrypt service in well-defined spots. In this respect, apart from the countries that own the systems, the global choice would probably be to combine GPS with GLONASS or BeiDou, rather than GPS with Galileo. On the other hand, GPS and Galileo can work on the same front-end elements and a GPS/Galileo receiver can share an RF channel between the two systems, whereas a GPS/GLONASS or GPS/BeiDou receiver requires independent RF channels.

It is expected that the worldwide GNSS market will continue to grow, though at a much less rapid rate than the internal markets for the systems in their own countries. There are some other additional issues that may affect growth. For example, the frequencies for BeiDou overlap with those for Galileo. Even though Galileo presented plans to use these frequencies first, International Telecommunications Union (ITU) regulations may create problems for Galileo in terms of frequency allocation if Galileo starts to use the band after BeiDou. The ITU normally gives priority in frequency allocation to the first nation that starts to broadcast in that band. The subsequent users are required to obtain permission prior to using that frequency. Uncertainty of this nature is always undesirable for application development.

10.3 From SDR to SDR we go

When we talk about GNSS software receivers today, we usually mean a receiver which resides in a PC. In Chapter 6, we discussed the concept of a software receiver in the GPS arena, in which all the signal processing is carried out in the software. We gave definitions for software-defined radio (SDR) and the software receiver in Chapter 6 that are substantially different.

The concept of SDR appeared in 1995. Joe Mitola, who introduced the concept to the scientific community, defines it as a class of reprogrammable and reconfigurable radios [4]. The first SDR, SPEAKeasy, a joint project of several branches of the US military, had started earlier, in 1992. SPEAKeasy Phase II [5],[6] was a receiver in which the baseband processing was on FPGA. It is no surprise that the first SDR looked pretty similar to the first conventional GPS receiver; indeed, a GPS receiver was originally designed as SDR with signal processing on digital hardware. In that sense, almost any GPS receiver is an SDR-based receiver. A. J. Van Dierendonck writes [7, p. 335]: "All GPS receivers in production today are *probably*[1] all true digital signal-processing receivers." The only exception to this is high-sensitivity mass-correlator receivers [8]. Petrovski and Tsujii [9] proposed that a GNSS receiver took the form of a front end and a baseband processor, thus simplifying classification, as a receiver implementation is confined only to a baseband processor implementation.

The new definition of "software receiver" for GPS in the narrower sense of completely existing as software, has sparked the interest of researchers and engineers worldwide by giving them instant, inexpensive, and flexible access to the inside of a receiver.

In Chapter 7, we discussed that, from an algorithmic point of view, for GPS there is almost no difference between the software approach in a narrow sense and the hardware approach. The only superficial difference lies in the baseband processor implementation. There is hardly any change in the algorithms. We have included FFT in the software approach, but it can be, and is, implemented in hardware receivers as well. This is because all GPS processing was previously developed for digital implementation. In this sense, SDR is a different issue. Originally, SDR was related to radio, and radio can be completely based on analog components. Therefore to move radio from analog to digital hardware, required signal processing to change completely.

Figure 10.2 shows the concept of an SDR receiver, as it is generally accepted today. The important thing is that programmable hardware is part of SDR.

In this chapter, we discuss prospective receivers in terms of the general SDR concept. Figure 10.3 shows a receiver classification based on the programmability of the baseband processor.

[1] The italics appear in the original.

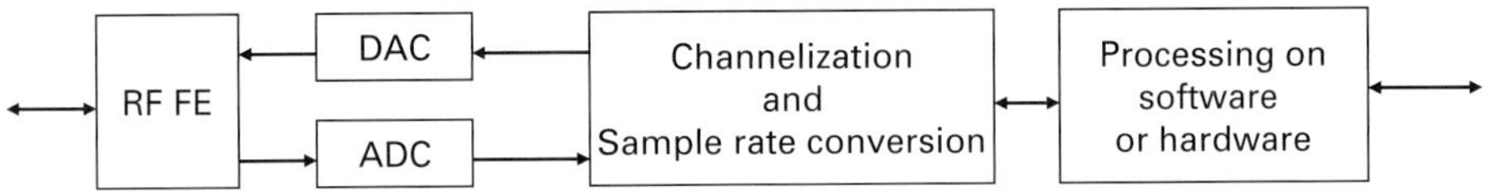

Figure 10.2 SDR receiver flowchart. After [5].

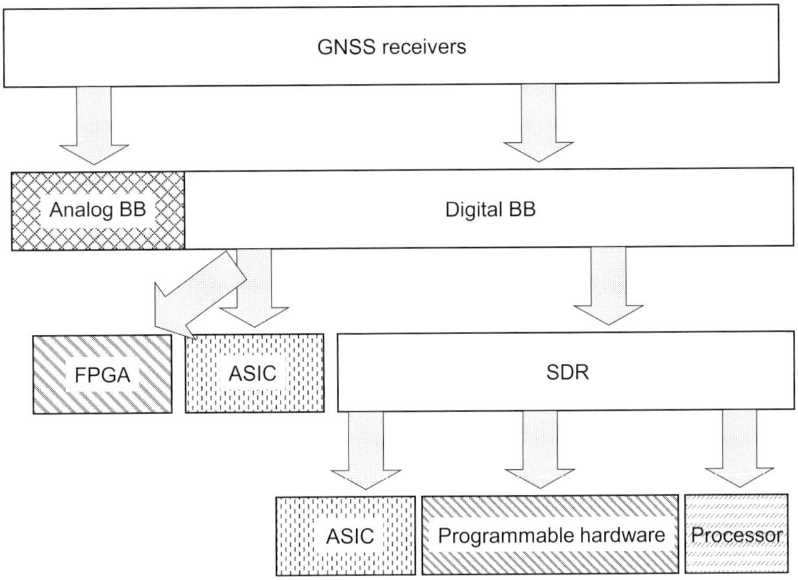

Figure 10.3 Receiver classification.

All GNSS receivers can be divided into two groups: those with digital and those with analog baseband processors, but there are hardly any in this latter category. A digital baseband processor can be implemented in programmable hardware (FPGA), in non-programmable hardware (ASIC), or on an embedded processor.[2] In general, even a receiver intended for mass production is prototyped first in FPGA.

Figure 10.4 depicts a simplified flowchart showing the development cycle for a receiver. First, a receiver is made on programmable hardware. If the potential market can justify a transition from FPGA to ASIC, the cost of which starts above one million dollars, then the cost of the device can be drastically reduced from hundreds of dollars to dollars; there are added bonuses that size and power consumption are reduced as well. For high-sensitivity receivers, ASIC is a necessity, because massive parallel correlation (not only important for high sensitivity, but also because they can work well with direct tiered code acquisition) in FPGA would be excessively bulky and expensive.

[2] The discussion from this point is based on certain definitions that may differ from vendor to vendor and that therefore should be understood in general terms.

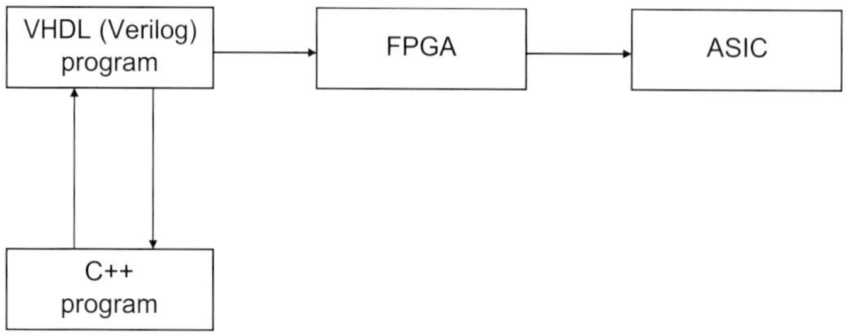

Figure 10.4 Digital hardware development flowchart.

As a result of putting the baseband processor in non-programmable hardware, the device may suffer from loss of flexibility, adaptability, and so on. However, even if the correlators are in ASIC, the device can still be programmable and considered as SDR.

Some cases of implementation in FPGA and ASIC can be referred to as SDR, and some cannot. SDR is defined by the following features [5]:

(1) reconfigurability, i.e. personality (see Section 10.7.2 for details) can be changed through reprogramming;
(2) flexibility, i.e. reconfiguration without changes in architecture;
(3) modularity, i.e. task encapsulation in separate blocks;
(4) scalability, i.e. the capability to add new models;
(5) future-proofability, i.e. it can accommodate new protocols;
(6) replicability, i.e. the capability to add channels by copying;
(7) cross-channel connectivity, i.e. the capability to share information between channels.

In this sense, all implementations in FPGA and some in ASIC may be considered as SDR. A massive-correlator receiver is maybe the only exception, due to the difficulties of providing the preceding SDR features to ASIC on the scale of hundreds of thousands of elements.

If after the receiver is released the FPGA is not accessible, then the receiver with FPGA cannot be classified as SDR. However, even if it has correlators implemented in ASIC, it still can be considered as SDR if it satisfies the above-mentioned conditions, in particular if it "derives its flexibility through software while using a static hardware platform" [5]. This means that if the receiver can control correlators through its software, it may have all the flexibility and programmability required to define it as SDR.

A generalized example of a GNSS receiver can be realized as shown in Figure 10.5. A receiver uses a microprocessor, for example an ARM9 processor, to control correlators, which in turn are realized as an IC with its parameters controlled by software. Therefore the receiver can be set up to work with either satellite system.

Currently, GNSS mass-market receivers are mostly constructed as two separated ASIC, analog and digital; for example, Fujitsu's MB15H156, MB87Q2040, uNav uN1008+, and uN2110. The receivers can also be implemented as a system in package (SiP), when a number of ICs are enclosed in a single module; for example, SiRFstarIII GSC3LT and Atmel ATR0635. Finally, they can be implemented as a system on chip (SoC), when all components are integrated into a single chip, for example the Taiwan MediaTek

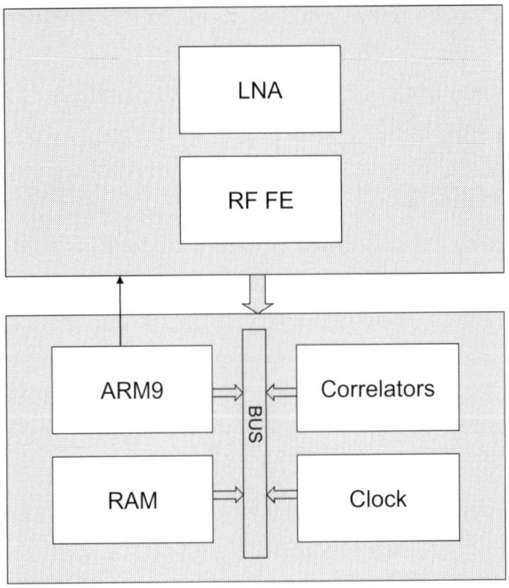

Figure 10.5 Generalized example of a GNSS receiver.

MT3336 host-based GPS solution and the STmicro STA2056. The chip may combine digital and RF functions; for example, the MT3336 host-based GPS solution, which includes on-chip CMOS RF and digital baseband, from MediaTek, and the STmicro STA2056. Almost all solutions (besides uNav) use the ARM 7 or ARM 9 processor.

The ARM 9 microprocessor is often used as the host processor in mobile devices. Theoretically, it is possible and attractive to use a host device processor, thus removing the processor from a receiver; however, almost all receivers use dedicated processors because utilizing a host processor is difficult due to GPS processing requiring a rather high interrupt rate (on the kilohertz scale), and this would interfere with other CPU functions. The integration, therefore, is possible only for pure snapshot receivers, which would work with GPS for only short intervals of time.

10.4 SA off, AGPS on, mass market open

In 2000, the United States turned selective availability (SA) off. Until that time, the precision of a GPS signal available to non-US military users was deliberately degraded by use of the SA feature. The author remembers the moment during the European GNSS conference in Edinburgh when "the much-anticipated but still-unexpected did happen" [10], as NASA Administrator Dan Goldin pre-empted the Department of Transportation officials and announced that SA was on its way out. His announcement made a sensation, and it heralded the beginning of GPS mass markets worldwide.

Soon after that, the Federal Communications Commission (FCC) Enhanced 9-1-1 mandate (E911) gave birth to AGPS technology, combining a cellular phone with a GPS receiver. For E911 applications, users expect location services to be available everywhere where a wireless phone call could be made, even inside buildings or in locations such as indoor-parking garages.

Indoor GPS receivers can even work on the signals that other GPS receivers consider to be parasites, i.e. on multipath signals. The technology for indoor positioning was developed by SnapTrack, Inc., a venture company of about 30 people in Silicon Valley, California. In 2000, Qualcomm Inc. bought SnapTrack for about $1 billion in stock. Qualcomm then patented assist information transmission over the cellular phone network.

The E911 requirements are a 100 m accuracy for 67% of calls and a 300 m accuracy for 95% of calls, and, for handset-based solutions, a 50 m accuracy for 67% of calls and a 150 m accuracy for 95% of calls [11]. These accuracy requirements came from the accuracy of snapshot positioning, which is necessary for such applications. Snapshot positioning means that the positioning algorithms are based on the code measurements available only from signal acquisition. This is because tracking would put additional constraints on the environment. The tracking would require uninterrupted signals for some period of time, which in general is not possible indoors, or even in a partially obstructed, a high-multipath, or similar environment. If measurement accuracy derived from tracking is limited by the signal bandwidth, the accuracy of the measurements derived from acquisition is often limited by the sampling rate. Increasing the sampling rate would put an additional burden on the processing engine and front end, potentially increasing the cost of the handset.

As a result, AGPS technology led to breakthroughs in several directions at once:

- instant positioning with TTFF less than 1 s;
- power savings, because a receiver operates in one-hundredth or one-thousandth of the time of a conventional operation;
- less interruption with cellular operation;
- high sensitivity.

Once they were combined with cellular phones, the numbers of GPS receivers increased dramatically, and are now counted in billions. A large proportion of the approximately seven billion units of mobile phones worldwide are equipped with GNSS. Numbers of Smartphones, developed by Google, Apple, Nokia, Samsung, Microsoft, and others, have reached one billion at the time of writing, with GPS functions available in most of them.

In January 2007, all restrictions (including those on allowed positioning accuracy) on the positioning service in Russia were lifted. This brought down one of the major barriers that limited GLONASS's commercialization in the past and opened up a huge new market in Russia for cellular phone applications and the like.

AGPS technology was then modified to work offline in self-assist mode. The task of adding AGPS functionality to a conventional mobile receiver can be divided into two parts as follows.

(1) The implementation of AGPS algorithms in a receiver –this is not the difficult part relatively, and is not at all expensive; modifications are concerned mostly with signal search and acquisition algorithms, unless massive parallel correlation is also required. Assist offline would require a slightly different set of assistance data.

(2) The preparation and delivery of the assistance information for the receiver. This part is fairly complicated and very expensive due to the required infrastructure. Some infrastructure can be reused.

Immediate assistance information can occur via (among others)

- cellular networks,
- TV signals [12],
- digital audio broadcasting,
- DGPS beacon systems,
- satellite systems,
- the Internet.

For assist offline or BGPS, we can avoid creating, or even using, a special infrastructure. Instead, we can implement the Internet capability into the mobile receiver so the receiver can use data available from global networks and services, which were developed for the geodetic community.

10.5 Convergence of mobile and geodetic applications

This is an important trend, which combines precise geodetic GNSS applications with mobile applications [13]. At first, this trend appeared in GIS applications, which combine some features from geodetic applications and some from mobile applications. Geodetic applications have always used very expensive high-end receivers and have enjoyed the luxury of using post-processing with unlimited processing time, precise ephemerides, and correction data from multiple reference stations. Today, these algorithms, data, models, and processing results are finding their place in mobile applications. This tendency, which was also discussed in Chapter 8, grows each year and will lead to the ultimate unification of geodetic and mobile applications in the future.

We continue with an example from Section 10.4, related to the preparation of assistance information for a receiver. The assistance GNSS data related to satellite orbits and clocks can be created in four ways.

(1) The provider creates and maintains his own GNSS network, and calculates ephemeris information, including orbit parameters and clocks.
(2) The provider uses global or regional networks, for instance IGS (global) or GSI (Japan), and calculates ephemerides using raw data.
(3) The provider uses global or regional services, which deliver ephemeris information.
(4) The provider uses his proprietary GNSS network to deliver broadcast ephemeris information.

The second and third methods lack integrity, and the fourth method does not support offline AGPS.

Self-assisted GNSS is the most complicated method in terms of receiver algorithms. It also has much higher requirements for memory and processor power.

The algorithms that use assist information are exactly the same as those for offline AGPS. However, we now need to implement additional algorithms in the receiver that would allow assist information to be calculated based on the previous ephemerides and geodetic models. The models describe various forces that are acting on the satellites.

These models include the gravitational fields due to the Earth, the Moon, and the planets, the shape of the Earth, tides, the Sun's radiation, satellite geometry, and so on.

Finally, the glue which allows us to bring these totally different (geodesy and mobile) applications together is the Internet.

10.6 Synergy of the Internet and GNSS

The Internet is what can hold mobile applications and geodetic services together. There is, however, more to it than that. We show in Figure 10.6 an example of the integration of mobile applications into the Internet. The device, which consists of an RF front end and a baseband processor in its minimal implementation, exists within the Internet, which works as a data link to the server on the one hand and to global services on the other. Thus we can combine, in one device, from a correctional point of view, AGPS, BGPS, DGPS, RTK, and VRS, because we can choose and supply any of the corrections required by those services in any combination.

10.6.1 Integration of a mobile device into the Internet

Mobile devices themselves can be integrated into the Internet. For example, Hada *et al.* [14] have described the integration of a vehicle into the Internet. The principles of such integration for vehicles can be summarized as follows:

each vehicle should be capable of almost instant positioning;
each vehicle should be capable of transmitting its position information;
positioning accuracy on a centimeter scale is required for collision avoidance;
vehicle equipment and ground infrastructure should be inexpensive.

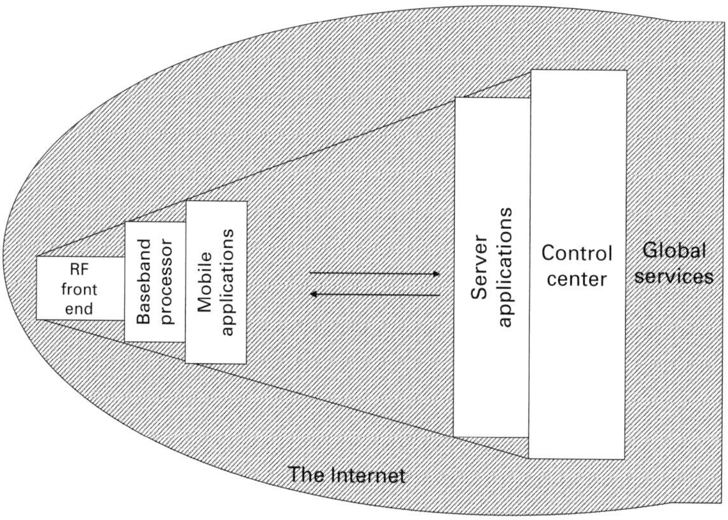

Figure 10.6 Example of integration of a mobile device on the Internet.

This approach already uses vehicle transmitted information to predict traffic jams and real-time weather conditions. Section 10.7 looks at how this approach, combined with the point of view of GPS as software, can bring us closer to a new GNSS paradigm.

10.6.2 The Internet as correction provider

The Internet can be seen as a source of orbit and clock information. The predicted ephemerides, which are used in BGPS, assist offline, and the like are available from the Internet at costs ranging from free to thousands of dollars. The reliable, accurate, and (importantly for many users) free information is available from the IGS (see Chapter 9 for details). This source of ephemeris information is where the technologies developed for mobile and geodetic applications are really merging via the Internet.

10.6.3 The Internet as data link

The Internet can be used as a medium to supply corrections, both AGPS and code and carrier differential.

All you need is to add to your device a front end, which costs a few dollars, and an Internet connection. This is enough for instant and precise (centimeter-scale) positioning.

Differential and VRS correction services via the Internet have many advantages over conventional correction services. The Internet-based service allows the creation of a low-cost infrastructure that can be used worldwide.

The Internet is of large enough capacity to accommodate all corrections, including AGPS, RTK, and VRS, and to provide reliable and stable mobile reception. A bi-directional communication between a server and a rover allows the transfer of only the required information through the optimum channel. A user can select a subset of the correction information that is required for a particular purpose, and the propagation agent can select the best server for this user.

The other advantage of such a service that it could easily grow. Everyone can create a new reference station in the Internet without a license, which would be necessary for a radio-transmitter. Eventually, however, issues of false information similar to those plaguing other Internet services will arise.

10.6.4 Improvement in GLONASS accuracy

In general, the real-time accuracy of GLONASS is noticeably less than that for GPS. The main source of accuracy degradation comes from broadcast ephemerides and clock parameters. The accuracy of precise GLONASS ephemerides, however, is within a few centimeters. This is, in particular, due to the regional character of the GLONASS ground segment in comparison to the GPS ground segment or the IGS network. The same type of problem may be anticipated for BeiDou. This was not such a big issue for geodetic users, who have access to precise ephemerides that are freely available on the Internet from, for example, the IGS. The Internet as a source of correction information can benefit GLONASS (and, in the future, BeiDou) users by providing more accurate ephemerides.

10.7 Towards a new GNSS paradigm

This look at GPS receivers as software has given us a new perspective. All GNSS receivers may be designed as SDR, because their physical layer behavior can be significantly altered through changes to the software, as "the term software radio generally refers to a radio that derives its flexibility through software while using a static hardware platform" [5].

This exploitation of the advantages of SDR can lead to the possibility of upgrading a receiver to receive new or modified signals, fix bugs, or add functionality after the receiver has been sent to a customer.

In the past, satellite transmitters were developed and sent into space. It was only after receivers had been developed that the total system started to function. This system, however, was static, and existed without changes being made for a long time.

We are now at a situation whereby all receivers, including those already sold to customers, can be connected through the Internet. This means that receivers can acquire corrections, assist information, upgrades, and bug fixes online.

And that is not all. A receiver's processing power can benefit enormously from being online by using cloud technology which involves other machines' processors to handle the most time-consuming or urgent procedures.

Some GNSS satellite transmitters, for example Galileo, are programmable while already in space. This allows us to change or adapt signal at any time, or to modify user equipment without recalling it or making it obsolete.

Online, receivers can be upgraded to a new signal design if necessary. In addition, GNSS simulators have also become software based and programmable on the fly. GNSS satellite transmitters can change their signal on the fly.

So, technology already allows for a system to be changed dynamically on both sides: satellites and user receivers, and in the space and user segments. This, however, comes at a price. And that price is privacy. Because of these advances, mobile device functionality may become degraded if the device is offline. In this respect, the combination of using online and offline positioning, which has been described in the preceding chapters, has become essential. By implementing all the methods for positioning without network described in Chapter 9, a receiver can be connected to the Internet periodically, for example every few days or once a week, and at the same time ensure that a user maintains privacy and retains almost all the advantages of AGPS, upgrades and bug fixing; the exception is sharing a processing load.

In a 2001 article about offline instant positioning technology [12], the author and colleagues stated that GPS is the second fastest growing field after the Internet. We also noted that these two areas were beginning to converge.

Now this convergence process has become more obvious, more essential, and more inevitable. Today, Internet integration may benefit the user even more, because of the SDR nature of GNSS receivers.

Let us sum up a few features of this new GNSS paradigm as SDR system rather than software GNSS.

10.7.1 Online updates and upgrades

It should be possible to upgrade and fix bugs in all GNSS receivers (not only PC based, which we traditionally call a software receiver in a narrow sense, but also SDR receivers which comprise most if not all receivers used today) using patches, even if the receiver incorporates FPGA and ASIC. If a receiver has FPGA or embedded processors, it can obviously be programmed, fixed, upgraded, and patched on the fly, similar to programs on your computer. But this is also the case for a receiver with ASIC correlators, which can be controlled through the software.

10.7.2 Programmable personality change

We can divide the mobile GNSS applications market into two main groups that have different requirements for availability and accuracy. The first group does not use carrier phase positioning such as RTK; the applications are focused on achieving high sensitivity, low power consumption, and low cost. The second group uses carrier measurements, and consequently has different and usually higher requirements for baseband and navigation processors. This increases not only positioning accuracy, but also power consumption and cost, especially for multi-GNSS.

The time required for signal acquisition is defined by signal power, receiver (more precisely antenna) dynamics, front-end clock stability, and by available assist information. Employing massive correlation decreases acquisition time dramatically. The number of correlators that it is possible to implement in cheap mobile hardware is in excess of hundreds of thousands.

Until recently, the capability to implement massive correlation in "hardware" receivers only limited the sensitivity of software receivers. Recently developed AMP (accelerated massive parallelism) technology (see Chapter 6) allows the use of graphic card hardware for massive correlation implementation in software receivers residing on a general processor.

A receiver can also be changed from one type to another via online reprogramming. We have shown how the iPRx receiver can function with the same front end as a snapshot receiver with high sensitivity and TTFF < 1s, *and* as an RTK receiver.

It would take only a change in antenna and reprogramming to turn a device from, for example, a high-sensitivity snapshot receiver to a fairly accurate GIS device.

10.7.3 Full set of online corrections

Following from Section 10.7.2, a receiver should be able to benefit from all corrections (rather than just a subset, or be capable of choosing a subset by itself), assistance information, code and carrier differential corrections, and VRS corrections.

The new GNSS paradigm combines in one device the advantages of AGPS, such as high sensitivity and instant positioning, with those of DGPS, or even RTK, such as high accuracy.

10.7.4 Application of cloud computing technology

A receiver can integrate processing power, corrections, updates, and tracking/fleet management where appropriate through cloud technology.

10.7.5 Third-party tools and services

Existing benefits arising from integration with online services in third-party applications include Internet-based maps. For example, a PC-based software receiver no longer needs mapping software of its own. The output from a receiver can be used by, for example, Google Maps or Google Earth, which are existing free online services. Figure 10.7 shows an example of the Google Earth output from an iPRx software receiver. These services can also be provided in real time. This is just one example; a receiver can access other third-party services in a similar way.

Some companies may even provide encapsulated indoor or high-sensitivity services, with not only software, but also assistance data and even processing on their machines, with many resources provided through cloud technology.

10.7.6 One for all and all for one

A "multi-mighty" receiver may not be needed as a front end can be reprogrammed from one system to another on the fly. For example, the iPRx receiver can be changed from working with GPS, Galileo, and GLONASS to working with GPS, Galileo, and BeiDou

Figure 10.7 Google Earth output from iPRx receiver.

by using other firmware and some software parameters. This allows for redundancy in a mobile device without sacrificing size, cost, and power consumption.

10.7.7 Offline operation

And what about areas outside the Internet? As we have shown, a receiver can use BGPS instead of AGPS, and PPP instead of RTK. In relation to the Internet, the following options are available for a mobile device.

10.7.7.1 Network position calculation

A snapshot of code phase data is delivered through the Internet from a mobile device to a server. The server calculates the position and either sends it back to the mobile device or uses it in another specified manner, for example it keeps it in memory or transfers it to another customer.

10.7.7.2 AGPS

Assistance data are provided to a mobile device via the Internet. The mobile device then can calculate its own position. A connection to the Internet at time of positioning is required. The calculated position can optionally be passed back to the Internet.

10.7.7.3 BGPS

Assistance data can be updated once or twice a week, or during a month in the future, for example at the same time as recharging the device batteries. The mobile device can then calculate its position without connection to the Internet.

References

[1] B. Parkinson, Introduction and heritage of NAVSTAR, the Global Positioning System, in *Global Positioning System: Theory and Applications, Vol. I*, B. W. Parkinson and J. J. Spilker, Eds. Washington, D.C.: American Institute of Aeronautics and Astronautics Inc., 1996.

[2] V. Engelsberg, I. Petrovski, and V. Babakov, Expert advice – GLONASS business prospects, *GPS World*, Mar. 2008, pp. 12–14. Available at http://www.nxtbook.com/nxtbooks/questex/gps0308/index.php?startid=12 (last accessed June 15, 2013).

[3] Y. Urlichich, V. Subbotin, G. Stupak, V. Dvorkin, A. Povalyaev, and S. Karutin, GLONASS. Developing strategies for the future, *GPS World*, pp. 42-49, Apr. 2011.

[4] J. Mitola III, The software radio architecture, *IEEE, Commun. Mag*. pp. 26–38, May 1995.

[5] J. Reed, *Software Radio: A Modern Approach to Radio Engineering*. Upper Saddle River, NJ: Prentice Hall PTR, 2002.

[6] W. Bonser, SPEAKeasy military software defined radio, *Proc. Int. Symp. on Advanced Radio Technologies*, Boulder, CO, 1998.

[7] A. J. Van Dierendonck, GPS receivers, in *Global Positioning System: Theory and Applications, Vol. I*, B. W. Parkinson and J. J. Spilker, Eds. Washington, D.C.: American Institute of Aeronautics and Astronautics Inc., 1996.

[8] F. van Diggelen, *A-GPS: Assisted GPS, GNSS, and SBAS*. Boston, MA: Artech House, 2009.

[9] I. Petrovski and T. Tsujii, *Digital Satellite Navigation and Geophysics: A Practical Guide with GNSS Signal Simulator and Receiver Laboratory*. Cambridge: Cambridge University Press, 2012.

[10] G. Gibbons, Deselecting unavailability, the ten year anniversary of the death of selective availability, *Inside GNSS*, p.12, June 2010. Available at http://www.insidegnss.com/node/2123 (last accessed Oct. 7, 2013).

[11] Federal Communications Commission, Guidelines for testing and verifying the accuracy of wireless E911 location systems, FCC OET Bulletin 71, Apr. 12, 2000.

[12] I. Petrovski, K. Sasano, B. Townsend, M. Ishii, and H. Torimoto, TV broadcast utilization for code and carrier phase differential GPS, *Proc. GNSS* 2001, Seville, Spain, May 2001.

[13] I. Petrovski, T. Tsujii, H. Hojo, First AGPS - Now BGPS. Instantaneous Precise Positioning Anywhere, *GPS World*, November 2008, vol.19., No.11, pp.42–48.

[14] H. Hada, I. Petrovski, H. Sunahara, *et al.*, DGPS and RTK positioning using the Internet, *GPS Solutions*, Vol. 4, no. 1, pp. 34–44, July 2000.

Part IV

Testing mobile devices

11 Testing equipment and procedures

Testing procedures are an essential part of the development, manufacture, and integration of GNSS receivers into mobile devices. The core instrument for testing mobile device GNSS functionality is a GNSS simulator, and this chapter discusses their use. Readers who may be interested in in-depth information about simulators and the principles of their operation and design, should consult [1], which is also accompanied by a bundled DIF signal simulator. This chapter describes GNSS testing equipment and procedures mostly following [2]–[5].

11.1 Testing equipment

11.1.1 Multi-channel simulator

Almost everything in the business surrounding GNSS can be related to either signal generation or signal processing. Simulators are used in the GNSS field mostly for testing. The methodological and research values of GNSS signal simulators are often overlooked. In [1], Petrovski and Tsujii looked at a simulator as a model of a real GNSS in order to understand better how a GNSS signal is generated and affected by various factors. Here we look at the simulator from a different perspective, as testing equipment only. Quite often, live signals from satellites are used to test GNSS devices. However, with live satellites a user has no access to the true data and no control over the signal. This leaves the simulator as an essential piece of test equipment.

The purpose of a GNSS simulator is to generate a GNSS signal as it would be received by a GNSS receiver from a number of satellites. During the development of the iP-Solutions iPRx receiver, we used extensively the Spirent GSS6300 Multi-GNSS generator and GSS6700 Multi-GNSS simulator for testing. Figures 11.1 and 11.2 show a test-bench setup in our office with Spirent simulators and the iPRx receivers under test. We used a multi-channel simulator to work on all positioning tasks with GPS. A multi-channel simulator provides a system-level environment incorporating multiple parameters into the simulated scenario. We also used a single-channel simulator to generate single and combined GPS/SBAS, Galileo signals. Single-channel simulators allow the optimization of the receiver's tracking and acquisition loops, rather than the whole system. In principle, this allows more control over specific parameters, such as a

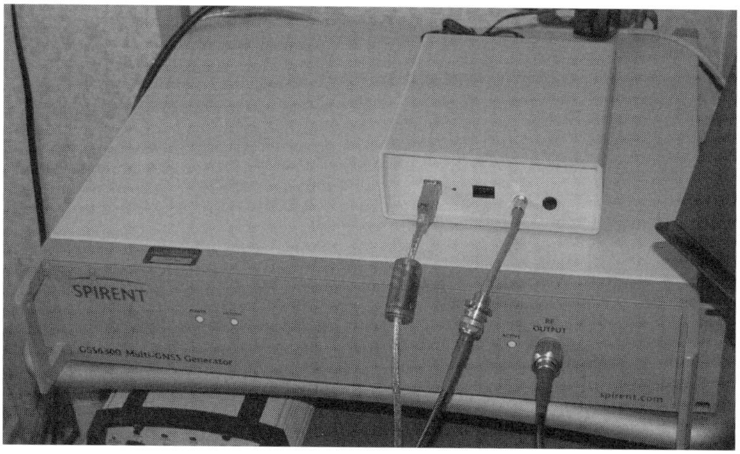

Figure 11.1 Spirent GSS6300 generator and iP-Solutions FEM front end/recorder.

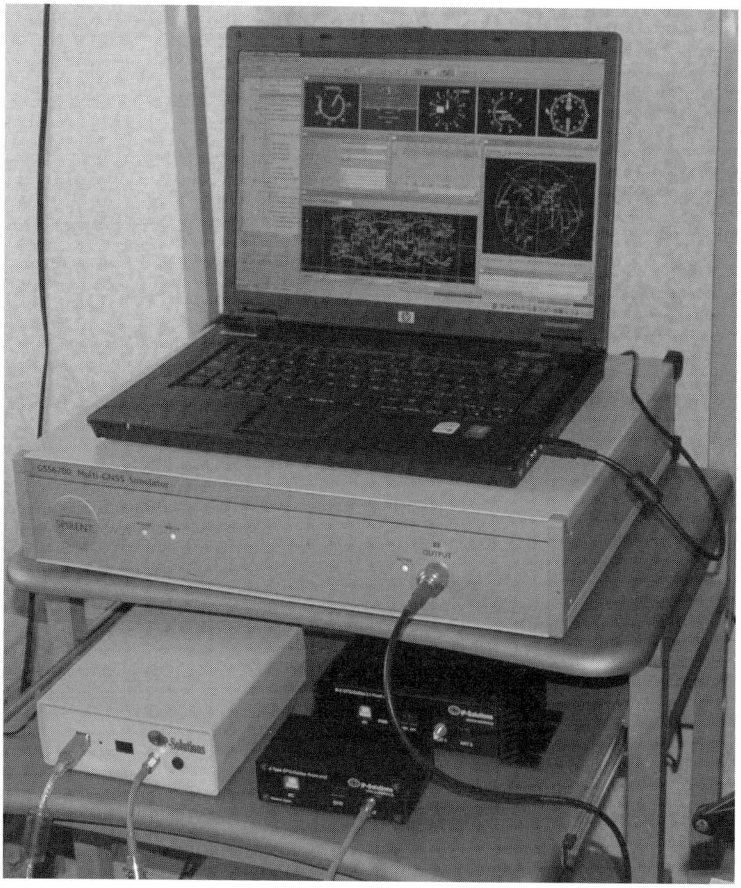

Figure 11.2 Spirent GSS6700 simulator and iP-Solutions FEM and E-Type front end/recorders.

Doppler profile. In contrast, a multi-channel simulator incorporates such variables as the satellite motion, vehicle dynamics, and clock parameters, and generally cannot be configured to fit a specific profile. It becomes indispensable when it comes to the dynamic accuracy tests that are needed for most GNSS receivers. The simulator provides a time-marked reference trajectory, which can then be compared with the trajectory estimated by the receiver.

In general, simulators provide a universal, high-level capability to cover the necessary tests from design to manufacture of GNSS equipment. Moreover, their use allows us to exclude most of the other testing equipment that might otherwise be necessary.

11.1.2 RPS: record and playback systems

Record and playback systems (RPS) record a GNSS signal to memory as a DIF signal, in a similar manner to that considered for bird and animal tracking in Chapter 9. Actually, any GNSS receiver has an RF signal converted to a DIF signal prior to signal processing in the baseband processor.

These DIF signals can be played back. First, they are converted to an analog signal by digital-to-analog converters, and then the IF analog signal is up-converted to the desired RF frequency. The resulting signal is filtered to remove images which were generated during the up-conversion process. The system is shown schematically in Figure 11.3. The resulting RF signal closely reproduces the original signal, and can be acquired, tracked, and used for positioning by conventional GNSS receivers. Figure 11.3(a) shows the signal being recorded to a memory. The signal in the memory can then be played back. It can be arranged in such a way that the signal is recorded on a standalone device, such as the one by Spirent systems (Figure 11.3(b)) , or recorded on a general PC, like for the iP-Solutions RF recorders (Figure 11.3(a)) . In both cases, the signal is played back through a simulator. Figure 11.4 shows the iP-Solutions recorder on top of the Fuji simulator, which is also used for playback. The Spirent RPS is shown in Figure 11.5.

Record and playback systems are appealing because they bring repeatability to tests with live satellites. However, RPS are inferior to simulators in terms of flexibility and control over the models and parameters.

It is important, however, to understand the limitations in some test applications of RPS as a function of their design. Most RPS, besides those designed for general purposes and which are very expensive, use 1 or 2 bit quantization in the recorder, similar to that used in most GNSS receivers. This does not affect most of the tests, but it does affect those tests related to interference for which the receiver under test relies on quantization with substantially wider bit resolution. In this case, a signal with a low bit resolution can be saturated much more quickly. The receiver under test may be able to find signal if more quantization levels are available, so it can distinguish the GNSS signal from the interference.

Having said that, RPS are indispensable in many situations. Consider, for example, a flight test. If we use the results from the flight test for receiver algorithm development, it may be essential to record the signal once and then use it with various algorithms over

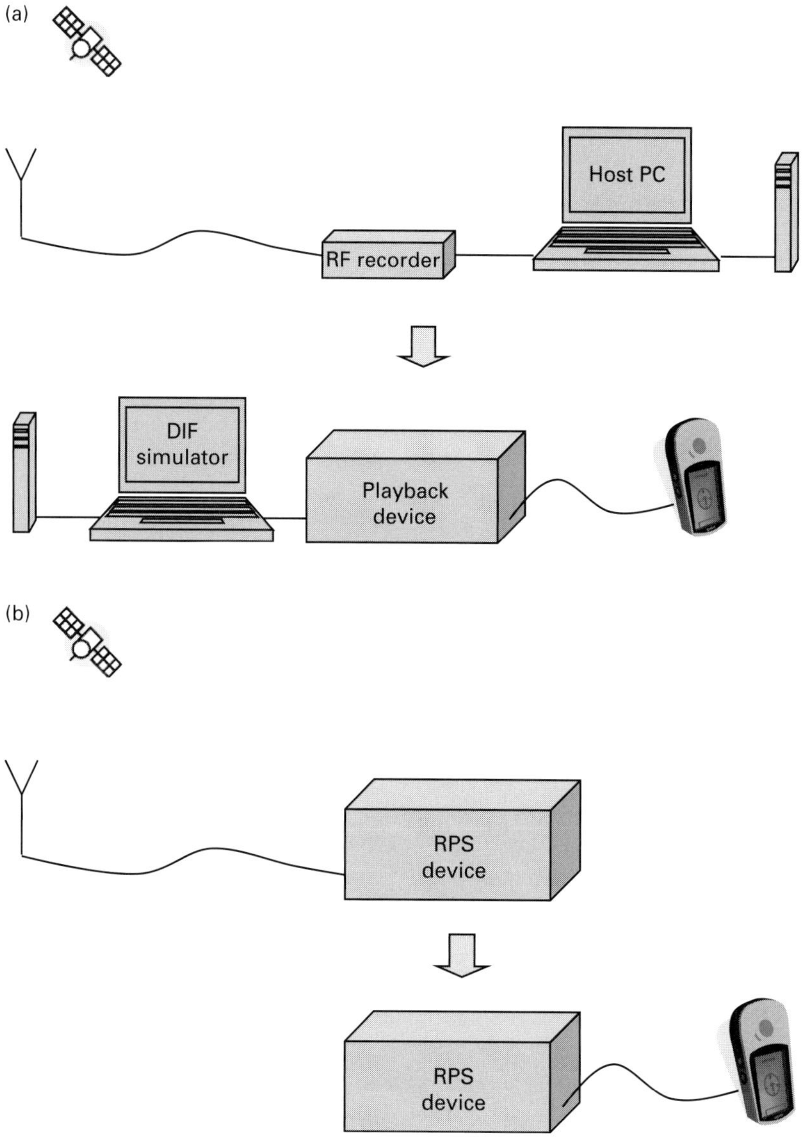

Figure 11.3 Record and playback system (RPS). (a) PC-based RPS; (b) Standalone RPS.

and over again, without repeating expensive and demanding tests every time significant changes are made to the algorithms.

Another example would be when the signal is experiencing some kind of unique effects, for example a maximum in solar activity, which occurs once in 11 years.

A third example is the measurement of a signal during solar eclipses, earthquakes, and other unique natural events. The signal is recorded and can be analyzed interactively, using the obtained results to modify the signal processing. This can provide us with unique tools for researching our planet.

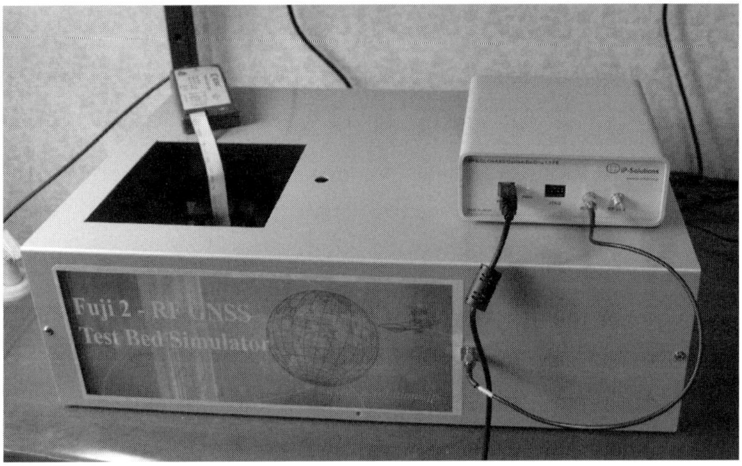

Figure 11.4 iP-Solutions RF recorder on top of a Fuji simulator.

Figure 11.5 Spirent RPS device.

11.2 Device life cycle

Tests accompany a GNSS receiver throughout its life, from its conception at the design stage, through manufacturing, its incorporation into a mobile device, and on to its role as serving a user. The life cycle of a GNSS receiver as part of a mobile device can be divided into a few stages, each of which require testing.

11.2.1 Research and development

The research and development (R&D) stage is the period when new algorithms and solutions emerge. There are two different types of tests conducted during R&D: basic tests, which are based on providing a valid reference signal, truth data, and repeatability, and advanced tests, which simulate situations seldom encountered under normal circumstances.

The reference signal and truth data comprise the basic elements of GNSS testing equipment that allow a user to tune up various functions of a receiver, for example navigation message decoding. The test equipment emits a reference signal and provides access to the signal structure and navigation message. This permits the verification of receiver algorithms to recover time synchronization and decode navigation messages.

Reference truth data allow the optimization of receiver accuracy and sensitivity, and assist in tuning up models to compensate for atmospheric effects, etc. All components of the GNSS error budget can be easily separated and dealt with in a simulator.

Repeatability is an essential feature at this stage. It is interesting to note that in general a software receiver realized on a general processor does not guarantee repeatability, although the latest Intel software libraries can bring about repeatability at a cost of efficiency. This is not an issue for most other receivers, when the processing timing is fixed in FPGA or ASIC flow, or even in embedded processors working under real-time operating systems (RTOS). Work with a real-time software receiver means that tests are repeatable, even with a simulator, only to a certain level. This is because there is always a number of processing threads which are executed simultaneously on a PC processor. In two sequential runs, a tracking thread running with a higher priority, for example, can give way to an acquisition thread after each different time delay, which may lead to the acquisition of satellites in each different time sequence, and so on.

In general, for R&D the simulation of specific situations, which include scenarios of satellite failure for receiver autonomous integrity monitoring (RAIM) algorithms testing, extensive ionospheric errors, and interference, is required. For mobile devices, tests that facilitate the development of advanced multipath-mitigation techniques are especially important.

11.2.2 Design

The design and validation stages describe the periods at which specific receiver designs are created and compared with a benchmark design or a given specification.

Validation tests may be required at each design stage as proof of a concept. They may include any or all imaginable tests, depending on the stage and specification.

In line with our receiver concept, many parameters are related to the RF front end rather than the receiver itself. These parameters may include the local oscillator's phase noise limits, the required noise figure, second- and third-order intermodulation, and filtering characteristics.

An important example of validation tests is in the chip development cycle. Chip development requires a large amount of initial investment, and it is impossible to change the design once completed; therefore, more elaborate performance validation tests are required using a prototype.

Positioning tests to validate the receiver performance in static positioning can be arranged with a geodetic reference point. The reference point must use a high-quality antenna located in a low-multipath environment. The true coordinates must be found using post-processing geodetic software in a similar way to that for reference station installation, which we have discussed in the preceding chapters.

11.2.3 Certification

Consumer testing and certification become more important as more GNSS come to market. The addition of the new systems, GLONASS, Galileo, and BeiDou, requires corresponding modifications to the test procedures. Certification is very important for equipment that must be endorsed for federally controlled applications, such as aviation, public fleet management, and E911-type services for mobile phone users. Both the multi-GNSS products and the equipment used to test them must be certified.

11.2.4 Production

Production includes the production of chips, modules, parts built by the OEM, and finally the end-user device.

Here we are interested in the tests involved in mobile device production. These tests may include pack sample tests for the components that are supplied by the third parties and for the final manufactured product, and qualification tests for any sensitive specifications that might be specifically required from the device.

Periodic pack sample tests are necessary to ensure that the incoming components and outgoing products meet their specifications. Production lines, both at the suppliers and in house, can encounter pre-planned or unexpected changes in technology. Slight changes in supplied materials can affect performance as well, affecting the quality of the product and its ability to meet a specification. Moreover, the specifications of GNSS systems themselves are changing, with new frequencies and signal designs coming on board. The various tests should continuously ensure the quality of product.

Production tests should also be easily adaptable for automation. Tests that can be useful at typical manufacturing facilities along the production chain include incoming components pack sample tests, qualification tests, and final production tests.

The final test is an important part of the manufacturing process for an end-user device manufacturer if it includes a receiver as an integrated part. In contrast to a validation test, which aims to demonstrate that particular specification parameters are being met for a particular component, a final test has to ensure that all components in the system are working together properly after they are integrated. The final tests may be organized as a complex functionality test in such a way that they combine, for example, a sensitivity test and positioning test in one run. Sensitivity, velocity, and acceleration profiles can be defined in a single scenario. If the receiver operates normally in that scenario, it is proved that the whole set of receiver performance specifications is fulfilled and meets the product acceptance requirements. The reference data used to check dynamic accuracy in this test are obtained from the simulation data files. Tests at this stage also include one for power consumption, which is measured using a wattmeter under normal receiver operation.

In many cases, it would seem to be enough just to ring up the circuits to test them. However, a handful of technical issues, such as those related to soldering, wiring, soldering material, impedance mismatch, etc., can cause extra signal losses. Such losses in RF may go undetected with ring-up tests, but still cause severe degradation in receiver parameters, or even equipment failure. In fact, ring-up tests cannot even properly ensure that antenna connections are sound. Problems also can arise due to unannounced or involuntary changes in the technological processes within the component or material supplier's facilities, undetected changes within the manufacturer's own production technologies and processes, and machinery failure or aging. The recommended reflow parameters in manufacturing can vary among components and therefore sometimes cannot be set to fit all of them precisely.

A particularly sensitive part of any mobile device with GNSS functionality is the clock, which is usually TCXO or voltage-controlled TCXO. The clock parameters can be affected by the assembly process. The clock can be supplied for assembly either inside a component module, on its own, or within a host device. In all of these cases, the clock should be tested after the final assembly. Almost all potential problems with the clock, including not only catastrophic parameter failure, such as drift, but also in many cases the complete operational failure, cannot be detected by means of the circuit ring-up test. An excess clock drift beyond the spec will affect the receiver sensitivity because of imposed limitations on coherent integration. It may also cause failure of the assist mode because of incorrectly chosen Doppler bins, or even affect the receiver's ability to acquire satellites due to a shifted frequency search area.

Further, the circuit ring-up test cannot check a baseband processor or a navigation processor. If the navigation function is performed as part of the upper-level system and not by a dedicated processor, the overall integration must be checked. This should include a stream of outgoing data from the baseband processor for positioning calculation and raw data, and possibly an incoming stream of assistance data for assisted acquisition.

If we have a software receiver using a host processor for acquisition and tracking as well as navigation, it becomes essential to test how the host processor copes with the extra tasks. Even in the case of a hardware receiver, the existence of multiple correlators leads to a significant load for a host processor from the measurements supplied by the multiple correlators. Moreover, the software or firmware for a host processor changes constantly and may affect the positioning tasks. All these types of error, however, can be spotted with one properly designed positioning test.

11.2.5 Consumer testing

Essentially, everybody who gets hold of a mobile device becomes involved in or affected by testing at one time or other. Naturally, when we talk about test procedures, we think of an engineer or scientist developing or validating a new GNSS device.

But even a consumer who walks out of a shop with a new GNSS unit can be viewed as being involved in a test: he or she switches on the unit and sees if it works. That is the test. A conference attendant browsing at an exhibition approaches a booth and asks the exhibitor for a demonstration of a GNSS device. The exhibitor is performing a test.

11.3 Test examples

There are many different tests required for a mobile device; these include testing its main functionalities, tests for electromagnetic compatibility (EMC), and so on. Here we consider only the tests that are specific to GNSS.

11.3.1 General tests

After the receiver tracking loops are designed, we need to check if they deliver the correct observations, i.e. code phase, carrier phase, Doppler, and carrier-to-noise ratio. The navigation message decoding function deciphers the satellite signal's data content. PVT (position, velocity, time) represents the calculation of those values for the receiver. The user interface communicates with the receiver operator by means of anything from a serial data stream to a full map display. All these components should be tested. Using the simulator, we know the true observations and can compare them directly with those measured epoch by epoch.

Such tests are in general possible to perform with live satellites, but this involves much greater effort and provides less usable results. For example, the antenna position would have to be surveyed, ephemerides predicted, and the predicted range to a satellite and Doppler shift calculated. This procedure, however, would not guarantee decisive results. For instance, satellite orbits can be predicted very precisely to satisfy the requirements of such tests, but satellite clock predictions may not be so precise, and these errors will in effect show up as incorrectly predicted orbits. Also, it is difficult to separate the receiver-hardware-related measurement biases from the satellite-hardware- and signal-propagation-related measurement biases.

When one is working on a navigation processor, one may wish to optimize ionospheric and tropospheric error compensation in the receiver. In the corresponding test, we can toggle the ionospheric and tropospheric errors in a simulator and the user can directly assess the accuracy and performance of the ionospheric and tropospheric correction algorithms implemented in the receiver.

While debugging the navigation processor, one can also compare side by side an encoded navigation message with one decoded by the receiver, and ensure the correctness of the decoding algorithm.

Some tests may be specific to particular receivers. In a 1 PPS accuracy test, for example, one compares the 1 PPS output from the tested receiver with either the 1 PPS pulse output from a standard GPS timing receiver or the 1 PPS output of the simulator. Then one calculates the variance of the error.

One should also be interested in measuring receiver sensitivity. It is possible in such tests to change the signal power level at predefined levels to see how the receiver will react to the change in terms of tracking, acquisition, and navigation. One also can simulate numerous non-standard situations and errors in the signal to see how a receiver reacts. The sensitivity tests should be done to check acquisition sensitivity, which is the measurement of a minimum power level at which a receiver is able to identify correctly a satellite signal. A tracking sensitivity test should be conducted for receivers other than snapshot receivers to measure the lowest level at which the receiver reliably tracks the satellite. Normally, sensitivity measurements are quite complicated and require additional signal attenuation and calibration. However, one can use a set of precompiled test scenarios usually available in a simulator.

Another test benefiting from a simulator determines a receiver's positioning accuracy in differential mode. This test can be performed with either a high-end reference receiver and live satellite signals or a simulator. Some simulators can provide differential corrections in the RTCM standard. For such tests, the antenna of a receiver under test is co-located with the antenna of the reference receiver. The positioning data logged from the tested receiver are then calculated and compared with the reference receiver data to determine the standard deviation of the position results, as in the standalone static test. When using a reference receiver, a zero base-line test is recommended.

Some other tests include special-date tests, such as those relating to end-of-week, end-of-year, week rollover, and leap-second events. Most of these types of tests are available as predefined scenarios with a simulator.

11.3.2 AGPS tests

AGPS functionality can be tested in different ways. In the first, a GNSS signal simulator simulates a GNSS signal for client and server (Figure 11.6(a)). In this case, the signal for the client device is additionally attenuated and the server provides assistance information. Such a system can also be augmented by a network simulator to test the whole system, including the cellular phone components and the server. A more complicated setup may also include interference [6]. However, the simulation for client-only devices can be simplified, as shown in Figure 11.6(b). In this case, the simulator provides an RF signal to the client device and generates assistance information for that device. The simulator must have an option to output assistance data, which should include satellite ephemerides. In contrast, more comprehensive testing for cellular phones may combine GNSS testing with a cellular network test bed (Figure 11.6(c)).

AGPS tests place some additional requirements on simulators with respect to the available signal power dynamic range and the signal level control resolution, which should be at least from 1 to 0.5 dB.

More complicated AGPS tests may involve interference testing, such as that described by Boulton [6]; see Figure 11.7. The described test analyzes the possible interference to a mobile device from LightSquared's base station. A similar test can be used in general to verify the level of interference from various sources. Such tests include cases of testing at the GPS sensitivity limits of the devices, representative of

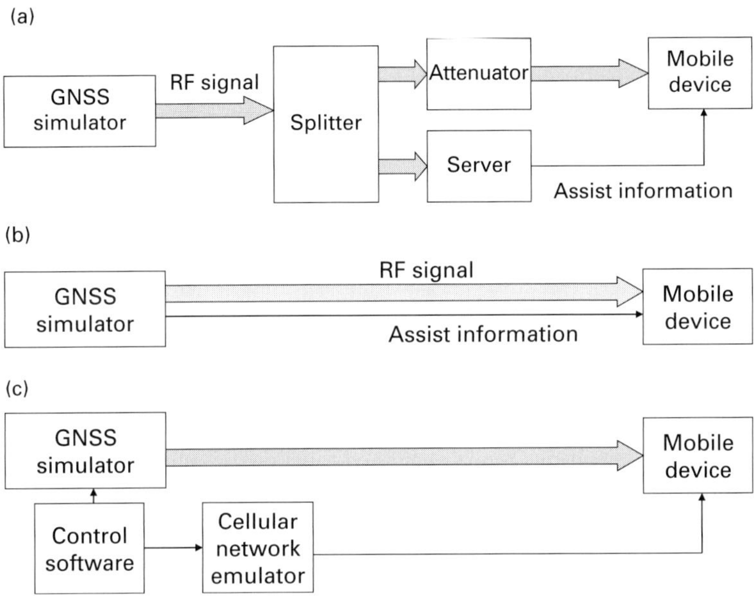

Figure 11.6 AGPS tests with (a) GNSS simulator and AGPS server, (b) GNSS simulator only, (c) GNSS simulator and cellular network emulator.

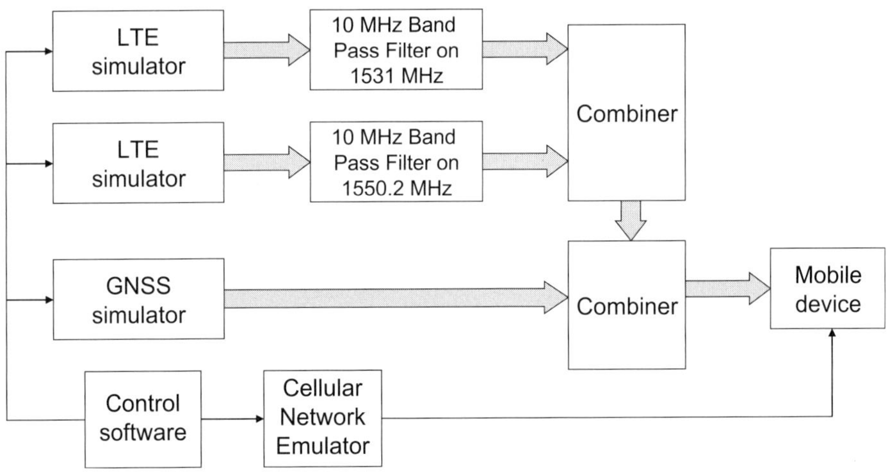

Figure 11.7 AGPS and interference test. After [6].

indoor or other highly obscured settings, testing at intermediate received GPS levels (with equal signal strength from all space vehicles) to evaluate performance in indoor, dense urban outdoor, or other environments with significant blockage and reflection of GPS signals, and testing at strong received GPS signal levels corresponding to outdoor usage with open-sky conditions.

11.3.3 Multi-GNSS test specifics

End-user products of today and tomorrow not only combine various GNSS, but also their manufacturers aim to market them in many countries worldwide, because applications are now global and truly international. So, when multi-GNSS equipment comes to a local market, the issue of local certification may become crucial for success there.

First of all, a multi-GNSS receiver requires a final test to include all GNSS implemented in the mobile device, because a proper GPS-only test cannot ensure the proper operation of GLONASS or BeiDou components. The RF hardware components may differ significantly, starting from system to system, and even if they are the same, the particular settings will still differ. As previously described, the bandwidth and frequencies for various GNSS signals are different, which leads to an implementation of different sets of filters and separated signal tracks. The correlators may also be different. Therefore, these separate tracks should be tested separately.

Moreover, the same components can work for one system and fail for another. For example, a dual GPS/GLONASS device can share the same clock, but the sampling rate for GLONASS is twice as high due to the wide bandwidth of FDMA (see Figure 5.23). Consequently, one can imagine a situation in which the clock will pass the GPS positioning test but fail when processing GLONASS signals in the field. The second argument for special multi-GNSS tests is that many multi-GNSS markets may require certification, including that of the manufacturing process. This certification would require a final test for each GNSS used in the product.

Testing equipment using multi-GNSS systems introduces additional issues that must be dealt with by receiver designers. Multi-GNSS implies multiple radio frequencies. Differences appear in the various GNSS time systems and coordinate reference systems. The required tests are nearly impossible to carry out with live satellite signals, but are possible with simulators, provided the corresponding functions are supported. There are many advantages in knowing the truth model and in controlling the signal environment. The list of possible tests is much longer than those previously listed, and most of the tests are either impossible, or less effective, with live satellites. One needs to ensure that a receiver can properly integrate multi-GNSS coordinate systems (including various associated time frames) and error budget models.

As discussed earlier, the various GNSS systems' coordinate frames seem to be fairly well integrated already, and no additional tests are required in this sense for a standard receiver. The time frames, however, are unlikely to be coordinated well enough, not only because they differ, but also because they are calculated in a different way – either as an assembly of clocks or a master clock. In relation to time frames, time references and the handling of differences caused by leap seconds for various GNSS must be covered by a special test, with a simulator taking the different time frames and how they relate to UTC via local UTC scales into account.

Another issue that should be considered for multi-GNSS design and testing is that relating to the underlying error models. An ionospheric model is of particular importance. For example, GPS uses the Klobuchar model whereas Galileo uses the NeQuick model. Although, on a receiver level, these models can be applied separately to the corresponding signals, in a simulator they must be combined in one common truth model, which is used to calculate code delay and phase advance (if implemented) in the simulator's signals. Consequently, broadcast ionospheric model parameters for all simulated signals should be derived from the same truth model.

Tests should ensure an optimal combination of the GNSS in either the coordinate or the measurement domain. The method of combination depends on the optimization criteria and on the signal statistics for the particular systems. What would be more advantageous to use in terms of accuracy if, for example, a receiver acquired GPS L1 and GLONASS L1 + L2 signals? Would it be GPS L1 or GPS L1 + GLONASS L1 + L2? What about if, for example, a receiver acquired GPS L1 + L5 and Galileo E1? Would it be better accuracy for GPS L1 + L2 + BeiDou L1 or for GPS L1 + L2 only? Would the decision be affected by mask angle or foliage? We can almost not imagine how to carry out such analyses without employing state-of-the-art simulation techniques.

A further item of interest is that interference will affect different systems in different ways. The new and future signals of GPS and Galileo are less prone to interference due to their design. GLONASS is less prone to interference due to being an FDMA system on top of CDMA. As a result of an application of the various underlying systems, undertaking interference tests of a multi-system receiver becomes a more complex task with many parameters to consider.

11.4 Case study: new paradigm SDR simulator

The iP-Solutions digital IF signal simulator, ReGen, was originally developed to test and validate algorithms and methods of GNSS signal processing for ARAMIS™ (adaptive receiver applied for monitoring ionospheric scintillation).

ARAMIS™ was developed for and in close cooperation with the Japanese Aerospace Exploration Agency (JAXA), with the purpose of operating onboard aircraft using tight and ultra-tight integration with the INS. At that time (and even now), no off-the-shelf simulators could meet the requirements for ARAMIS™ test conditions, so we had to develop a DIF signal simulator in house. The requirements included specific scintillation models and ionospheric bubble simulation.

ARAMIS™ is a software receiver, and it was a perfect choice to use a signal simulator that can generate a digitized baseband signal directly, thus eliminating two front ends, which in this case would introduce additional noise.

The ReGen architecture allows us to customize or change implemented models easily.

The most critical functionality is related to scintillation generation, which applies phenomenological and split-beam scintillation models [1]. The other important function is the simulation of spatially correlated ionospheric errors, in particular via the ionospheric gradient model. Figure 11.8 shows the ReGen panel. The other important feature of ReGen is its ability to simulate electron depletion areas in the ionosphere. Figure 11.8(b) shows a simulated GPS constellation (the darker icons indicate visible

(a)

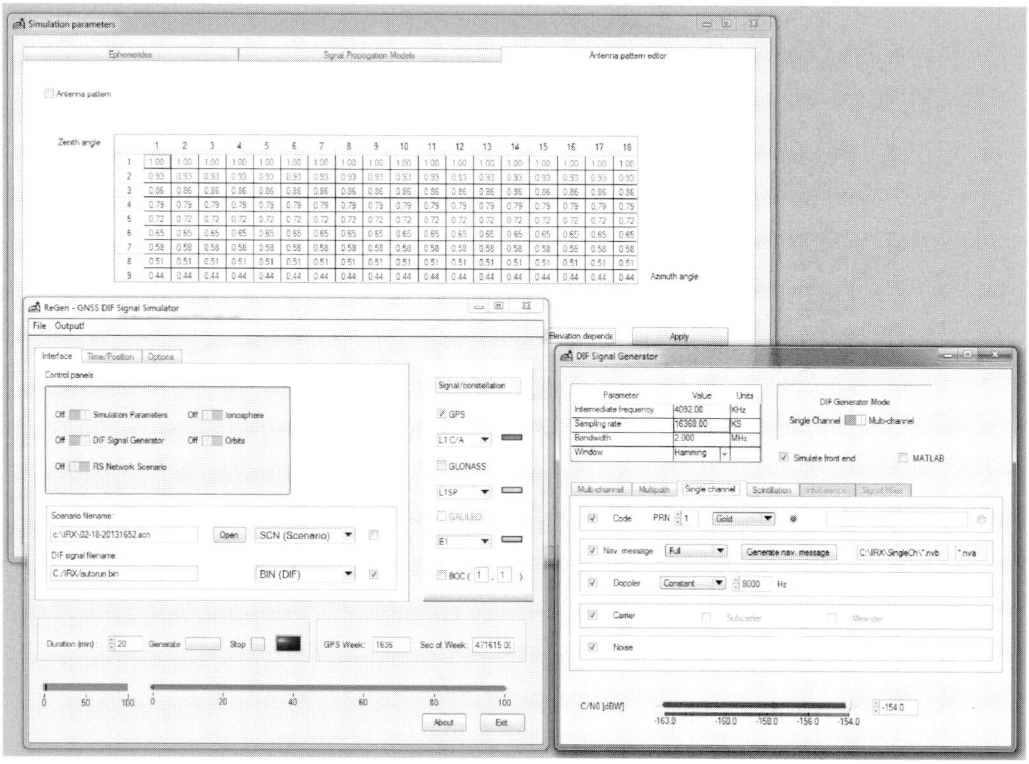

(b)

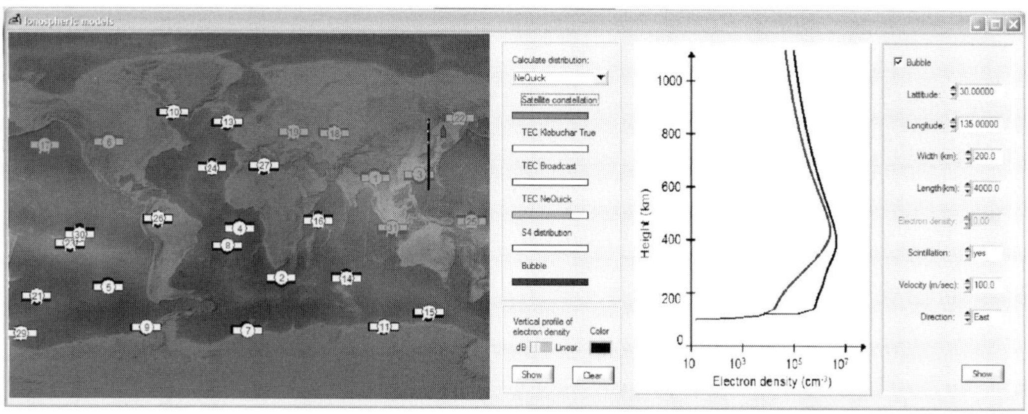

Figure 11.8 iP-Solutions ReGen. (a) GUI with antenna and signal configuration panels. (b) Ionospheric model panel. (c) Reference station panel.

satellites) on top of the NeQuick ionospheric model, The panel allows also the simulation and visualization of an electron depletion area or bubble. It also shows a TEC vertical profile for the simulated model at two different epochs at the receiver location. Figure 11.8(c) shows the panel for network simulation using a spatially correlated

(c)

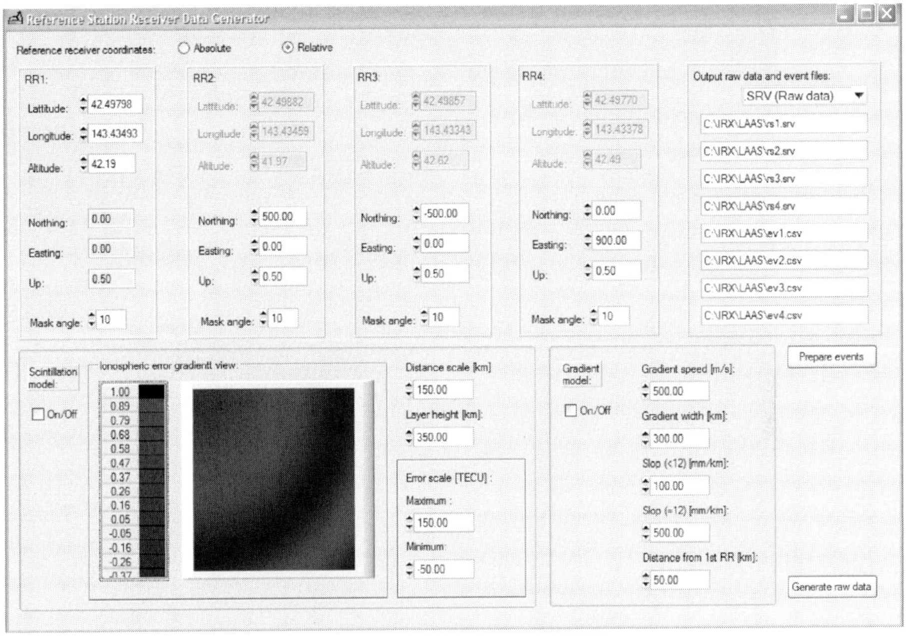

Figure 11.8 *(cont.)*

ionospheric model. This simulation was made in order to facilitate software testing for the local area augmentation system (LAAS) control center.

This type of simulator, together with a software receiver, makes an excellent educational tool. It can be used on a PC in a class working with digitized signals at no additional expense. It also allows us to explore unique possibilities, not only to have all access to GNSS on our desk, but also to investigate signal generation, error budget, and signal processing thoroughly. That is why iP-Solutions and JAXA decided to make academic versions of the software signal generator and software receiver. They are available free of charge with the Cambridge University Press book *Digital Satellite Navigation and Geophysics* [1].

In addition, when augmented by a USB receiver front end and playback device for a DIF signal generator, this software creates a complete GNSS receiver–simulator system. Thus, a previously developed DIF playback system has been upgraded to an RF simulator called Fuji. This simulator is changeable through the software interface, though it is not a software simulator in the narrow GNSS sense, but rather it is like modern satellite transmitters, where the signal can be reprogrammed on the fly. DIF playback can also be used on its own with conventional receivers. Figure 11.9 shows the 2 bit DIF signal being sent to a complex programmable logic device (CPLD) device through micro test clips, emulating the signals from the RF front end. The third clip, which would supply the front-end clock in this case, was substituted with a clock input from the simulator. The RF front end was powered down in order not to supply any outputs to the CPLD.

(a)

(b)

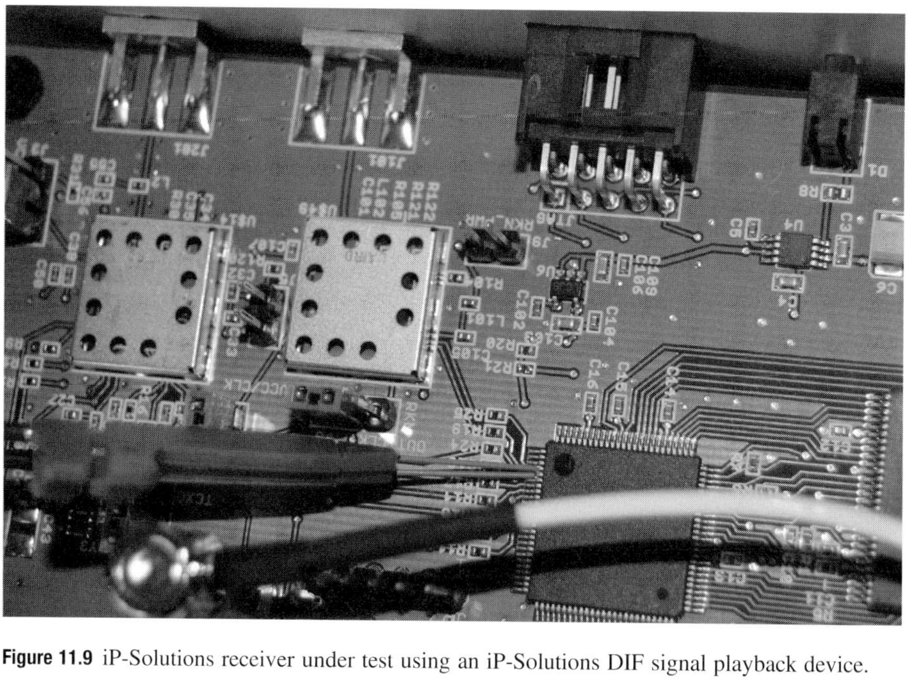

Figure 11.9 iP-Solutions receiver under test using an iP-Solutions DIF signal playback device.

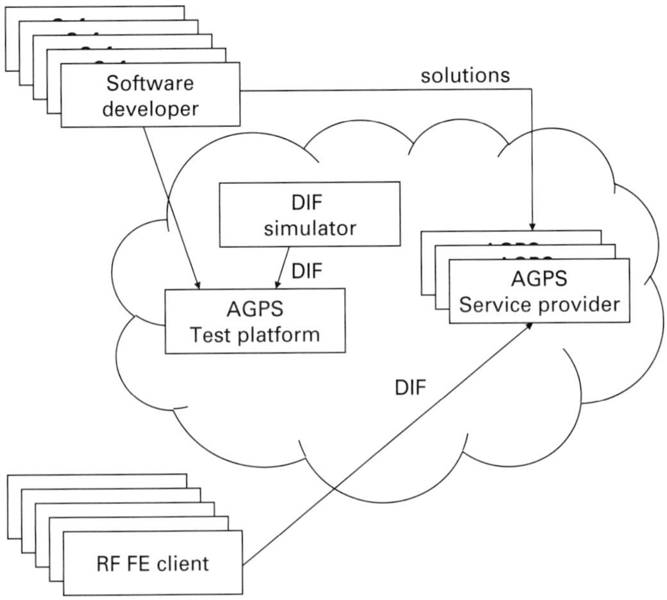

Figure 11.10 Simulation on the cloud service.

Although on the one hand the Fuji simulator (see Figure 11.4) has the same functionality as a normal high-end RF simulator, its design as an SDR simulator allows for unprecedented flexibility on the part of the user. We again need to underline the difference between an SDR receiver and a software receiver, which we have discussed in the preceding chapters. The Fuji simulator's functionality is facilitated by the programmable hardware (FPGA), but the hardware can be changed from the software layer. This is one of the features of an SDR device. Most conventional GNSS receivers today are in fact SDR receivers. On the transmitting side, some of the new generation satellite transmitters are also SDR, in the sense that they can be reprogrammed on the fly to change the signal. In this sense, the Fuji simulator is a new generation simulator, which gives the user the flexibility to change not only the underlying models and parameters of the signal, but also the signal structure itself.

The by-product of this new design, we observe that the DIF signal fits neatly into the new GNSS paradigm discussed in Chapter 10.

Mobile devices can output chunks of RF signal to be processed not in their embedded baseband processors, but by services in the cloud, including selectable third-party services (Figure 11.10). The services can compete on who provides better sensitivity and shorter TTFF, who covers specific regions, and so on. Device manufacturers can free themselves from this functionality and leave it up to the user, providing only the necessary RF hardware, which is limited to the front end, as depicted in Figure 11.11. This allows a reduction in weight, cost, and power consumption of the device, putting more weight on the outside services, see Figure 11.10, for the concept. This would require the development of services for which the proper simulation will be not an RF

Figure 11.11 iP-Solutions front end.

signal, but chunks of a DIF signal supplied through the cloud (Figure 11.10). The ReGen DIF simulator is an example of a simulator that can provide a simulation service for developers through the cloud in real time.

References

[1] I. Petrovski and T. Tsujii, *Digital Satellite Navigation and Geophysics: A Practical Guide with GNSS Signal Simulator and Receiver Laboratory*, Cambridge: Cambridge University Press, 2012.

[2] I. Petrovski, B. Townsend, and T. Ebinuma, Testing multi-GNSS equipment, systems, simulators and the production pyramid, *Inside GNSS Mag.*, July–Aug. 2010. Available at http://www.insidegnss.com/auto/julaug10-Petrovski.pdf (last accessed Oct. 7, 2013).

[3] I. Petrovski and T. Ebinuma, Everything you always wanted to know about GNSS simulators but were afraid to ask, *Inside GNSS*, Sept. 2010, pp. 48–58. Available at http://www.insidegnss.com/auto/sep10-Petrovski.pdf (last accessed Oct. 7, 2013).

[4] I. Petrovski, T. Tsujii, J-M. Perre, B. Townsend, and T. Ebinuma, GNSS simulation: a user's guide to the galaxy, *Inside GNSS*, pp. 36–45, Oct. 2010. Available at http://www.insidegnss.com/auto/oct10-petrovski.pdf (last accessed Oct. 7, 2013).

[5] I. Petrovski, T. Tsujii, and H. Hojo, First AGPS – now BGPS. Instantaneous precise positioning anywhere, *GPS World*, vol. 19, no. 11, pp. 42–48, Nov. 2008.

[6] P. Boulton, GPS interference testing, *Inside GNSS*, pp. 32–45, July/Aug. 2011. Available at http://www.insidegnss.com/auto/julyaug11-Butler.pdf (last accessed Oct. 7, 2013).

Index